*Soil Fertility
and Fertilizers*

K deficiency in corn. Chlorosis and necrosis of edges of lower leaves, as K is translocated to newer, developing leaves. Midrib usually remains green.

K deficiency in soybean. Chlorosis and necrosis of lower leaf edges where tissue along veins and base of leaf remain green.

K deficiency in alfalfa. Small white spots occur along leaf margins, although yellowing of leave edges can also occur. Normal plant is on the right.

S deficiency in corn. Plant is stunted with light green and/or yellow leaves. Although usually occurring on newer leaves, symptoms can be observed on the entire plant. S deficiency symptoms can be confused with N deficiency.

S deficiency in wheat. Chlorotic newer leaves are observed, as S is not translocated from older to newer leaves as readily as N.

S deficiency in soybean. Plant is stunted with light green and/or yellow newer leaves.

JOHN L. HAVLIN
NORTH CAROLINA STATE UNIVERSITY

JAMES D. BEATON
POTASH & PHOSPHATE INSTITUTE OF CANADA

DALE
NSTITUTE

SON
NSTITUTE

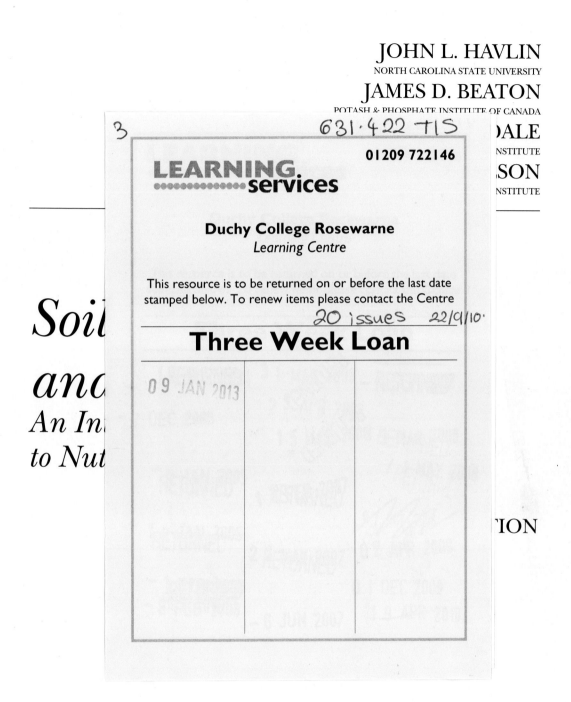
Soil
and
An In
to Nut

TION

Prent   H
UPPER

Library of Congress Cataloging-in-Publication Data

Soil fertility and fertilizers : an introduction to nutrient
   management / Samuel L. Tisdale . . . [et al.].—6th ed.
     p.   cm.
   Includes bibliographical references and index.
   ISBN 0-13-626806-4
   1. Fertilizers.   2. Soil fertility.   3. Crops—Nutrition.
   I. Tisdale, Samuel L.
   S633.S715   1999
   631.4'22—dc21   99-38911
                                  CIP

Publisher: *Charles Stewart*
Production Editor: *Lori Harvey*
Production Liaison: *Eileen M. O'Sullivan*
Managing Editor: *Mary Carnis*
Director of Manufacturing and Production: *Bruce Johnson*
Production Manager: *Marc Bove*
Marketing Manager: *Melissa Bruner*
Editorial Assistant: *Jennifer Stagman*
Cover Designer: *Liz Nemeth*
Formatting/page make-up: *Carlisle Communications, Ltd.*
Printer/Binder: *Courier Westford*

 © 1999 by Prentice-Hall, Inc.
Simon & Schuster/A Viacom Company
Upper Saddle River, New Jersey 07458

Earlier editions copyright 1956, 1966, 1975, 1985,
and 1993 by Macmillan Publishing Company, a
division of Macmillan, Inc.

Printed in the United States of America

10 9 8 7 6 5 4 3 2 1

ISBN 0-13-626806-4

Prentice-Hall International (UK) Limited, *London*
Prentice-Hall of Australia Pty. Limited, *Sydney*
Prentice-Hall Canada Inc., *Toronto*
Prentice-Hall Hispanoamericano, S.A., *Mexico*
Prentice-Hall of India Private Limited, *New Delhi*
Prentice-Hall of Japan, Inc., *Tokyo*
Simon & Schuster Asia Pte. Ltd., *Singapore*
Editora Prentice-Hall do Brasil, Ltda., *Rio de Janeiro*

Contents

CHAPTER 8

Micronutrients

245

CHAPTER 9

Soil Fertility Evaluation

300

CHAPTER 10

Fundamentals of Nutrient Management

358

CHAPTER 11

Nutrients, Water Use, and Other Interactions

406

Preface

Soil Fertility and Fertilizers: An Introduction to Nutrition Management, 6e, was first published in 1956 under the title *Soil Fertility and Fertilizers*. Although this sixth edition has been substantially revised to reflect rapidly advancing knowledge and technologies in plant nutrition and nutrient management, the outstanding contributions of Dr. Samuel L. Tisdale (1918–89) and Dr. Werner L. Nelson (1914–92) will always be remembered and appreciated.

The importance of soil fertility and plant nutrition to the health and survival of all life cannot be understated. As human populations continue to increase, human disturbance of the earth's ecosystem to produce food and fiber will place greater demand on soils to supply essential nutrients. Therefore, it is critical that we increase our understanding of the chemical, biological, and physical properties and relationships in the soil-plant-atmosphere continuum that control nutrient availability.

The soil's native ability to supply sufficient nutrients has decreased with the higher plant productivity levels associated with increased human demand for food. One of the greatest challenges of our generation will be to develop and implement soil, crop, and nutrient management technologies that enhance plant productivity and the quality of the soil, water, and air. If we do not improve and/or sustain the productive capacity of our fragile soils, we cannot continue to support the food and fiber demand of our growing population.

To the Student

The goal of this book is to impart to the student a thorough understanding of plant nutrition, soil fertility, and nutrient management so that she or he can (1) describe the influence of soil biological, physical, and chemical properties and interactions on nutrient availability to crops; (2) identify plant nutrition–soil fertility problems and recommend proper corrective action; and (3) identify soil and nutrient management practices that maximize productivity and profitability while maintaining or enhancing the productive capacity of the soil and quality of the environment.

The specific objectives are to (1) describe how plants take up or absorb plant nutrients and how the soil system supplies these nutrients; (2) identify and describe plant-nutrient deficiency symptoms and methods used to quantify nutrient problems; (3) describe how soil organic matter, cation exchange capacity, soil pH, parent material, climate, and human activities affect nutrient availability; (4) evaluate nutrient and soil amendment materials on the basis of content, use, and effects on the soil and the crop; (5) quantify, using basic chemical principles, application rates of nutrients and amendments needed to correct plant nutrition problems in the field; (6) describe nutrient response patterns, fertilizer use efficiency, and the economics involved in fertilizer use; and (7) describe and evaluate soil and nutrient management practices that either impair or sustain soil productivity and environmental quality.

To the Teacher

Motivate your students to learn by showing them how the knowledge and skills gained through the study of soil fertility will be essential for success in their careers. Use teaching methodologies that enhance their critical thinking and problem-solving skills. In addition to understanding the qualitative soil fertility and plant-nutrition relationships, students must know how to quantitatively evaluate nutrient availability and nutrient management. Environmental protection demands that nutrients be added in quantities and by methods that maximize crop productivity and recovery of the added nutrients.

Since some of the examples used in this text may not be representative of your specific region, frequently integrate additional field examples from your region to illustrate the qualitative and quantitative principles. Strongly reinforce the reality that production agriculture, sustainability, and environmental quality are compatible provided soil, crop, and nutrient management technologies are used properly. Develop in your students the desire and discipline to expand beyond this text through reading and self-learning. Demand of your students what will be demanded of them after they graduate—to think, communicate, cooperate, and solve problems from an interdisciplinary perspective.

An instructors' manual is available from the publisher and provides qualitative and quantitative questions pertinent to each chapter. Instructors should utilize these questions as learning aids to help students gain confidence with the material and to prepare for exams. Answers to each question and complete solutions to quantitative calculations are provided.

We hope your students find the text a valuable resource throughout their careers. Please feel free to provide suggestions for enhancing the effectiveness of the text as a teaching and learning aid.

Acknowledgments

We wish to thank the following reviewers for their helpful comments on the manuscript:

Mark M. Alley, Virginia Polytechnic Institute and State University; James R. Brown, University of Missouri at Columbia; Richard H. Fox, Pennsylvania State University; John H. Grove, University of Kentucky; Gordon V. Johnson, Oklahoma State University; Jay W. Johnson, Ohio State University; Bill Raun, Oklahoma State University; H. M. Reisenauer, University of California-Davis; and Joe Toudeton, Auburn University.

Most of the photos included in the color plates (see inside book cover) are the property of one of the authors; however, several were borrowed from colleagues. We have received permission from the following persons to use these photos in this edition of *Soil Fertility and Fertilizers:*

Dr. Gordon Miner; Dr. Jim Shelton; Dr. Bobby Wells; Dr. Jeff Jacobson; and Dr. David Whitney.

John L. Havlin
James D. Beaton

CHAPTER 1

Introduction

During most of our existence on earth, hunting and gathering have procured food. As populations grew, organized agricultural systems were developed to ensure food security (Table 1.1). All of the systems shown in Table 1.1 exist in various parts of the globe today, since the entire population has not advanced at the same rate. As a result, famine has been a reality in much of the underdeveloped regions that exhibit the highest population growth rates and rely on inefficient and unproductive farming methods. In contrast, developed countries utilizing modern agricultural technologies are generally self-sufficient in food production and provide the majority of food exports.

Compared with just 50 years ago, the incidence of famine around the world has decreased three- to fourfold (Fig. 1.1). Unfortunately, these data do not include people severely undernourished or exhibiting various levels of nutrient deficiencies. Estimates of chronic or seasonal undernourishment exceed 500 million people. Recent projections of world food security show that the number of people at risk of hunger in Africa and Asia will likely increase over the next 50 years, although the percentage of population at risk of hunger in each country will decrease (Table 1.2). In contrast, substantial improvements in food security are projected for Latin America and Southeast Asia. Despite improvements in food

TABLE 1.1 Capability of Agricultural Systems to Produce Food and Support Population

Agricultural System	Cultural Stage or Time	Cereal Yield (t/ha)	World Population (millions)	Hectares per Person
Hunting and gathering	Paleolithic		7	
Shifting agriculture	Neolithic (10,000 years ago)	1	35	40.0
Medieval rotation	500–1450 A.D.	1	900	1.5
Livestock farming	Late 1700s	2	1,800	0.7
Modern agriculture	Twentieth century	4	4,200	0.3
	Twenty-first century	6	12,000	0.1

SOURCE: McCloud, *Agron J.,* 67:1 (1975).

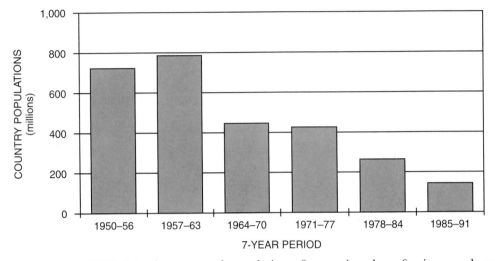

FIGURE 1.1   Average total population of countries where famine was documented, 1950–91.

security in some regions, by 2060, 640 million people, or 9% of the population in developing countries, will be at risk of famine or undernourishment.

The importance of advancing agricultural productivity throughout the world is obvious. The world's population will exceed 6 billion people by 2000, 9 billion by 2025, and 11 billion by 2050. About 95% of the projected increase in population will occur in developing countries, primarily in Africa. By 2060 increases in global food production are projected to be 100 to 300% of 1980 levels (Table 1.3). Are sufficient land resources available to expand agriculture production to meet the food demands expected in 2060? In developing countries, for example, 3.59 billion people have access to 6,495 million hectares of land, or 0.55 persons per hectare (Table 1.4). Depending on the level of inputs used, this land area could support between 5.6 and 33.2 billion people. In 2060 the population in these regions is projected to be 9 billion people; thus, sufficient land area exists to meet potential food demands. However, considerations for the risks to biodiversity, wildlife habitat, and soil and environmental quality will reduce available land area for cultivation. Estimates of only 20 to 30% increases in new land brought under production illustrate the dependence on increasing productivity per unit land area. Advances in agricultural production technologies must occur to enhance productivity per hectare and ensure food security.

In the United States, the amount of farmland has remained relatively stable at approximately 1 billion acres, with about 500 million acres in cropland and pas-

TABLE 1.2   Projected Number of People at Risk of Hunger*

Region	1980	2000	2020	2040	2060
			Million		
Developing	501 (23)	596 (17)	717 (14)	696 (11)	461 (9)
Africa	120 (26)	185 (22)	292 (21)	367 (19)	415 (18)
Latin America	36 (10)	40 (8)	39 (6)	33 (4)	24 (3)
South and Southeast Asia	321 (25)	330 (17)	330 (13)	232 (8)	130 (4)
West Asia	27 (18)	41 (16)	55 (14)	64 (12)	72 (11)

*Numbers in parentheses show percentages of population.

SOURCE: Compiled by Economic Research Service (1996).

TABLE 1.3   Projected Global Production of Food Commodities through 2060

Commodity	1980	2000	2020	2040	2060
	Million Tons*				
Wheat	441	603	742	861	958
Rice	249	368	480	586	659
Coarse grains	741	1,022	1,289	1,506	1,669
Animal products	82	108	138	164	184
Dairy	470	613	750	877	997
Protein feed	36	52	64	76	85

*Wheat, rice, and coarse grain in million tons; animal products in million tons carcass weight; dairy products in million tons whole milk equivalent; protein feed in million tons protein equivalent.
SOURCE: Compiled by Economic Research Service (1996).

TABLE 1.4   Developing Country Population-Supporting Capacities

Location	Total Land Area (Million Hectares)	Population in 2000 (Millions)	Persons per Hectare in 2000	Potential Population-Supporting Capacity in 2000 (Persons per Hectare)	
				Low Inputs	High Inputs
Total	6,495	3,590	0.55	0.86	5.11
Africa	2,878	780	0.27	0.44	4.47
Southwest Asia	677	265	0.39	0.27	0.48
South America	1,770	393	0.22	0.78	6.99
Central America	272	215	0.79	1.07	4.76
Southeast Asia	898	1,937	2.16	2.74	7.06

SOURCE: Compiled by Economic Research Service (1996).

ture. Although the number of farms and producers has decreased dramatically, with a concomitant increase in farm size, the total land area used for food and fiber production will not likely increase (Figs. 1.2 and 1.3). Therefore, substantial increases in productivity per acre will be needed if the United States is to remain a major food exporter, which is essential to attain world food security.

Historically in the United States, crop yields have increased greatly over the last half-century (Table 1.5). This remarkable achievement is directly related to the development of agricultural technology over the last 60 years (Fig. 1.4). The

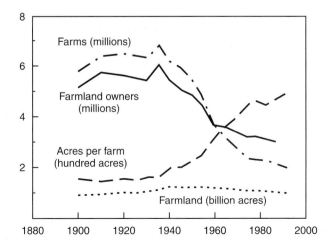

FIGURE 1.2   Farms, farmland, farm owners, and average acres per farm, 1900–92.   *USDA-ERS, based on Census of Agriculture, 1954 and 1992.*

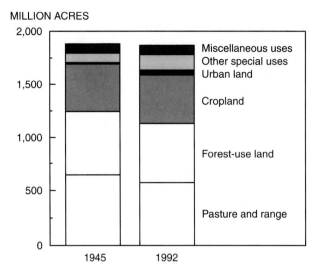

FIGURE 1.3   Major uses of land in the contiguous 48 states. *USDA-ERS, 1996.*

TABLE 1.5   Average Yields of Major Crops in the United States, 1950–92

	Corn (bu/a)	Wheat (bu/a)	Soybean (bu/a)	Alfalfa (tn/a)
1950	37.6	14.3	21.7	2.1
1964	62.1	26.2	22.8	2.4
1972	96.9	32.7	28.0	2.9
1982	113.2	35.5	31.5	3.4
1992	128.7	41.4	37.0	3.6

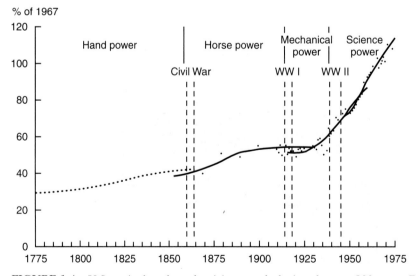

FIGURE 1.4   U.S. agricultural productivity growth during the past 200 years. *Farrell,*
Productivity in U.S. Agriculture, *ESS Report no. AGESS810422, USDA, 1981.*

MILLION NUTRIENT TONS

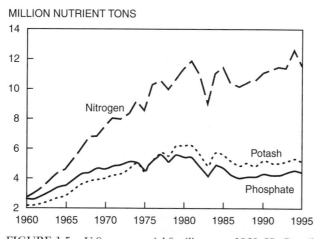

FIGURE 1.5   U.S. commercial fertilizer use, 1960–95. *Compiled by ERS from Tennessee Valley Authority, 1994 and earlier issues; Association of American Plant Food Control Officials, 1995.*

principal factors contributing to higher crop yields include development of improved varieties and hybrids, nutrient and pest management, soil and water conservation, and cultural practices. Development and increased use of fertilizer and pesticides are directly related to increased crop productivity in the United States (Figs. 1.5 and 1.6). Since 1980 concerns for environmental quality have resulted in the development and adoption of improved input management technologies that have stabilized nutrient and pesticide use.

Grower adoption of many agronomic technologies developed since 1950 has increased crop productivity (Table 1.6). These data demonstrate that the technologies contributing most to the increase in average corn yields from 32 to 100 bu/a were N fertilization, breeding/genetics, weed control, and other cultural practices. In contrast, decreased use of manure and declining organic matter

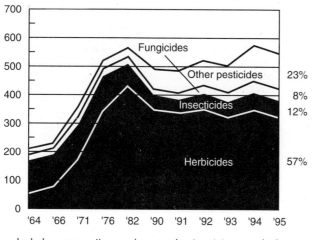

Includes corn, cotton, soybeans, wheat, potatoes, and other vegetables, and citrus, apples, and other fruit (about 67% of U.S. cropland).

FIGURE 1.6   Total pesticide use on major crops, 1964–95. *USDA-ERS, 1996.*

TABLE 1.6   Sources of Corn Yield Increases with Changing Production Practices in Minnesota, 1930–79

Cultural Practice or Yield Limiting Factor	*bu/a*	*kg/ha*	*% Net Gain/Loss*
		Contribution to 1979 Yield	
Pre-1930 yield levels	32.1	2,012	—
Productivity Gains/Losses			
Hybrids			
Double crosses	5.9	371	9
Three-way crosses	0.5	28	1
Single crosses	3.7	235	6
Genetic gain	29.1	1,825	43
Fertilizer N	31.9	2,003	47
Plant population	14.4	905	21
Herbicides	15.5	975	23
Row spacing	2.8	173	4
Planting date	5.8	364	8
Drilling vs. hill drop	5.1	322	8
Fall plowing	3.6	224	5
Rotations			
Soybeans	7.7	484	11
Alfalfa/clovers	−2.2	−136	−3
Sweet clover	−5.1	−318	−7
Interference effect	−4.6	−291	−7
Manure	−10.1	−633	−15
Organic matter	−9.1	−571	−13
Insects			
Corn borer	−3.5	−220	−5
Corn rootworm	−2.3	−145	−3
Soil erosion	−6.5	−345	−8
Unidentified negative factors	−15.5	−975	−23
Net gain	68.1	4,275	—
1977–79 yield level	100.2	6,287	—

SOURCE: Cardwell, *Agron J.,* 74:984 (1982).

(OM) contributed to yield loss. Evaluation of the gains and losses in yield shows that if losses had not occurred, yields would have been 158 bu/a instead of 100 bu/a. Elimination of fertilizer would decrease yields 40 to 90%, depending on the crop, soil, and climatic region. If fertilizer were not used, about 30 to 40% more land would be needed. Therefore, growers must take advantage of technologies that increase productivity, as well as those that minimize productivity loss.

Yield Limiting Factors

Obtaining the maximum production potential of a particular crop depends on the environment during the growing season and the skill of the producer in identifying and eliminating or minimizing those factors that reduce yield potential.

TABLE 1.7   Factors Affecting Crop Yield Potential

Climate Factors	Soil Factors	Crop Factors
Precipitation	Organic matter	Crop species/variety
Quantity	Texture	Planting date
Distribution	Structure	Seeding rate and geometry
Air temperature	Cation exchange capacity	Row spacing
Relative humidity	Base saturation	Seed quality
Light	Slope and topography	Evaportranspiration
Quantity	Soil temperature	Water availability
Intensity	Soil management factors	Nutrition
Duration	Tillage	Pests
Altitude/latitude	Drainage	Insects
Wind	Others	Diseases
Velocity	Depth (root zone)	Weeds
Distribution		Harvest efficiency
CO_2 concentration		

More than 50 factors affect crop growth and yield potential (Table 1.7). Although the producer cannot control many of the climate factors, most of the soil and crop factors can and must be managed for maximum productivity.

Two major factors affecting the upper limit of yield potential are (1) the amount of moisture available during the growing season and (2) the length of the growing season. For maximum yield potential, crop plants must utilize a high percentage of the available solar energy. Based on available solar energy in the United States, the maximum potential yield for most crops far exceeds current yield levels. For example, maximum potential yields are on the order of 490 to 580 bushels of corn, 140 to 225 bushels of soybeans, and 180 to 230 bushels of wheat per acre.

For high yields, controllable and uncontrollable factors must operate in unison, because many of them are interrelated. Most of the factors influencing yield potential interact with each other to either increase or decrease plant growth and/or yield. Many of these interactions, especially as they influence plant response to nutrients, are discussed in Chapter 11. The challenge of a producer or consultant is to accurately identify all yield limiting factors and eliminate or minimize the influence of all those that can be managed. The importance of this principle was identified by Justus von Leibig in 1862. Leibig stated his *Law of the Minimum* as follows:

> Every field contains a maximum of one or more and a minimum of one or more nutrients. With this minimum, be it lime, potash, phosphoric acid, magnesia, or any other nutrient, the yields stand in direct relation. It is the factor that governs and controls . . . yields. Should this minimum be lime . . . yield . . . will remain and be no greater even though the amount of potash, silica, phosphoric acid, etc . . . be increased a hundred fold.

All successful agricultural producers use this important principle, either knowingly or unknowingly. For example, a producer may have planted the correct variety at the optimum time and population and applied all of the optimum nutrients using the most efficient methods, and still not attained maximum yield potential, because plant available water was the most limiting factor (Fig. 1.7).

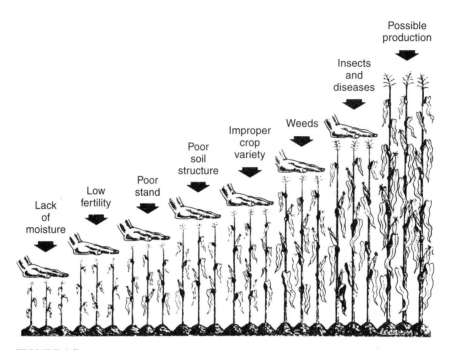

FIGURE 1.7   Leibig's Law of the Minimum states that the most limiting factor deter-mines yield potential. Producers should minimize or eliminate the most limiting factor first, then the second most limiting factor, and so forth. Only in this manner can maxi-mum yield potential be achieved.

Thus, until the producer minimizes water as a limiting factor to yield potential, yield response to management of any other factor(s) influencing yield potential will be substantially less than if plant available water were nonlimiting. Figure 1.8 graphically illustrates Leibig's Law of the Minimum.

From this discussion, it is apparent that sufficient nutrient availability is re-quired to realize maximum plant growth and yield potential. Before thoroughly discussing the complex soil chemical, biological, and physical factors that influ-ence the supply of nutrients to plants, as well as nutrient management strategies for optimizing crop productivity, a brief review of the nutrients required for plant growth is necessary.

Elements in Plant Nutrition

A mineral element is considered essential to plant growth and development if the element is involved in plant metabolic functions and the plant cannot complete its life cycle without the element. Usually the plant exhibits a visual symptom in-dicating a deficiency in a specific nutrient, which normally can be corrected or prevented by supplying that nutrient. Visual nutrient deficiency symptoms can be caused by many other plant stress factors; therefore, caution should be exercised when diagnosing deficiency symptoms. The following terms are commonly used to describe nutrient levels in plants:

Deficient: when the concentration of an essential element is low enough to severely limit yield and distinct deficiency symptoms are visible. Extreme defi-

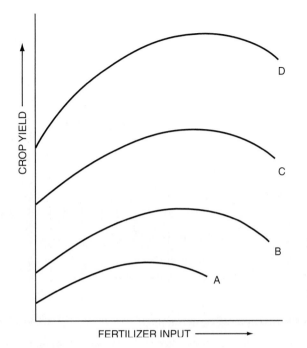

FIGURE 1.8   Yield response to nitrogen (N) fertilizer with water, phosphorus (P), and seeding rate limiting yield potential (A); with only P and seeding rate limiting yield potential (B); with only seeding rate limiting yield potential (C); and with no manageable factors limiting yield potential (D).

ciencies can result in plant death. With moderate or slight deficiencies, symptoms may not be visible, but yields will still be reduced.

Critical range: the nutrient concentration in the plant below which a yield response to added nutrient occurs. Critical levels or ranges vary among plants and nutrients but occur somewhere in the transition between nutrient deficiency and sufficiency.

Sufficient: the nutrient concentration range in which added nutrient will not increase yield but can increase nutrient concentration. The term *luxury consumption* is often used to describe nutrient absorption by the plant that does not influence yield.

Excessive or toxic: when the concentration of essential or other elements is high enough to reduce plant growth and yield. Excessive nutrient concentration can cause an imbalance in other essential nutrients, which can also reduce yield.

The general relationship between nutrient concentration in plant tissue and plant yield is shown in Figure 1.9. Yield is severely affected when a nutrient is deficient, and when the nutrient deficiency is corrected, growth increases more rapidly than nutrient concentration. Under severe deficiency, rapid increases in yield with added nutrient can cause a small decrease in nutrient concentration. This is called the *Steenberg effect* (Fig. 1.9) and results from dilution of the nutrient in the plant by the rapid plant growth. When the concentration reaches the critical range, plant yield is generally maximized. Nutrient sufficiency occurs over a wide concentration range, wherein yield is unaffected. Increases in nutrient concentration above the critical range indicate that the plant is absorbing nutrients

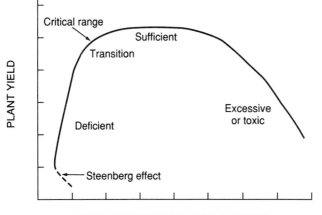

FIGURE 1.9   Relationship between essential plant nutrient concentration and plant growth or yield. As nutrient concentration increases toward the critical level, plant yield increases. Above the critical level the plant contains sufficient levels for normal growth and can continue to absorb nutrients without increasing yield (luxury consumption). Excessive absorption of a nutrient or element can be toxic to the plant and reduce yield or cause plant death.

above that needed for maximum yield. This *luxury consumption* is common in most plants. Elements absorbed in excessive quantities can reduce plant yield directly through toxicity or indirectly by reducing concentrations of other nutrients below their critical ranges.

Sixteen elements are considered essential to plant growth; their relative concentrations in plants are listed in Table 1.8. Carbon (C), hydrogen (H), and oxygen (O) are the most abundant elements in plants. The photosynthetic process in green leaves converts CO_2 and H_2O into simple carbohydrates from which amino acids, sugars, proteins, nucleic acid, and other organic compounds are synthesized. Carbon, H, and O are not considered mineral nutrients. The supply of CO_2 is relatively constant. The supply of H_2O rarely limits photosynthesis directly but does so indirectly through the various effects resulting from moisture stress.

The remaining 13 essential elements are classified as macronutrients and micronutrients, and the classification is based on their relative abundance in plants (Table 1.8). The macronutrients are N, P, potassium (K), sulfur (S), calcium (Ca), and magnesium (Mg). Compared with the macronutrients, the concentrations of the seven micronutrients—iron (Fe), zinc (Zn), manganese (Mn), copper (Cu), boron (B), chlorine (Cl), and molybdenum (Mo)—are very small. Five additional elements—sodium (Na), cobalt (Co), vanadium (Va), nickel (Ni), and silicon (Si)—have been established as essential micronutrients in *some* plants. Micronutrients are often referred to as *minor* elements, but this label does not mean that they are less important than macronutrients. Micronutrient deficiency or toxicity can reduce plant yield just as macronutrient deficiency or toxicity does.

Although aluminum (Al) is not an essential plant nutrient, its concentration in plants can be high when soils contain relatively large amounts of Al in soil solution. In fact, plants absorb many nonessential elements, and more than 60 ele-

TABLE 1.8   Relative and Average Plant Nutrient
Concentrations

Plant Nutrient	Relative Concentration	Average Concentration*
H	60,000,000	6.0%
O	30,000,000	45.0%
C	30,000,000	45.0%
N	1,000,000	1.5%
K	400,000	1.0%
Ca	200,000	0.5%
Mg	100,000	0.2%
P	30,000	0.2%
S	30,000	0.1%
Cl	3,000	100 ppm (0.01%)
Fe	2,000	100 ppm
B	2,000	20 ppm
Mn	1,000	50 ppm
Zn	300	20 ppm
Cu	100	6 ppm
Mo	1	0.1 ppm

*Concentration expressed by weight on a dry matter basis.

ments have been identified in plant materials. When plant material is burned, the remaining *plant ash* contains all of the essential and nonessential mineral elements except C, H, O, N, and S, which are burned off as gases.

The plant content of mineral elements is affected by many factors, and their concentration in crops varies considerably. Shown in Table 9.1 are the contents of some of the mineral elements in common crop plants. Although the latest available data have been used, the figures should be regarded only as averages. Soil, climate, crop variety, and management factors exert considerable influence on plant composition and in individual cases may cause appreciable variation from the values in the table.

Plant nutrient concentration data are valuable to successful nutrient management programs and can be used to help establish nutrient recommendations. Because many biological and chemical reactions occur with nutrients in soils, the quantity of nutrients absorbed by plants does not equal the quantity applied. Proper management can maximize the proportion of nutrient absorbed by the plant. As plants absorb nutrients from the soil, complete their life cycle, and die, the nutrients in the plant residue are returned to the soil. These plant nutrients are subject to the same biological and chemical reactions as fertilizer nutrients. Although this cycle varies somewhat among nutrients, understanding nutrient dynamics in the soil-plant-atmosphere system is essential to successful nutrient management.

The remaining chapters will detail our current knowledge of soil fertility and nutrient management. Use of this knowledge to identify nutrient availability problems and provide economically and environmentally sound nutrient management recommendations will be essential to a world with a secure food supply.

Selected References

RUSSEL, D. A., AND G. G. WILLIAMS. 1977. History of chemical fertilizer development. *SSSAJ*, 41:260–65.

VIETS, F. G. 1977. A perspective on two centuries of progress in soil fertility and plant nutrition. *SSSAJ*, 41:242–49.

EPSTEIN, E. 1972. *Mineral Nutrition of Plants: Principles and Perspectives.* John Wiley & Sons, New York.

MENGEL, K., AND E. A. KIRKBY. 1987. *Principles of Plant Nutrition.* International Potash Institute, Bern, Switzerland.

RÖMHELD, V., AND H. MARSCHER. 1991. Function of micronutrients in plants. In J. J. MORTREDT et al. (Eds.), *Micronutrients in Agriculture.* No. 4. Soil Science Society of America, Madison, Wisc.

CHAPTER 2

Basic Soil-Plant Relationships

The interaction of numerous physical, chemical, and biological properties in soils controls the availability of plant nutrients. Understanding these processes enables us to manage selected soil properties to optimize nutrient availability and plant productivity. The purpose of this chapter is to provide a review of ion exchange reactions in soils, ion movement in soil solution, and ion uptake by plants.

Nutrient supply to plant roots is a very dynamic process (Fig. 2.1). Plants absorb nutrients (cations and anions) from the soil solution and release small quantities of ions such as H^+, OH^-, and HCO_3^- (reactions 1 and 2). Changes in ion concentrations in soil solution are "buffered" by ions adsorbed on the surfaces of soil minerals (reactions 3 and 4). Ion removal from solution causes partial desorption of the same ions from these surfaces. Soils contain minerals that can dissolve to resupply soil solution with many ions (reactions 5 and 6). Likewise, increases in ion concentration in soil solution resulting from fertilization or other inputs can cause some minerals to precipitate.

Soil microorganisms remove ions from soil solution and incorporate them into microbial tissues (reaction 7). When microbes or other organisms die, they release nutrients to the soil solution (reaction 8). Microbial activity produces and

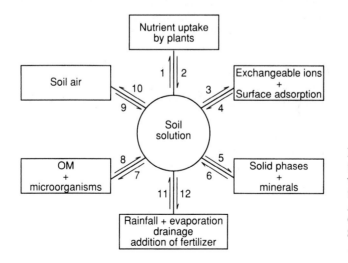

FIGURE 2.1 Relationships between the various components of the dynamic soil system. *Adapted from Lindsay,* Chemical Equilibria in Soils, *Wiley Interscience, 1979.*

decomposes organic matter or humus in soils. These dynamic processes are very dependent on adequate energy supply from organic C (i.e., crop residues), inorganic ion availability, and numerous environmental conditions. Plant roots and soil organisms utilize O_2 and respire CO_2 through metabolic activity (reactions 9 and 10). As a result, CO_2 concentration in the soil air is greater than in the atmosphere. Diffusion of gases in soils decreases dramatically with increasing soil water content.

Numerous environmental factors and human activities can influence ion concentration in soil solution, which interacts with the mineral and biological processes in soils (reactions 11 and 12). For example, adding P fertilizer to soil initially increases the $H_2PO_4^-$ concentration in soil solution. With time, the $H_2PO_4^-$ concentration will decrease with plant uptake, $H_2PO_4^-$ adsorption on mineral surfaces, and P mineral precipitation.

All of these processes and reactions are important to the availability of plant nutrients; however, depending on the specific nutrient, some processes are more important than others. For example, microbial processes are more important to N and S availability than mineral surface exchange reactions, whereas the opposite is true for K, Ca, and Mg. Obviously, these processes are complex, and only their general description and importance to plant nutrient availability are presented. (See references at the end of the chapter for more detail).

Ion Exchange in Soils

Ion exchange in soils occurs on surfaces of clay minerals, inorganic compounds, organic matter, and roots (Fig. 2.2). The specific ion associated with these surfaces depends on the kinds of minerals present and the solution composition. Ion exchange is a reversible process by which a cation or anion adsorbed on the solid phase is exchanged with another cation or anion in the liquid phase. If two solid phases are in contact, ions may also be exchanged between their surfaces. Cation exchange is generally considered to be more important, since the anion exchange capacity (AEC) of most agricultural soils is much smaller than the cation exchange capacity (CEC). Ion exchange reactions in soils are very important to plant nutrient availability. Therefore, it is essential that we understand the nature of the solid constituents and the origin of their surface charge.

Cation Exchange

Solid materials in soils comprise about 50% of the volume, with the remaining volume occupied by water and air. The solid portion is made up of inorganic (mineral) material and organic matter in various stages of decay and humification. The inorganic material consists of sand, silt, and clay. In some soils, coarse fragments are present in varying amounts. The clay fraction primarily consists of layer silicate minerals made up of various combinations of silica tetrahedra and aluminum octahedra (Fig. 2.3). The structure of a silica tetrahedra is one Si^{4+} cation bonded to four O^{2-} ions, whereas the aluminum octahedra is one Al^{3+} cation bonded to six OH^- anions. The long *chains* or *layers* of tetrahedra and octahedra are bonded together to form the *layer silicates*.

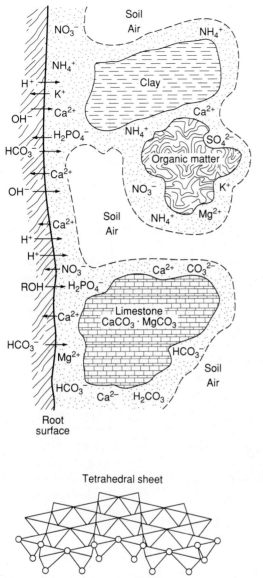

FIGURE 2.2   Diagram of the mineral and organic exchange surfaces in soils.

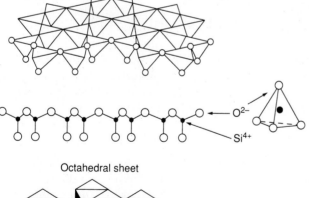

Tetrahedral sheet

O^{2-}

Si^{4+}

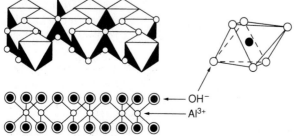

Octahedral sheet

OH^-

Al^{3+}

FIGURE 2.3   Chemical structure of silica tetrahedra, Al octahedra, and the tetrahedral and octahedral sheets.   *Adapted from Sposito, The Chemistry of Soils, Oxford University Press, 1989.*

Layer silicate clay minerals in soils are of three general classes: 1:1, 2:1, and 2:1:1. The 1:1 clays are composed of one silica sheet and one alumina sheet. Kaolinite is the most important 1:1 clay mineral (Fig. 2.4). The 2:1 clays are composed of an aluminum octahedral layer between two silica tetrahedral layers. Examples of the 2:1 clays are smectites (montmorillonite), mica (illite), and vermiculite (Fig. 2.5). Muscovite and biotite mica are 2:1 primary minerals that are often abundant in silt and sand fractions.

Chlorites are 2:1:1 layer silicates commonly found in soils. This clay mineral consists of an interlayer hydroxide sheet in addition to the 2:1 structure referred to previously.

The major source of negative charge associated with layered silicates arises from replacement of either the Si^{4+} or Al^{3+} cations with cations of lower charge. Cation replacement in minerals is called *isomorphic substitution* and occurs predominately in the 2:1 minerals, with very little substitution in the 1:1 minerals. Isomorphic substitution occurred during the formation of these mineral thousands of years ago and is largely unaffected by present environmental conditions.

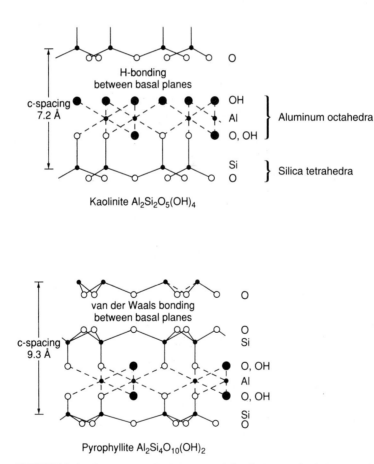

FIGURE 2.4 Structures of a 1:1 mineral, kaolinite, and a 2:1 mineral, pyrophyllite. No isomorphic substitution occurs in the tetrahedral or octahedral layers. *Bear (Ed.), Chemistry of the Soil, ASC Monograph Series No. 160, 1964.*

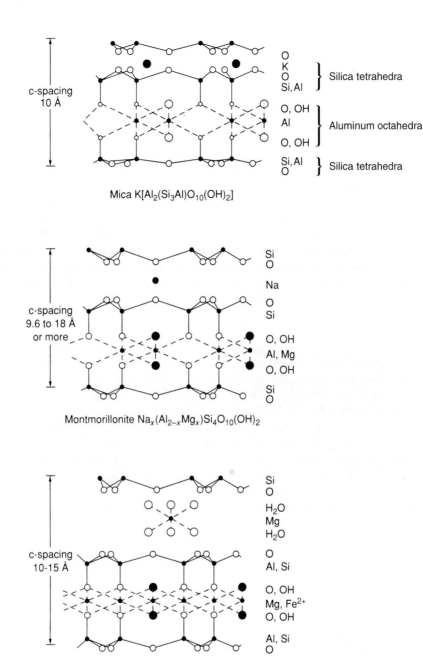

FIGURE 2.5 Structures of mica, montmorillonite, and vermiculite, all 2:1 minerals. Isomorphic substitution occurs in the tetrahedral and octahedral layers. *Bear (Ed.)*, Chemistry of the Soil, *ACS Monograph Series No. 160, 1964.*

TABLE 2.1   Common Layer Silicate Minerals in Soils

Clay Mineral	Layer Type	Layer Charge	c-Spacing (Å)	CEC (meq/100 g)	pH-Dependent Charge
Kaolinite	1:1	0	7.2	1–10	High
Mica (Illite)	2:1	1.0	10	20–40	Low
Vermiculite	2:1	0.8	10–15	120–150	Low
Montmorillonite	2:1	0.4	Variable	80–120	Low
Chlorite*	2:1:1	1.0	14	20–40	High
Organic matter				100–300	High

*Chlorite is a 2:1:1 mineral with a Mg hydroxide interlayer.

In mica, substitution of Al^{3+} for one out of every four Si^{4+} cations in the tetrahedral layer results in an increase of one negative charge. In montmorillonite, Mg^{2+} or Fe^{2+} replaces some of the Al^{3+} in the octahedral layers, again resulting in an increase of one negative charge for each substitution. Compare the unsubstituted 2:1 pyrophyllite mineral in Figure 2.4 with the isomorphic substitution in the 2:1 mica and montmorillonite minerals in Figure 2.5. In the 2:1 vermiculite, isomorphic substitution occurs in both the octahedral and tetrahedral layers.

The location of the isomorphic substitution (tetrahedral, octahedral, or both) imparts specific properties to the clay minerals that affect the quantity of negative surface charge (Table 2.1). For example, isomorphic substitution in the tetrahedral layer locates the negative charge closer to the mineral surface compared with octahedral substitution. The high negative surface charge combined with the unique geometry of the tetrahedral layers allows K^+ cations to neutralize the negative charge between two 2:1 layers (Fig. 2.5). The resulting mica mineral exhibits a lower c-spacing, and the mineral is considered "collapsed," with very little of the surface negative charge available to attract cations. Thus, mica has a lower CEC than montmorillonite because the interlayer surfaces are not exposed (Table 2.1).

The negative charge associated with isomorphic substitution is uniformly distributed over the surface of the clay minerals and is considered a *permanent charge* in that it is unaffected by solution pH (Fig. 2.6). Another source of negative

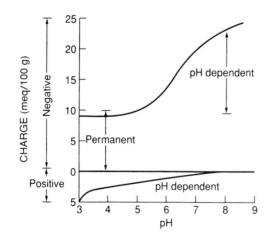

FIGURE 2.6   Permanent and pH-dependent charge associated with clay minerals.   *Guenzi (Ed.)*, Pesticides in Soil and Water, *ASA, 1974.*

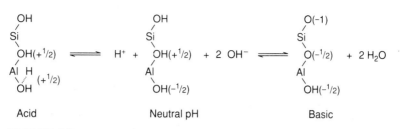

FIGURE 2.7   The pH-dependent charge associated with broken edges of kaolinite.

charge on clay minerals is associated with the *broken edges* of the layer silicates (Fig. 2.7). The quantity of negative or positive charge on the broken edges depends on the pH of the soil solution. The broken edge charge is called a *pH-dependent* charge. Under acidic conditions (low pH) the broken edge is positively charged because of the excess H^+ ions associated with the exposed Si-OH and Al-OH groups (Fig. 2.7). As soil solution pH increases, some of the H^+ ions are neutralized, and the negative charge on the broken edge increases. Increasing the pH above 7 results in a nearly complete removal of H^+ ions on the Si-OH and Al-OH groups, which maximizes the negative charge associated with the broken edge. The increase in negative charge with increasing pH is shown in Figure 2.6. Only about 5 to 10% of the negative charge on 2:1 clays is pH dependent, whereas 50% or more of the charge developed on 1:1 clay minerals can be pH dependent.

Another source of pH-dependent charge is associated with soil organic matter (Fig. 2.8). Most of the negative charge originates from the dissociation of H^+ from carboxylic acid ($—COOH \leftrightarrow —COO^- + H^+$) and phenolic acid ($—C_6H_4OH \leftrightarrow —C_6H_4O^- + H^+$) groups. As pH increases, some of these H^+ ions are neutralized, increasing the negative charge on the surface of these large molecules.

The CEC of a soil represents the total quantity of negative charge available to attract cations in solution. It is one of the most important chemical properties of soils and strongly influences nutrient availability. The CEC is expressed in terms of milliequivalents of negative charge per 100 g of oven-dried soil (meq/100 g).[1] The CEC also represents the total meq/100 g of cations held on the negative sites. The meq unit is used instead of mass because CEC represents the total charges involved: since the specific cations associated with the CEC will vary, it is more meaningful to simply quantify the total charges involved.

The definitions of equivalents and equivalent weight are as follows:

- **Atomic weight:** weight in grams of 6×10^{23} atoms of the substance. One *mole* of substance is 6×10^{23} atoms, molecules, ions, compounds, and so on; therefore, units of atomic weight are grams per mole (g/mole).
- **Equivalent weight:** quantity (mass) of a substance (e.g., cation, anion, or compound) that will react with or displace 1 gram of hydrogen (H^+), which equals Avogadro's number of charges (+ or −). This is equal to the weight in

---

[1] The SI unit system is used by the scientific community. Thus, meq/100 g becomes cmol/kg in SI units, representing the centimoles (cmol) of charge per kilogram of soil. The conversion is 1 meq/100 g soil = 1 cmol/kg. We use meq/100 g in this text because most soil-testing laboratories in the United States use meq/100 g for CEC measurement.

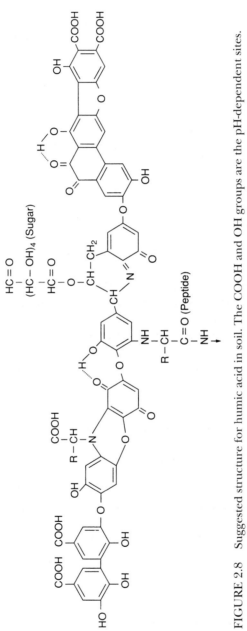

FIGURE 2.8 Suggested structure for humic acid in soil. The COOH and OH groups are the pH-dependent sites. *Mortvedt, Giordano, and Lindsay (Eds.), Micronutrients in Agriculture, ASA, 1972.*

grams of 6×10^{23} charges; therefore, units of equivalent weight are grams per equivalent (g/eq).

The definitions of atomic weight and equivalent weight are very similar:

$$\text{atomic wt} = \text{grams per } 6 \times 10^{23} \text{ ions or molecule}$$
$$\text{equivalent wt} = \text{grams per } 6 \times 10^{23} \text{ charges } (+ \text{ or } -)$$

The use of equivalents in soil chemistry and fertility is a convenient way to express quantities of exchangeable ions in soils. In *cation exchange* problems atomic weight and equivalent weight are related:

$$\text{equivalent wt of } A = \text{atomic wt of } A/\text{valence of } A. \text{ For example,}$$
$$\text{equivalent wt of } K^+ = \frac{39 \text{ g/mole}}{1 \text{ eq/mole}} = 39 \text{ g/eq}$$
$$\text{equivalent wt of } Ca^{2+} = \frac{40 \text{ g/mole}}{2 \text{ eq/mole}} = 20 \text{ g/eq}$$
$$\text{equivalent wt of } Al^{3+} = \frac{27 \text{ g/mole}}{3 \text{ eq/mole}} = 9 \text{ g/eq}$$

If a soil contains 1 mole of Ca^{2+} ions (6×10^{23} ions), then there are $2 \times (6 \times 10^{23}$ charges), or 12×10^{23} charges, since each Ca^{2+} ion has two charges. The definition of equivalent weight, the weight of 6×10^{23} charges, means that 1 equivalent of Ca^{2+} weighs 20 g/eq, or 20 g/6×10^{23} charges. Recall that 1 mole of Ca^{2+} weighs 40 g/m, or 40 g/6×10^{23} ions.

The use of equivalents to express concentrations or quantities of nutrients in soils is very convenient because of the nature of exchange reactions. If Ca^{2+} replaces K^+ on the exchange, then each Ca^{2+} cation can replace two K^+ cations, but one equivalent of Ca^{2+} replaces one equivalent of K^+ or one equivalent of any other cation. Thus,

1 equivalent A = 1 equivalent B, where A and B are ions, compounds, and so forth.

This concept is crucial to understanding and quantifying many chemical reactions in soil fertility.

The equivalent weight of a compound is determined by knowing the reaction the compound is involved in. For example,

$$CaCO_3 + 2HCl \rightarrow Ca^{2+} + 2Cl^- + H_2O + CO_2$$

What is the *equivalent weight* of $CaCO_3$ in this reaction?

Answer: 1 mole of $CaCO_3$ neutralizes 2 moles of HCl; therefore,

$$\text{equivalent wt} = \frac{\text{molecular weight}}{2} = \frac{100}{2} = 50 \text{ g/eq}$$

The CEC of common clay minerals and organic matter is given in Table 2.1. Soils with predominately 2:1 colloids have a higher CEC than soils with mainly 1:1 minerals.

The CEC is strongly affected by the nature and amount of mineral and organic colloid present in the soil. Soils with large amounts of clay and organic matter have a higher CEC than sandy soils low in organic matter. Examples of CEC values for different soil textures are as follows:

Sands (light colored)	3–5 meq/100 g
Sands (dark colored)	10–20
Loams	10–15
Silt loams	15–25
Clay and clay loams	20–50
Organic soils	50–100

The most common cations associated with CEC are listed in Table 2.2. Except for Al^{3+}, most of the exchangeable cations are plant nutrients. In acidic soils the principal cations are Al^{3+}, H^+, Ca^{2+}, Mg^{2+}, and K^+ (Table 2.2). In neutral and basic soils the predominant cations are Ca^{2+}, Mg^{2+}, K^+, and Na^+. Cations are held on the exchange sites with different adsorption strengths; therefore, the ease with which cations can be replaced or exchanged with other cations also varies. For most minerals, the strength of cation adsorption, or *lyotropic series,* is

$$Al^{3+} > Ca^{2+} > Mg^{2+} > K^+ = NH_4^+ > Na^+$$

The properties of the cations determine the strength of adsorption or ease of desorption. First, the strength of adsorption is directly proportional to the charge on the cations. The H^+ ion is unique because of its very small size and high charge density; thus, its adsorption strength is between Al^{3+} and Ca^{2+}. Second, the adsorption strength for cations with similar charge is determined by the size or radii of the hydrated cation (Table 2.2). As the size of the hydrated cation increases, the distance between the cation and the clay surface increases. Larger hydrated cations cannot get as close to the exchange site as smaller cations, resulting in decreased strength of adsorption.

TABLE 2.2   Cation and Anions Associated with Exchange Capacity of Soils

Element	Atomic Weight (g/mole)	Equivalent* Weight (g/eq)	Ionic Radii	
			Nonhydrated (nm)	Hydrated (nm)
Cations				
Al^{3+}	27	9	0.051	
H^+	1	1		
Ca^{2+}	40	20	0.099	0.96
Mg^{2+}	24	12	0.066	1.08
K^+	39	39	0.133	0.53
NH_4^+	18	18	0.143	0.56
Na^+	23	23	0.097	0.79
Anions				
$H_2PO_4^-$	97	97		
SO_4^{2-}	96	48		
NO_3^-	62	62		
Cl^-	35	35		
OH^-	17	17		

*g/eq or mg/meq.

Determination of CEC

A conventional method of CEC measurement is to extract a soil sample with neutral 1 N ammonium acetate (NH_4OAc). All of the exchangeable cations are replaced by NH_4^+ ions, and the CEC becomes saturated with NH_4^+. If this NH_4^+ saturated soil is extracted with a solution of a different salt, say 1 N KCl, the K^+ ions will replace the NH_4^+ ions. If the soil-KCl suspension is filtered, the filtrate will contain the NH_4^+ ions that were previously adsorbed by the soil. The quantity of NH_4^+ ions in the leachate is a measure of the CEC.

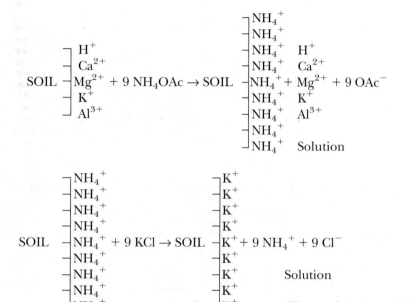

For example, suppose that the concentration of NH_4^+ in the filtrate was 270 ppm (20 g of soil extracted with 200 ml of KCl solution). The CEC is calculated as follows:

$$270 \text{ ppm } NH_4^+ = 270 \text{ mg } NH_4^+/l$$
$$(270 \text{ mg } NH_4^+/l) \times (0.2 \text{ l}/20 \text{ g soil}) = 2.7 \text{ mg } NH_4^+/g \text{ soil}$$
$$(2.7 \text{ mg } NH_4^+/g \text{ soil})/(18 \text{ mg } NH_4^+/meq) = 0.15 \text{ meq CEC/g soil}$$
$$0.15 \text{ meq CEC/g soil} \times 100/100 = 15 \text{ meq}/100 \text{ g soil}$$
$$CEC = 15 \text{ meq}/100 \text{ g}$$

The equivalent weight of NH_4^+ is given in Table 2.2.

Base Saturation

One of the important properties of a soil is its base saturation, which is defined as the percentage of total CEC occupied by basic cations (Ca^{2+}, Mg^{2+}, K^+, and Na^+). To illustrate, suppose that the following ions were measured in the 200-ml NH_4OAc extract obtained from leaching the 20 g of soil in the previous example:

$$Ca^{2+} = 100 \text{ ppm}$$
$$Mg^{2+} = 30 \text{ ppm}$$
$$K^{+} = 78 \text{ ppm}$$
$$Na^{+} = 23 \text{ ppm}$$

The equivalent weights of the cations are found in Table 2.2. The following calculations are used to express the cation concentrations in CEC units and determine the base saturation.

$$Ca^{2+} = 100 \text{ ppm} = 100 \text{ mg/l} \times (0.2 \text{ l/20 g soil})/(20 \text{ mg/meq}) \times 100/100$$
$$= 5 \text{ meq } Ca^{2+}/100 \text{ g}$$
$$Mg^{2+} = 30 \text{ ppm} = 30 \text{ mg/l} \times (0.2 \text{ l/20 g soil})/(12 \text{ mg/meq}) \times 100/100$$
$$= 2.5 \text{ meq } Mg^{2+}/100 \text{ g}$$
$$K^{+} = 78 \text{ ppm} = 78 \text{ mg/l} \times (0.2 \text{ l/20 g soil})/(39 \text{ mg/meq}) \times 100/100$$
$$= 2 \text{ meq } K^{+}/100 \text{ g}$$
$$Na^{+} = 23 \text{ ppm} = 23 \text{ mg/l} \times (0.2 \text{ l/20 g soil})/(23 \text{ mg/meq}) \times 100/100$$
$$= 1 \text{ meq } Na^{+}/100 \text{ g}$$
$$\text{Total} = 10.5 \text{ meq bases}/100 \text{ g}$$

$$\text{Base saturation \%} = (\text{total bases/CEC}) \times 100$$
$$= [(10.5 \text{ meq}/100 \text{ g})/(15 \text{ meq}/100 \text{ g})] \times 100$$
$$= 70\%$$

The % saturation with any cation may be calculated in a similar fashion. For example, from the preceding data, % Mg saturation = (2.5 meq Mg/10.5 meq CEC) × 100 = 23.8% Mg.

As a general rule, the degree of base saturation (BS %) of normal uncultivated soils is higher for arid- than for humid-region soils. In humid regions, the BS % of soils formed from limestones or basic igneous rocks is greater than that of soils formed from sandstones or acidic igneous rocks.

The availability of nutrients such as Ca^{2+}, Mg^{2+}, and K^{+} to plants increases with increasing BS %. For example, a soil with 80% BS would provide cations to growing plants far more easily than the same soil with a BS of only 40%. The relation between BS % and cation availability is modified by the nature of the soil colloids. As a rule, soils with large amounts of organic or 1:1 clays can supply nutrient cations to plants at a much lower BS % than soils high in 2:1 clays.

Base saturation is related to soil pH (Fig. 2.9). As the percentage of Ca^{2+}, Mg^{2+}, and K^{+} on the CEC increases, the pH increases. In this example pH 5.5 equals about 50% BS and pH 7.0 equals 90% BS. Although the shape of the curve varies slightly among different soils, the relationship can be helpful in evaluating lime requirements for acidic soils. For example, assume that a soil has a pH of 5.5 and a CEC of 20 meq/100 g. The grower needs to lime the soil to pH 6.5 for optimum production. Using Figure 2.9, the initial BS at pH 5.5 is about 50%. At pH 6.5 the BS is estimated to be 75%. The following calcula-

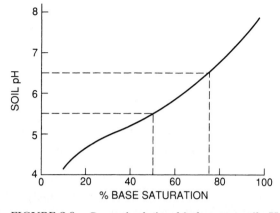

FIGURE 2.9   General relationship between soil pH and base saturation.

tion is used to estimate the lime ($CaCO_3$) required to raise soil pH from 5.5 to 6.5 (BS from 50 to 75%):

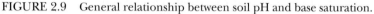

$$\text{Final BS} = 75\% = (0.75) \times 20 \text{ meq CEC}/100 \text{ g} = 15 \text{ meq}/100 \text{ g}$$
$$\text{Initial BS} = 50\% = (0.50) \times 20 \text{ meq CEC}/100 \text{ g} = \underline{10 \text{ meq}/100 \text{ g}}$$
$$\text{Total bases needed to raise pH} = 5 \text{ meq}/100 \text{ g}$$

$$CaCO_3 \text{ required* (kg/ha)} = 5 \text{ meq } CaCO_3/100 \text{ g} \times 50 \text{ mg } CaCO_3/\text{meq}$$
$$= 250 \text{ mg } CaCO_3/100 \text{ g soil}$$
$$= 0.25 \text{ g } CaCO_3/100 \text{ g}$$

$$0.25 \text{ g } CaCO_3/100 \text{ g} = X \text{ kg } CaCO_3/2 \times 10^6 \text{ kg soil}$$
$$X = 5,000 \text{ kg } CaCO_3/\text{ha} - 15 \text{ cm}$$

Anion Exchange

Anions in soil solution are also subject to adsorption to positively charged sites on clay mineral surfaces and organic matter. The positive charges responsible for adsorption and exchange of anions originate in the broken bonds, primarily in the alumina octahedral sheet, exposing OH groups on the edges of clay minerals (Fig. 2.7). Anion exchange may also occur with OH groups on the hydroxyl surface of kaolinite. Displacement of OH ions from hydrous Fe and Al oxides is considered to be an important mechanism for anion exchange, particularly in highly weathered soils of the tropics and subtropics.

The AEC increases as soil pH decreases (Fig. 2.6). Further, anion exchange is much greater in soils high in 1:1 clays and those containing hydrous Fe and Al oxides than it is in soils with predominately 2:1 clays. Montmorillonitic minerals usually have an AEC of less than 5 meq/100 g, whereas kaolinites can have an AEC as high as 40 meq/100 g at pH 4.7. The pH of most productive soils in North America is usually too high for full development of AEC.

*One hectare (ha) of soil to a depth of 15 cm weighs about 2×10^6 kg.
One acre (a) of soil to a depth of 6 in. weighs about 2×10^6 lb. = acre furrow slice (afs).

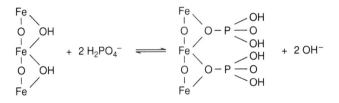

FIGURE 2.10   Chemisorption of phosphate ($H_2PO_4^-$) to iron hydroxide [$Fe(OH)_3$] minerals in soils.   *Bohn et al.*, Soil Chemistry, *John Wiley & Sons, 1979.*

Anions such as Cl^- and NO_3^- may be adsorbed, although not to the extent of $H_2PO_4^-$ and SO_4^{2-}. The order of adsorption strength is $H_2PO_4^- > SO_4^{2-} > NO_3^- = Cl^-$. In most soils $H_2PO_4^-$ is the primary anion adsorbed, although some acidic soils also adsorb significant quantities of SO_4^{2-}.

The mechanisms for anion retention in soils are much more complex than the simple electrostatic attractions involved in most cation exchange reactions. Anions may be retained by soil particles through specific adsorption or chemisorption reactions that are nonelectrostatic (Fig. 2.10).

Buffering Capacity

Plant nutrient availability depends on the concentration of nutrients in solution but, more importantly, on the capacity of the soil to maintain the concentration. The buffering capacity represents the ability of the soil to resupply an ion to the soil solution. The buffering capacity involves all of the solid components in the soil system; thus, the ions must also exist in soils as solid compounds or adsorbed to cation-anion exchange sites (Fig. 2.1). For example, when H^+ ions in solution are neutralized by liming, H^+ will desorb from the exchange sites. The solution pH is thus buffered by exchangeable H^+ and will not increase until significant quantities of exchangeable acids have been neutralized. Similarly, as plant roots absorb or remove nutrients such as K^+, exchangeable K^+ is desorbed to resupply solution K^+. With some nutrients, such as $H_2PO_4^-$, solid P minerals dissolve to resupply or buffer the solution $H_2PO_4^-$ concentration.

Soil buffer capacity (BC) can be described by the ratio of the concentrations of absorbed (ΔQ) and solution (ΔI) ions:

$$BC = \frac{\Delta Q}{\Delta I}$$

Figure 2.11 illustrates the quantity (Q) and intensity (I) relationships between two soils. Soil A has a higher BC than soil B, as indicated by the steeper slope ($\Delta Q/\Delta I$). Thus, increasing the concentration of adsorbed ion increases the solution concentration in soil B much more than that in soil A, indicating that $BC_A > BC_B$. Alternatively, decreasing the solution concentration by plant uptake decreases the quantity of ion in solution much less in soil A than in soil B.

The BC in soil increases with increasing CEC, organic matter, and other solid constituents in the soil. For example, the BC of montmorillonitic, high organic matter soils is greater than that of kaolinitic, low organic matter soils. Since CEC increases with increasing clay content, fine-textured soils will exhibit higher BC

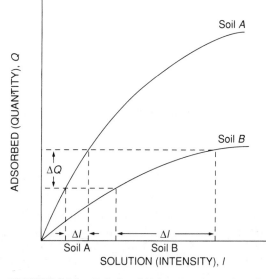

FIGURE 2.11 Relationship between quantity of adsorbed nutrient and concentration of the nutrient in solution (intensity). BC ($\Delta Q/\Delta I$) of soil A is greater than that of soil B.

than coarse-textured soils. If exchangeable K^+ decreases (e.g., as a result of plant uptake), the capacity of the soil to buffer solution K^+ is reduced. The nutrient will likely become deficient, and K^+ fertilizer will be needed to increase exchangeable K^+. Addition of P fertilizer will increase anion-exchangeable $H_2PO_4^-$, but, more importantly, some $H_2PO_4^-$ will precipitate as solid P compounds that contribute to the BC of P in soils. Thus, it is apparent that BC is a very important soil property that strongly influences nutrient availability.

Root Cation Exchange Capacity

Plant roots exhibit a CEC ranging from 10 to 30 meq/100 g in monocotyledons such as grasses and 40 to 100 meq/100 g in dicotyledons such as legumes (Table 2.3). The exchange properties of roots are attributable mainly to carboxyl groups (—COOH), similar to the exchange sites on humus (Fig. 2.8), and account for 70 to 90% of root CEC.

Legumes and other plant species with high CEC values tend to absorb divalent cations such as Ca^{2+} preferentially over monovalent cations, whereas the reverse occurs with grasses. These cation exchange properties of roots help to explain why, in grass-legume pastures on soils containing less than adequate K^+, the grass survives but the legume disappears. The grasses are considered to be more effective absorbers of K^+ than are the legumes.

TABLE 2.3   CEC of Roots

Species	CEC meq/100 g Dry Root
Wheat	23
Corn	29
Bean	54
Tomato	62

Movement of Ions from Soils to Roots

For ions to be absorbed by plant roots, they must come into contact with the root surface. There are generally three ways in which nutrient ions in soil may reach the root surface: (1) root interception, (2) mass flow of ions in solution, and (3) diffusion of ions in the soil solution. The relative importance of these mechanisms in providing nutrients to plant roots is shown in Table 2.4. The contribution of diffusion was estimated by the difference between total nutrient needs and the amounts supplied by interception and mass flow.

Root Interception

The importance of root interception as a mechanism for ion absorption is enhanced by the growth of new roots throughout the soil and perhaps also by mycorrhizal infections. As the root system develops and exploits more soil, soil solution and soil surfaces retaining adsorbed ions are exposed to the root mass, and absorption of these ions occurs by a contact exchange mechanism. Ions such as H^+ attached to the surface of root hairs may exchange with ions held on the surface of clays and organic matter because of the intimate contact between roots and soil particles. The ions held by electrostatic forces at these sites tend to oscillate within a certain volume (Fig. 2.12). When the oscillation volumes of two ions overlap, the ions exchange places. In this way Ca^{2+} on a clay surface could then presumably be absorbed by the root and utilized by the plant.

The quantity of nutrients that can come in direct contact with plant roots is the amount in a volume of soil equal to the volume of roots. Roots usually occupy 1% or less of the soil; however, roots growing through soil pores with higher than average nutrient content would contact a maximum of 3% of the available soil nutrients.

Root interception of nutrients can be enhanced by mycorrhiza, a symbiotic association between fungi and plant roots. The beneficial effect of mycorrhiza is greatest when plants are growing in infertile soils. The extent of mycorrhizal in-

TABLE 2.4   Relative Significance of the Principal Ways in Which Plant Nutrient Ions Move from Soil to the Roots of Corn

Nutrient	Amount of Nutrient Required for 150 bu/a of Corn (lb/a)	Percentage Supplied by		
		Root Interception	Mass Flow	Diffusion
Nitrogen	170	1	99	0
Phosphorus	35	3	6	94
Potassium	175	2	20	78
Calcium	35	171	429	0
Magnesium	40	38	250	0
Sulfur	20	5	95	0
Copper	0.1	10	400	0
Zinc	0.3	33	33	33
Boron	0.2	10	350	0
Iron	1.9	11	53	37
Manganese	0.3	33	133	0
Molybdenum	0.01	10	200	0

SOURCE: Barber, *Soil Bionutrient Availability*, John Wiley & Sons, New York (1984).

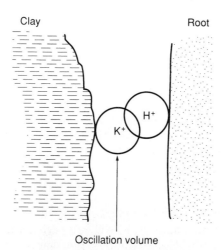

FIGURE 2.12 Conceptual model for root interception or contact exchange of nutrients between ions on soil and root exchange sites. Overlapping oscillation volumes cause exchange of H^+ on the root with K^+ on the clay mineral surface.

fection also is enhanced under conditions of slightly acidic soil pH, low P, adequate N, and low soil temperatures. The hyphal threads of mycorrhizal fungi act as extensions of plant root systems, resulting in greater soil contact. (See Figure 10.10 for a diagrammatic representation of a mycorrhizal infected root.)

The two major groups of mycorrhizas are ectomycorrhizas and endomycorrhizas. The ectomycorrhizas occur mainly in the tree species of the temperate zone but can also be found in semiarid zones. The endomycorrhizas are more widespread. The roots of most agronomic crops have vesicular arbuscular mycorrhiza. The fungus grows into the cortex. Inside the plant cells small structures known as *arbuscules,* considered to be the site of transfer of nutrients from fungi to host plants, are formed. The positive effect of inoculation of English oaks with ectomycorrhizas is shown in Figure 2.13.

The increased nutrient absorption is partly due to the larger nutrient-absorbing surface provided by the fungi. This area has been calculated to be up to 10 times that of uninfected roots. Fungal hyphae extend up to 8 cm into the soil surrounding the roots, thus increasing the absorption of nutrients such as P that do not diffuse readily to the roots (Table 2.5). Enhanced P uptake is the primary cause of improved plant growth from mycorrhiza. This improvement in growth may lead to more rapid uptake of other elements.

Mass Flow

Movement of ions in the soil solution to root surfaces by mass flow is an important factor in supplying nutrients to plants (Table 2.4). Mass flow occurs when plant nutrient ions and other dissolved substances are transported in the flow of water to the root that results from transpirational water uptake by the plant. Mass flow can also take place in response to evaporation and percolation of soil water.

The amounts of nutrients reaching roots by mass flow are determined by the rate of water flow or the water consumption of plants and the average nutrient concentrations in the soil water. Mass flow supplies an overabundance of Ca^{2+}

FIGURE 2.13 Differential growth responses of 16-week-old English oaks (*Quercus robur* L.) inoculated (left) with *Pisolithus tinctorius* (Pers.) Coker and Couch, an ectomycorrhizal former, and uninoculated (right). *Garrett, School of Forestry, Fisheries, and Wildlife, University of Missouri, Columbia, Mo.*

TABLE 2.5   Effect of Inoculation of Endomycorrhiza and of Added P on the
Content of Different Elements in Corn shoots (µg)

| Element Added | Content in Shoots (µg) | | | |
| | No P | | 25 ppm P Added | |
	No Mycorrhiza	Mycorrhiza	No Mycorrhiza	Mycorrhiza
P	750	1,340	2,970	5,910
K	6,000	9,700	17,500	19,900
Ca	1,200	1,600	2,700	3,500
Mg	430	630	990	1,750
Zn	28	95	48	169
Cu	7	14	12	30
Mn	72	101	159	238
Fe	80	147	161	277

SOURCE: Lambert et al., *J. Soil Sci.* 43:976 (1979).

and Mg^{2+} in many soils, as well as most of the mobile nutrients, such as NO_3^-, Cl^-, and SO_4^{2-} (Table 2.4). As soil moisture is reduced (increased soil moisture tension), water movement to the root surface slows down. The movement of nutrients by mass flow is reduced at low temperatures because the transpirational demands of plants is substantially less at low temperatures than high temperatures. In addition, the transport of ions in the flow of water evaporated at the soil surface diminishes at low soil temperatures.

Diffusion

Diffusion occurs when an ion moves from an area of high concentration to one of low concentration. Most of the P and K move to the root by diffusion (Table 2.4). As plant roots absorb nutrients from the surrounding soil solution, the nutrient concentration at the root surface decreases compared with the "bulk" soil solution concentration (Fig. 2.14). Therefore, a nutrient concentration gradient is established that causes ions to diffuse toward the plant root. A high plant requirement for a nutrient results in a large concentration gradient, favoring a high rate of ion diffusion from the soil solution to the root surface.

Many soil factors influence nutrient diffusion in soils; the most important one is the magnitude of the diffusion gradient. The following equation (known as *Fick's law*) describes this relationship:

$$dC/dt = De\ A\ dC/dX$$

where dC/dt = rate of diffusion (change in concentration C with time)
      dC/dX = concentration gradient (change in concentration with distance)
          De = effective diffusion coefficient (defined later)
          A = cross-sectional area through which the ions diffuse

The diffusion equation shows that the rate of nutrient diffusion (dC/dt) is directly proportional to the concentration gradient (dC/dX). As the difference in

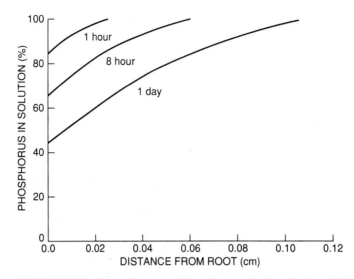

FIGURE 2.14   The influence of P uptake on the distribution of P in the soil solution as a function of distance from the root surface.

nutrient concentration between the root surface and the bulk solution increases, the rate of nutrient diffusion increases. Also, increasing the cross-sectional area for diffusion increases dC/dt. The diffusion rate is proportional to the diffusion coefficient, De, which controls how far nutrients can diffuse to the root. For a given spacing between roots, De determines the fraction of nutrients in the soil that can reach the roots during a specific period of plant growth. De is described as follows:

$$De = Dw\, \theta\ (1/T)\ (1/b)$$

$$\text{where } Dw = \text{diffusion coefficient in water}$$
$$\theta = \text{volumetric soil water content}$$
$$T = \text{tortuosity factor}$$
$$b = \text{soil BC}$$

This relationship shows that as soil moisture content (θ) increases, De increases, which results in an increase in the diffusion rate, dC/dt. As the moisture content of the soil is lowered, the moisture films around the soil particles become thinner and the diffusion of ions through these films becomes more tortuous. Transport of nutrients to the root surface is most effective at a soil moisture content corresponding to field capacity. Therefore, raising θ reduces tortuosity, or the diffusion path length, which in turn increases dC/dt.

Tortuosity (T) is also related to soil texture. Nutrients diffusing in finer-textured soils experience a more tortuous path to the root surface. As T increases with increasing clay content, $1/T$ decreases, which reduces the diffusion coefficient and thus dC/dt. Also, ions diffusing through soil moisture in clay soils are much more likely to be attracted to adsorption sites on the clay than in a sandy soil.

The diffusion coefficient in soil (De) is directly related to the diffusion coefficient for the same nutrient in water (Dw). Inherent in the Dw term is a temperature factor such that increasing temperature increases Dw, De, and then dC/dt. The diffusion coefficient is inversely related to the soil BC, b. Increasing the BC of the soil decreases De, which in turn decreases the rate of nutrient diffusion. Therefore, decreasing the BC by increasing the nutrient concentration in solution increases De, which then increases dC/dt. Increasing the solution concentration also increases the diffusion gradient, dC/dX, which contributes to the increased rate of diffusion.

Uptake of ions at the root surface, which is responsible for creating and maintaining diffusion gradients, is strongly influenced by temperature. Within the range of about 10 to 30°C, an increase of 10°C usually causes the rate of ion absorption to go up by a factor of 2 or more. Diffusion of nutrient ions is slow under most soil conditions and occurs over very short distances in the vicinity of the root surface. Typical average distances for diffusion to the root are 1 cm for N, 0.02 cm for P, and 0.2 cm for K. The mean distance between corn roots in the top 15 cm of soil is about 0.7 cm, indicating that some nutrients would need to diffuse half of this distance, or 0.35 cm, before they would be in a position to be absorbed by the plant root.

Roots do not absorb all nutrients at the same rate. Thus, certain ions may build up at the root surface, especially during periods of rapid absorption of water. This situation results in a phenomenon known as *back diffusion*, in which the concentration gradient, and hence the movement of certain ions, will be away from the root surface and back toward the soil solution. Nutrient diffusion away from the root is much less common than diffusion toward the root; however, higher concentrations of some nutrients in the rhizosphere can affect the uptake of other nutrients.

The importance of diffusion and mass flow in supplying ions to the root surface depends on the ability of the solid phase of the soil to supply the liquid phase with these ions. Solution concentrations of ions are influenced by the nature of the colloidal fraction of the soil and the degree to which these colloids are saturated with cations. For example, the ease of replacement of Ca from colloids by plant uptake varies in this order: peat > kaolinite > illite > montmorillonite. An 80% Ca-saturated 2:1 clay provides the same percentage Ca^{2+} release as a 35% Ca-saturated kaolinite or a 25% Ca-saturated peat.

Mass flow and diffusion processes are also important in fertilizer management. Soils that may exhibit low diffusion rates because of high BC, low soil moisture, or high clay content may require the application of immobile nutrients near the roots to maximize nutrient availability and plant uptake.

Ion Absorption by Plants

Plant uptake of ions from the soil solution can be described by *passive* and *active* processes, where ions passively move to a "boundary" through which ions are actively transported to organs in plant cells that metabolize the nutrient ions. Solution composition or ion concentrations outside and inside of the boundary are controlled by different processes, each essential to plant nutrition and growth.

Passive Ion Uptake

A considerable fraction of the total volume of the root is accessible for the passive absorption of ions. The *outer* or *apparent free space*, where the diffusion and exchange of ions occur, is located in the walls of the epidermal and cortical cells of the root and in the film of moisture lining the intercellular spaces (Fig. 2.15). Cell walls of the cortex are the principal locale of the outer space. This extracellular space is outside of the outermost membrane, the Casparian strip, which is a barrier to diffusion and exchange of ions.

Ions in soil solution enter the root tissue through diffusion and ion exchange processes. The concentration of ions in the apparent free space is normally less than the bulk solution concentration; therefore, diffusion occurs with the concentration gradient, from high to low concentration. Interior surfaces of cells in the cortex are negatively charged, attracting cations. Cation exchange readily occurs along the extracellular surfaces and explains why cation uptake usually exceeds anion uptake. To maintain electrical neutrality, the root cells release H^+, decreasing soil solution pH near the root surface.

Diffusion and ion exchange are passive processes because uptake into the outer space is controlled by ion concentration (diffusion) and electrical (ion exchange) gradients. These processes are nonselective and do not require energy produced from metabolic reaction within the cell (Table 2.6). Passive uptake occurs outside of the Casparian strip and plasmalemma, which are the boundary membranes, or barriers to diffusion and ion exchange.

Extracellular spaces exist in the mesophyll cells of leaves where ions are able to diffuse and exchange. Most of the nutrient ions reach the outer space of leaves via the xylem from the roots. Mineral ions in rain, irrigation water, and foliar applications penetrate leaves through the stomata and cuticles to reach the interior of leaves, where they become available for absorption by mesophyll cells.

The movement of ions from roots to shoots is determined by the rates of water absorption and transpiration, suggesting that mass flow may be important in the movement of ions.

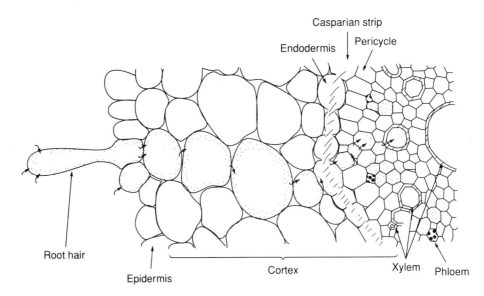

FIGURE 2.15   Cross section of a plant root. Site of passive uptake is the apparent free space, which is outside of the Casparian strip in the cortex.   *Epstein*, Mineral Nutrition of Plants, *Wiley Interscience, 1972.*

TABLE 2.6   Characteristics of Ion Movement into Inner and Outer Cell Spaces

Outer	Inner
Diffusion and exchange adsorption	Ion-binding compounds or carriers
Uptake occurs quickly	Uptake occurs slowly
Ions stoichiometrically exchangeable	Ions essentially nonexchangeable
Not highly selective	Specific with regard to site and entry
Nonmetabolic	Dependent on aerobic metabolism
Ions in solution or adsorbed in outer space	Ions in vacuoles and partly in cytoplasm

SOURCE: Gauch, *Annu. Rev. Plant Physiol.*, 8:31 (1957). Reprinted with permission of the author and the publisher, Annual Reviews, Inc., Palo Alto, Calif.

Active Ion Uptake

The membrane that provides the boundary between the apparent free space and the interior contents of the cell is the plasmalemma (Fig. 2.16). Ions that were passively absorbed occupy spaces between cells; however, the plasmalemma prevents passive transport of nutrients into the cell. Since ion concentrations are greater inside than outside of the cell, transport of ions across the plasmalemma is strictly against an electrochemical gradient. Therefore, ion transport across the plasmalemma into the cytoplasm requires energy derived from cell metabolism. Other organs within the cell are also surrounded by impermeable membranes. For example, the tonoplast is the barrier membrane for the vacuole, which regulates cell water content and serves as the reservoir for inorganic ions, sugars, and amino acids.

The ion-carrier mechanism involves a metabolically produced substance that combines with free ions. This ion-carrier complex can then cross membranes and

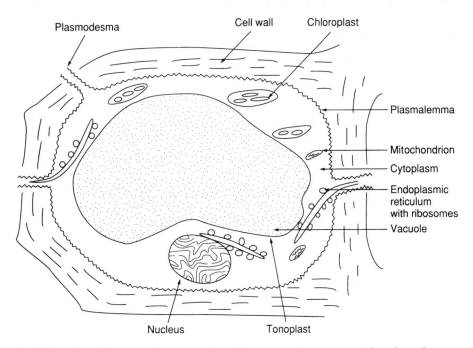

FIGURE 2.16   Diagram of a plant cell. Active ion uptake occurs at the plasmalemma. *Mengel and Kirkby*, Principles of Plant Nutrition, *IPI, 1987.*

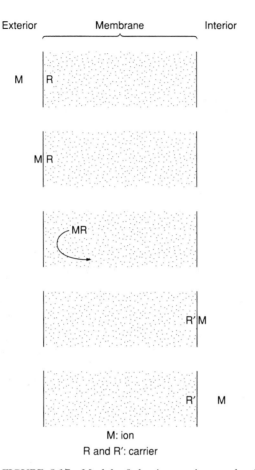

FIGURE 2.17   Model of the ion-carrier mechanism for ion transport across the plasmalemma.   *Epstein,* Mineral Nutrition of Plants, *Wiley Interscience, 1972.*

other barriers not permeable to free ions. After the transfer is accomplished, the ion-carrier complex is broken, the ion is released into the inner space of the cell, and the carrier is believed in some cases to be restored (Fig. 2.17).

Two different mechanisms are involved in the transport of ions into the inner space. For some ions a mechanism 1 operates at very low concentrations, while at high concentrations above about 1 mM a mechanism 2, with different properties, comes into play. The nature of the ion-carrier compounds is not known, but it is likely that they are directly connected with proteins or are themselves proteins.

Active ion transport is also a selective process such that specific ions are transported, or "carried," across the plasmalemma by specific carrier mechanisms. Although K, rubidium (Rb), and cesium (Cs) compete for the same carrier, they do not compete with elements such as Ca, strontium (Sr), and barium (Ba). The last three elements do, however, compete among themselves for another carrier. Selenium (Se) will compete with SO_4^{2-} but not with $H_2PO_4^{-}$ or with other monovalent anions. Interestingly, $H_2PO_4^{-}$ and HPO_4^{2-} apparently have separate carriers and do not compete with one another for entry into the inner space.

TABLE 2.7   Differential Response of Soybean Varieties to Fe Stress

Cultivar	Yield (g Dry Weight)		Fe Concentration in Tops (ppm)	
	Quinlan	Millville	Quinlan	Millville
North				
Amsoy 71	1.17	1.50	32	38
Hodgson	1.55	1.96	43	43
South				
Forrest	1.07	1.60	20	22
Bragg	1.46	1.87	39	45

SOURCE: Brown and Jones, *Agron, J.,* 69:401 (1977).

Many aspects of absorption, transport, and utilization of mineral nutrients in plants are under genetic control. Genotypes within a species may differ in the rate of absorption and translocation of nutrients, efficiency of metabolic utilization, tolerance to high concentrations of elements, and other factors. The differential tolerance of soybean varieties to Fe stress is exhibited in Table 2.7. The Bragg cultivar was able to absorb sufficient Fe to grow satisfactorily, whereas the Forrest cultivar developed severe Fe deficiencies. In addition to genetically controlled differences, the morphology of roots can significantly influence the uptake of nutrients. Some varieties are better able to exploit soil for nutrients and moisture because of larger or more finely branched root systems.

Selected References

BOHN, H. L., B. L. MCNEAL, and G. A. O'CONNOR. 1979. *Soil Chemistry.* John Wiley & Sons, New York.

EPSTEIN, E. 1972. *Mineral Nutrition of Plants: Principles and Perspectives.* John Wiley & Sons, New York.

MENGEL, K., and E. A. KIRKBY. 1987. *Principles of Plant Nutrition.* International Potash Institute, Bern, Switzerland.

RENDIG, V. V., and H. M. TAYLOR. 1989. *Principles of Soil–Plant Relationships.* McGraw-Hill, New York.

TAN, K. H. 1982. *Principles of Soil Chemistry.* Marcel Dekker, New York.

CHAPTER 3

Soil Acidity and Alkalinity

General Concepts

In aqueous systems, an acid is a substance that donates H^+ to some other substance. Conversely, a base is any substance that accepts H^+. An acid, when mixed with water, ionizes into H^+ and the accompanying anions, as represented by the dissociation of acetic acid (CH_3COOH) or hydrochloric acid (HCl):

$$CH_3COOH \leftrightharpoons CH_3COO^- + H^+$$

$$HCl \leftrightharpoons Cl^- + H^+$$

Dissociation of H^+ in a strong acid such as HCl is 100%, whereas only 1% H^+ dissociation occurs in a weak acid such as CH_3COOH.

The H^+ ions, or *active* acidity, increase with the strength of the acid. The undissociated acid is considered *potential* acidity. The total acidity of a solution is the sum of the active and potential acid concentrations. For example, suppose that the active and potential acidity are 0.099 M and 0.001 M,* respectively. The total acid concentration is 0.100 M, and since the H^+ activity is nearly equal to the total acidity, this would be a strong acid.

With weak acids, the H^+ activity is much less than the potential acidity. For example, a 0.100-M weak acid that is 1% dissociated means that the H^+ activity is $0.1 \times 0.01 = 0.001$ M.

Pure water undergoes slight self-ionization:

$$H_2O \leftrightarrow H^+ + OH^-$$

The H^+ actually attaches to another H_2O molecule to give

$$H_2O + H^+ \leftrightarrow H_3O^+$$

---

*M = molarity (moles/liter)

Since both H^+ and OH^- are produced, H_2O is both a weak acid and a weak base. The concentration of H^+ (or H_3O^+) and OH^- in pure H_2O, not in equilibrium with atmospheric CO_2, is 10^{-7} M. The product of H^+ and OH^- concentration, as shown in the following equation, is the dissociation constant for water, or K_w.

$$[H^+][OH^-] = [10^{-7}][10^{-7}] = 10^{-14} = K_w$$

The pH of H_2O in equilibrium with atmospheric CO_2 is about 5.7 because of the following reaction:

$$H_2O + CO_2 \xrightleftharpoons{H_2CO_3} H^+ + HCO_3^-$$

Adding an acid to H_2O will increase $[H^+]$, but $[OH^-]$ would decrease because K_w is a constant 10^{-14}. For example, in a 0.1-M HCl solution, the $[H^+]$ is 10^{-1} M; thus the $[OH^-]$ is

$$K_w = [H^+][OH^-] = 10^{-14}$$

$$[10^{-1}][OH^-] = 10^{-14}$$

$$[OH^-] = 10^{-13}M$$

The $[H^+]$ in solution can be conveniently expressed using pH and is defined as follows:

$$pH = \log 1/[H^+] = -\log[H^+]$$

Each unit increase in pH represents a 10-fold decrease in $[H^+]$ (Table 3.1). A solution with $[H^+] = 10^{-5}$ M has a pH of 5.0.

Solutions with pH < 7 are acidic, those with pH > 7 are basic, and those with pH = 7 are neutral. The pH represents the H^+ concentration in solution and does not measure the undissociated or potential acidity. For example, the pH of completely dissociated 0.1-M HCl is 1.0, while the pH of 0.1-M CH_3COOH, a weak acid, is 3.0. Similarly, the pH of 0.1-M NaOH, a strong base, is 13.0, while the pH of 0.1-M NH_4OH, a weak base, is 11.0.

TABLE 3.1   Relationship between pH and $[H^+]$ Concentration

Conc. of H_3O^+ (mol/l)	pH	Conc. of H_3O^+ (mol/l)	pH
10^{-1}	1	10^{-8}	8
10^{-2}	2	10^{-9}	9
10^{-3}	3	10^{-10}	10
10^{-4}	4	10^{-11}	11
10^{-5}	5	10^{-12}	12
10^{-6}	6	10^{-13}	13
10^{-7}	7	10^{-14}	14

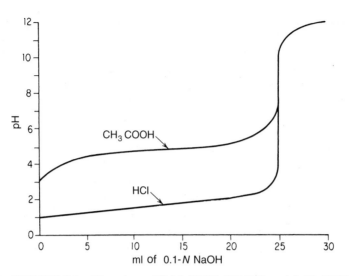

FIGURE 3.1   Titration of 0.10-N CH$_3$COOH and 0.10-N HCl with 0.1-N NaOH.

When acids and bases are combined, both are neutralized, forming a salt and water:

$$HCl \quad + \quad NaOH \longleftrightarrow H_2O + Na^+ + Cl^-$$

$$H^+ \quad Cl^- \qquad Na^+ \qquad OH^-$$

If a given quantity of acid is titrated with a base and the pH of the solution is determined at intervals during the titration, a curve is obtained by plotting pH against the amounts of base added (Fig. 3.1). Titration curves for strong and weak acids differ markedly. The neutralization reaction of HCl with NaOH is given in the previous equation, and that of CH$_3$COOH with NaOH is

$$CH_3COOH \quad + \quad NaOH \longleftrightarrow H_2O + CH_3COO^- + Na^+$$

$$CH_3COO^- + \quad H^+ \quad Na^+ \quad OH^-$$

Buffers

Buffers or buffer systems can maintain the pH of a solution within a narrow range when small amounts of acid or base are added. *Buffering* defines the resistance to a change in pH (see Chapter 2). An example of a buffer system is CH$_3$COOH and CH$_3$COONa:

$$CH_3COOH \rightleftarrows H^+ + CH_3COO^-$$

$$CH_3COONa \rightleftarrows Na^+ + CH_3COO^-$$

For example, a solution containing $1\text{-}M$ CH_3COOH and $1\text{-}M$ CH_3COONa has a pH of 4.6, compared with pH 2 for CH_3COOH alone. Adding the highly dissociated CH_3COONa to CH_3COOH increases the CH_3COO^- concentration, which shifts the equilibrium to form the undissociated CH_3COOH. The pH remains at 4.6 even with a 10-fold dilution with H_2O; however, dilution of $1\text{-}M$ CH_3COOH would raise the pH to 3.0. Thus, adding CH_3COONa to CH_3COOH buffers the solution pH.

If 10 ml of $1\text{-}M$ HCl is added to the CH_3COOH/CH_3COONa buffer solution, the pH will decrease to only 4.5, because the additional H^+ will shift the $CH_3COOH \leftrightarrow H^+ + CH_3COO-$ equilibrium to the left and the decreased CH_3COO^- will be replaced by the CH_3COO^- from the dissociation of CH_3COONa (equilibrium shift to the right in $CH_3COONa \leftrightarrow CH_3COO^- + Na^+$).

Conversely, if 10 ml of $1\text{-}M$ NaOH is added, the OH^- neutralizes H^+ to form water. Because of the large supply of undissociated CH_3COOH, the equilibrium shifts to the right, replacing the neutralized H^+; thus, the pH increases to only 4.7.

Soils behave like buffered weak acids, with the CEC of humus and clay minerals providing the buffer for soil solution pH.

Soil Acidity

About 25 to 30% of the soils in the world are classified as acidic and represent some of the world's most important food-producing regions (Fig. 3.2). In the United States most of the acidic soils occur in the east and northwest regions, where annual precipitation usually exceeds 24 in.

Sources of Soil Acidity

Soil OM, or humus, contains reactive carboxylic and phenolic groups that behave as weak acids releasing H^+ (Fig. 2.8). The soil OM content varies with the environment, vegetation, and soil; thus, its contribution to soil acidity varies accordingly. In peat and muck soils and in mineral soils containing large amounts of OM, organic acids contribute significantly to soil acidity.

CLAY AND OXIDE MINERALS   Clay minerals such as kaolinite (1:1) and montmorillonite (2:1) can buffer soil pH. The dissociation of H^+ from broken edges of clay minerals and Al and Fe oxide surfaces contributes to pH buffering in soil. High-clay- and/or high-OM-content soils exhibit greater buffering capacity than sandy and/or low-OM soils. The buffering capacity associated with clay minerals, Al and Fe oxides, and OM are as follows:

On Al and Fe oxides:

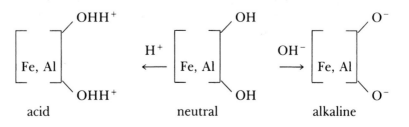

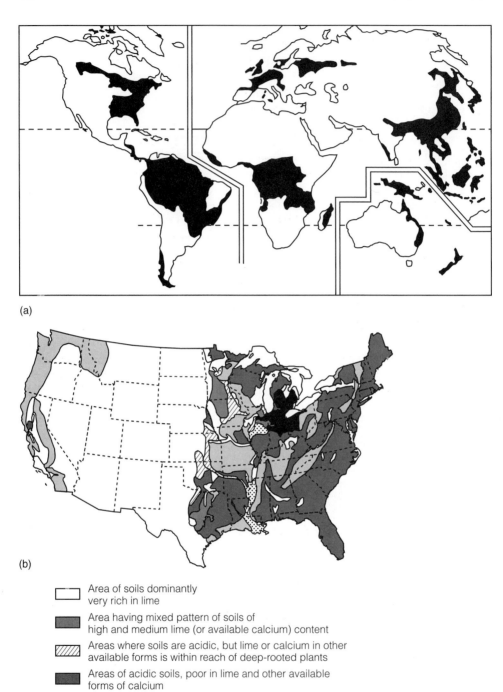

(a)

(b)

☐ Area of soils dominantly
very rich in lime

■ Area having mixed pattern of soils of
high and medium lime (or available calcium) content

▨ Areas where soils are acidic, but lime or calcium in other
available forms is within reach of deep-rooted plants

■ Areas of acidic soils, poor in lime and other available
forms of calcium

▨ Area where soils and parent materials are dominantly fairly
rich in lime, though surface soil is somewhat acidic in places

▨ Area having mixed pattern of soils of medium and
low lime content

■ Areas having mixed pattern of high, medium, and
low lime content

FIGURE 3.2   Major regions in the world (a) and the United States (b) with predominantly acidic soils.

On clay minerals:

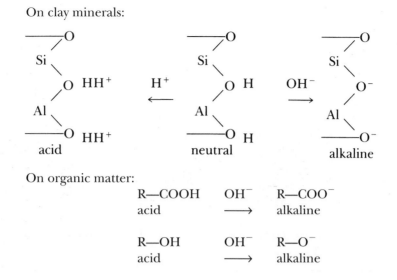

On organic matter:

$$R—COOH \quad OH^- \quad R—COO^-$$
acid $\quad\quad\quad \longrightarrow \quad$ alkaline

$$R—OH \quad OH^- \quad R—O^-$$
acid $\quad\quad\quad \longrightarrow \quad$ alkaline

AL AND FE POLYMERS The Al^{3+} ions displaced from the CEC are hydrolyzed to hydroxyaluminum complexes. Hydrolysis of Al^{3+} liberates H^+ and lowers pH unless there is a source of OH^- to neutralize H^+. Each successive step occurs at a higher pH. Figure 3.3 illustrates the range in pH wherein the various Al hydrolysis species predominate. The following reactions illustrate the Al hydrolysis:

$$Al^{3+} + H_2O \rightarrow Al(OH)^{2+} + H^+$$

$$Al(OH)^{2+} + H_2O \rightarrow Al(OH)_2^{+} + H^+$$

$$Al(OH)_2^{+} + H_2O \rightarrow Al(OH)_3^{0} + H^+$$

$$Al(OH)_3^{0} + H_2O \rightarrow Al(OH)_4^{-} + H^+$$

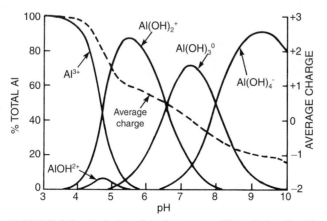

FIGURE 3.3 Relationship between pH and the distribution and average charge of soluble Al species.

If a base is added to a soil, H^+ will be neutralized first. When more of the base is added, the Al^{3+} hydrolyzes, with the production of H^+ in amounts equivalent to the Al^{3+} present. It should be noted that insoluble $Al(OH)_3$ will precipitate at $pH > 6.5$ whenever the $Al(OH)_3$ solubility product is exceeded.

Like H_2O, Al^{3+} can function as either an acid or a base, as illustrated in the following equations:

Al as a base:

$$Al(OH)_3 + H^+ \rightleftarrows Al(OH)_2{}^+ + H_2O$$

$$Al(OH)_2{}^+ + H^+ \rightleftarrows Al(OH)^{2+} + H_2O$$

$$Al(OH)_2{}^{2+} + H^+ \rightleftarrows Al^{3+} + H_2O$$

$$Al(OH)_3 + 3H^+ \rightleftarrows Al^{3+} + 3H_2O$$

Al as an acid:

$$Al^{3+} + OH^- \rightleftarrows Al(OH)^{2+}$$

$$Al(OH)^{2+} + OH^- \rightleftarrows Al(OH)_2{}^+$$

$$Al(OH)_2{}^+ + OH^- \rightleftarrows Al(OH)_3$$

Al as an anion:

$$Al(OH)_3 + OH^- \rightleftarrows Al(OH)_4{}^-$$

The hydroxy Al ions combine to form large, multicharged polymers. Polymerization is favored in the presence of clay surfaces. The mechanism for polymer formation is the sharing of OH^- groups by adjacent Al^{3+} ions, as represented in the following equations:

$$\left[(H_2O)Al\begin{smallmatrix}OH\\ \\H_2O\end{smallmatrix}\right]^{2+} + \left[\begin{smallmatrix}H_2O\\ \\HO\end{smallmatrix}Al(H_2O)_4\right]^{2+} \rightarrow \left[(H_2O)_4Al\begin{smallmatrix}OH\\ \\OH\end{smallmatrix}Al(H_2O)_4\right]^{4+} + 2H_2O$$

$$\downarrow$$

$$\left[(H_2O)_4Al\begin{smallmatrix}O\\ \\O\end{smallmatrix}Al(H_2O)_4\right]^{2+} + 2H^+$$

The Al polymers have high positive charge and are essentially nonexchangeable. CECs of `soil can be affected by the adsorption of the positively charged Al polymers. At high pH, the CEC is increased because $Al(OH)_3$ precipitates and positively charged Al polymer formation is reduced or is nonexistent (see the increasing negative charge with increasing pH in Fig. 3.3). Decreasing the soil pH will increase positively charged Al polymer formation, and adsorption to clay surfaces will decrease CEC (see the increasing positive charge in Fig. 3.3).

Fe hydrolysis is similar to that of Al.

$$Fe^{3+} + H_2O \leftrightharpoons Fe(OH)^{2+} + H^+$$

Although this reaction is more acidic than Al hydrolysis, the acidity is buffered by Al hydrolysis reactions. Thus, Fe hydrolysis has little effect on soil pH until most of the soil Al has reacted.

Al and Fe polymers occur as amorphous or crystalline colloids, coating the clay and other mineral surfaces. They are also held between the lattices of expanding clay minerals, preventing collapse of these lattices as water is removed during drying.

SOLUBLE SALTS   Acidic, neutral, or basic salts in the soil solution originate from mineral weathering, OM decomposition, or addition as fertilizers and manures. The cations of these salts will displace adsorbed Al^{3+} in acidic soils and thus decrease soil solution pH. Divalent cations have a greater effect on lowering soil pH than monovalent metal cations (see the lyotropic series in Chapter 2, page 22).

Band-applied fertilizer will result in a high soluble-salt concentration in the affected soil zone, which will decrease pH through Al hydrolysis. With high rates of band-applied fertilizer in soils with pH < 5.0–5.5, this acidification may be detrimental to plant growth.

CARBON DIOXIDE   In calcareous soils the pH is influenced by the partial pressure of CO_2 in the soil atmosphere. The pH of a soil containing free $CaCO_3$ in equilibrium with atmospheric CO_2 is 8.5; however, increased CO_2 in soil air causes the pH to decrease to 7.3–7.5. Decomposition of organic residues and root respiration increase CO_2 in soil air, which combines with water to produce H^+ and lower pH according to the following:

$$H_2O + CO_2 \leftrightarrow H_2CO_3$$

$$H_2CO_3 \leftrightarrow H^+ + HCO_3^-$$

Factors Affecting Soil Acidity

Increasing soil acidity in crop production systems is caused by (1) use of commercial fertilizers, especially NH_4^+ sources that produce H^+ during nitrification; (2) crop removal of cations in exchange for H^+; (3) leaching of cations being replaced first by H^+ and subsequently by Al^{3+}; and (4) decomposition of organic residues. Natural acidification of soils is enhanced with increasing rainfall since rain has a pH of 5.7 or less, depending on pollutants such as SO_2, NO_2, and others.

ACIDITY AND BASICITY OF FERTILIZERS   Fertilizer materials vary in their soil reaction pH. Nitrate sources carrying a basic cation should be less acid forming than NH_4^+ sources. Compared with P fertilizers, those containing or forming NH_4^+ exhibit greater effect on soil pH (Table 3.2).

Phosphoric acid released from dissolving P fertilizers such as triple superphosphate (TSP) and monoammonium phosphate (MAP) can temporarily acidify localized zones at the site of application. The former material will reduce pH to as low as 1.5, whereas the latter will decrease pH to approximately 3.5; however, the quantity of H^+ produced is very small and has little long-term effect on pH. Diammonium phosphate (DAP) will initially raise the pH of the soil to about 8, unless the initial soil pH is greater than the pH of the fertilizer (see Chapter 5). Acidity produced by the nitrification of the NH_4^+ in DAP will offset this initial pH increase.

The method for estimating the acidity or basicity of fertilizers is based on the following assumptions:

1. The acidifying effect of fertilizer is caused by all of the contained S and Cl^-, one-third of the P, and one-half of the N.
2. The presence of Ca, Mg, K, and Na in the fertilizer will slightly increase or cause no change in soil pH.
3. Half of the fertilizer N is taken up as NO_3^- and accompanied by equivalent amounts of K^+, Ca^{2+}, Mg^{2+}, or Na^+. Uptake of the other half is associated with H^+ as the counterion or exchanged for HCO_3^- from plant roots.

This method assumes that plant growth reduces the potential or theoretical amount of acidity produced by nitrification of NH_4^+ fertilizers because of unequal absorption of cations and anions by the crop.

Significant amounts of NH_4^+ utilized directly by plants, with resultant acidification of soil close to roots, is not considered. Accordingly, 1.8 lb of pure $CaCO_3$ would be required to neutralize the acidity resulting from the addition of each pound of NH_4-N (Table 3.2). When cation-anion uptake effects are not considered, 3.57 lb of $CaCO_3$ are needed to neutralize the acidity produced per pound of NH_4-N nitrified to NO_3^-.

REMOVAL OF BASIC CATIONS   Since solutions containing salts must be electrically neutral (equal positive and negative charges), water leaching below the root zone will contain both cations and anions. Every pound of N as NO_3^- leached from the soil takes with it 3.57 lb of $CaCO_3$ or its equivalent in cations.

Crop uptake of cations can either reduce or increase the soil acidity produced by nitrification of NH_4^+ from fertilizers, crop and animal wastes, or OM. These variations are explained by differences in N and excess bases (EB) taken up by the plant. EB are defined as total cation (Ca^{2+}, Mg^{2+}, K^+, and Na^+) uptake minus total anion (Cl^-, SO_4^{2-}, NO_3^-, and $H_2PO_4^-$) uptake. Plants with an EB/N ratio below 1.0 decrease the acidity formed by nitrification, whereas those with a ratio above this value increase acidity. Only a few crops (e.g., buckwheat and spinach) have values slightly above 1.0. Cereal and grass crops have average ratios of 0.43 and 0.47, respectively, meaning that only 43 and 47%, respectively, of N uptake is acid forming.

TABLE 3.2 Soil Acidity Produced by N Fertilizers

N Source	Nitrification Reaction	Residual Soil Acidity				Official Value*
		Maximum		Minimum		
		Acid Residue	$CaCO_3$ Equiv.	Acid Residue	$CaCO_3$ Equiv.	
			kg $CaCO_3$/kg of N		— kg $CaCO_3$/kg of N —	
Anhydrous ammonia	$NH_3(g) + 2O_2 \rightarrow H^+ + NO_3^- + H_2O$	H^+ NO_3^-	$50/14 = 3.6$	None	0	1.8
Urea	$(NH_2)_2CO + 4O_2 \rightarrow$ $2H^+ + 2NO_3^- + CO_2 + H_2O$	$2H^+$ $2NO_3^-$	$100/28 = 3.6$	None	0	1.8
Ammonium nitrate	$NH_4NO_3 + 2O_2 \rightarrow$ $2H^+ + 2NO_3^- + H_2O$	$2H^+$ $2NO_3^-$	$100/28 = 3.6$	None	0	1.8
Ammonium sulfate	$(NH_4)_2SO_4 + 4O_2 \rightarrow$ $4H^+ + 2NO_3^- + SO_4^{2-} + 2H_2O$	$4H^+$ $2NO_3^-$ SO_4^{2-}	$200/28 = 7.2$	$2H^+$ SO_4^{2-}	$100/28 = 3.6$	5.4
Monoammonium phosphate	$NH_4H_2PO_4 + O_2 \rightarrow$ $2H^+ + NO_3^- + H_2PO_4^- + H_2O$	$2H^+$ NO_3^- $H_2PO_4^-$	$100/14 = 7.2$	H^+ $H_2PO_4^-$	$50/14 = 3.6$	5.4
Diammonium phosphate	$(NH_4)_2HPO_4 + O_2 \rightarrow$ $3H^+ + 2NO_3^- + H_2PO_4^- + H_2O$	$3H^+$ $2NO_3^-$ $H_2PO_4^-$	$150/28 = 5.4$	H^+ $H_2PO_4^-$	$50/28 = 1.8$	3.6

*Value adopted by the Association of Official Analytical Chemists (Pierre, 1934).

SOURCE: Adams, *Soil Acidity and Liming*, no. 12, p. 234. ASA (1984).

TABLE 3.3   Effect of N Sources and Lime on Soil Profile Acidity in Two 2-Year Cotton-Corn
Rotation Experiments in Alabama after 32 Years, 1930–62

| | Soil pH by Soil Depth after 32 Years* | | | | | | | |
| | Dothan ls | | | | Lucedale sl | | | |
N and Lime Treatment†	0–15 cm	15–30 cm	30–45 cm	45–60 cm	0–15 cm	15–30 cm	30–45 cm	45–60 cm
$(NH_4)_2SO_4$	4.8	4.8	4.3	4.3	4.8	4.8	4.9	4.9
$(NH_4)_2SO_4$ + lime‡	6.2	5.4	4.7	4.6	6.0	5.5	5.3	5.1
$NaNO_3$	5.7	5.5	5.1	5.0	5.8	5.5	5.3	5.1
$NaNO_3$ + lime‡	6.4	6.4	6.3	6.3	6.8	6.6	5.8	5.2

*Initial surface soil pH was 6.0 at both sites.

†Nitrogen rate was 40 kg/ha during 1930 to 1945 and 53 kg/ha during 1946 to 1962.

‡Basic slag was lime source; it was applied annually at a rate of 500 kg/ha during 1930 to 1945 and at a rate of 800 kg/ha during 1946 to 1962.

SOURCE: Adams, *Soil Acidity and Liming,* no. 12, p. 238. ASA (1984).

LONG-TERM EFFECT   Soil acidity does not develop in a year or two, but the extent and rate of pH decline vary among soils. A problem might develop in 5 years on a sandy soil or 10 years on a silt loam soil but might take 15 years or more on a clay loam. Soil acidity is an increasing problem because lime use has been decreasing, while N use and yields have been increasing. An example of long-term use of $NaNO_3$ and $(NH_4)_2SO_4$ with and without lime is shown in Table 3.3. With $(NH_4)_2SO_4$ soil pH in the surface 15 cm decreased 1.6–1.8 units after 32 years compared with $(NH_4)_2SO_4$ plus lime. With $NaNO_3$ pH declined only 0.7–1.0 pH units, in contrast with $NaNO_3$ plus lime.

In long-term studies in Kansas, surface soil pH decreased more than 2 pH units (6.5 to 4.1 pH) with 40 years of 224-kg N/ha applied annually as NH_4NO_3 to bromegrass (Fig. 3.4). With only 20 years of N fertilization, surface soil pH decreased only 1 pH unit (6.5 to 5.5 pH).

The Soil as a Buffer

Soil behaves like a weak acid that will buffer the pH. In acidic soils adsorbed Al^{3+} will maintain an equilibrium with Al^{3+} in the soil solution, which hydrolyzes to produce H^+.

$$\text{clay} \begin{bmatrix} Al^{3+} \\ Ca^{2+} \\ Mg^{2+} \\ K^+ \\ Al^{3+} \end{bmatrix} \rightleftarrows Al^{3+} + H_2O \rightleftarrows AlOH^{2+} + H^+$$

If the H^+ is neutralized by adding a base and the Al^{3+} in solution precipitates as $Al(OH)_3$, more exchangeable Al^{3+} will desorb to resupply solution Al^{3+}. Thus, the pH remains the same or is buffered. As more base is added, the preceding reaction continues, with more adsorbed Al^{3+} neutralized and replaced on the CEC with the cation of the added base. As a result, soil pH gradually increases.

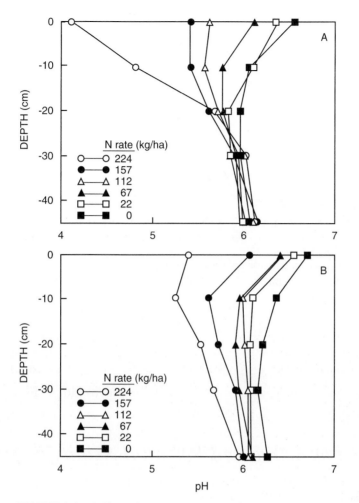

FIGURE 3.4   Soil pH decreases dramatically with increasing N rate and years of application. Soil fertilized annually since 1946 (A) and fertilized annually from 1946 to 1965 (no N applied since 1965). Soils were sampled in 1985.   *Schwab et al., SSSAJ, 53:1412–1417, 1990.*

The reverse of the preceding reaction also occurs. As acid is continually added, OH^- in the soil solution is neutralized. Gradually, the $Al(OH)_3$ dissolves, which increases Al^{3+} in the solution and subsequently on the CEC. As the reaction continues, soil pH continuously but slowly decreases as the Al^{3+} replaces adsorbed basic cations.

The total amount of clay and OM in a soil and the nature of the clay minerals determine the extent to which soils are buffered. Soils containing large amounts of clay and OM are highly buffered and require larger amounts of lime to increase the pH than soils with a lower BC. Sandy soils with small amounts of clay and OM are poorly buffered and require only small amounts of lime to effect a given change in pH. Soils containing mostly 1:1 clays (Ultisols and Oxisols) are generally less buffered than soils with principally 2:1 clay minerals (Alfisols and

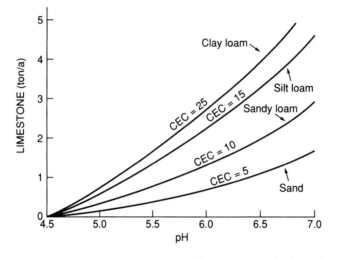

FIGURE 3.5   Approximate tons of limestone required to raise the pH of a 17-cm (7-in.) layer of soil of four textural classes with typical cation exchange capacities (CEC) in milliequivalents per 100 g of soil.   *Modified from Christenson and Doll,* Ext. Bul. *471, 1979; and Hanson,* Science and Technology Guide 9102, Agronomy, *11, 1977.*

Mollisols). For example, an increasing lime requirement with an increasing clay content and CEC is shown in Figure 3.5.

Determination of Active and Potential Acidity in Soils

Active Acidity

Currently, the most accurate and widely used method involves measuring pH in a saturated paste or a more diluted soil-water mixture with a pH meter and glass electrode. Soil pH is a useful indicator of the presence of exchangeable Al^{3+} and H^+. Exchangeable H^+ is present at pH < 4, while exchangeable Al^{3+} occurs predominantly at pH 4–5.5. Al polymers occur in the pH range 5.5–7.0.

Increasing the dilution of the soil from saturation to 1:1 to 1:10 soil to water ratios increases the measured pH compared with the pH of a saturated paste. To minimize differences in salt concentration among soils, some laboratories dilute the soil with 0.01 $CaCl_2$ instead of water. Adding Ca^{2+} decreases the pH compared with soil diluted with water. Changes in measured pH with dilution and added salt are generally small, ranging between 0.1 and 0.5 pH unit.

Potential Acidity

Soil pH measurements are excellent indicators of soil acidity or basicity; however, as an indicator of active acidity, soil pH does not measure potential acidity. Quantifying potential soil acidity requires titrating the soil with a base, which can be used to determine the lime requirement or quantity of $CaCO_3$ needed to increase the pH to a desired level. Thus, the lime requirement of a soil is related not only to the soil pH but also to its BC or CEC (Fig. 3.5). High-clay and/or high-OM soils have higher BCs and have a high lime requirement, whereas coarse-textured soils with little or no clay and OM have a low BC and a low lime requirement.

An example liming calculation for two soils with CEC = 20 meq/100 g and 10 meq/100 g is shown in the following:

$$\text{Clay soil: CEC} = 20 \text{ meq}/100 \text{ g}$$

$$\text{Initial pH} = 5.0 \text{ and } \% \text{ BS} = 50\%$$

$$\text{Final pH} = 6.5 \text{ and } \% \text{ BS} = 80\%$$

Need to neutralize 30% of acids on CEC (80 − 50%); thus

$$(0.30)\left(\frac{20 \text{ meq/CEC}}{100 \text{ g soil}}\right) = \frac{6.0 \text{ meq acids}}{100 \text{ g soil}} = \frac{6.0 \text{ meq CaCO}_3}{100 \text{ g soil}}$$

Therefore, the quantity of pure $CaCO_3$ needed is

$$\frac{6.0 \text{ meq CaCO}_3}{100 \text{ g soil}}\left(\frac{50 \text{ mg CaCO}_3}{\text{meq}}\right) = \frac{300 \text{ mg CaCO}_3}{100 \text{ g soil}}$$

Thus,

$$\frac{300 \text{ mg CaCO}_3}{100 \text{ g soil}}\left(\frac{1 \text{g}}{1,000 \text{ mg}}\right) = \frac{0.3 \text{ g CaCO}_3}{100 \text{ g soil}} = \frac{0.30 \text{ lb CaCO}_3}{100 \text{ lb soil}}$$

$$\frac{0.3 \text{ lb CaCO}_3}{100 \text{ lb soil}}\left(\frac{2 \times 10^4}{2 \times 10^4}\right) = \frac{6,000 \text{ lb CaCO}_3}{\text{afs}}$$

Assuming the same initial and final pH and % BS, the $CaCO_3$ required for a sandy loam soil (CEC = 10 meq/100 g) would be only 3,000 lb $CaCO_3$/afs.

Determining the Lime Requirement of Soils

The lime requirement of a soil can be determined by several different methods. Titrating the soil with an acid (HCl) or a base [$Ca(OH)_2$] will subsequently decrease or increase the soil pH, respectively (Fig.3.6). After equilibration, pH is determined and the values are plotted against the meq of acid or base added. From these data it is simple to determine the amount of lime to be added.

For example, increasing pH from 5.7 to 6.5 requires adding 1.0 meq base/100 g soil (Fig. 3.6). Thus, the quantity of pure $CaCO_3$ needed to increase pH would be

$$\frac{1.0 \text{ meq CaCO}_3}{100 \text{ g soil}}\left(\frac{50 \text{ mg CaCO}_3}{\text{meq}}\right) = \frac{50 \text{ mg CaCO}_3}{100 \text{ g soil}}$$

$$= 1,000 \text{ lb CaCO}_3 /\text{acre-6 in.}$$

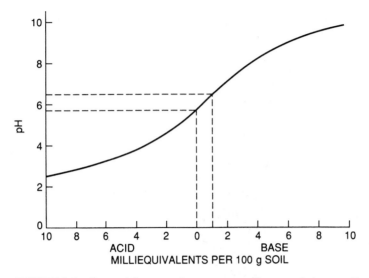

FIGURE 3.6   General lime requirement curve for a sandy loam soil.

The common methods to estimate lime requirement are based on the change in pH of a buffered solution added to a soil. When an acidic soil is added to a buffered solution, the buffer pH is depressed in proportion to the original soil pH and its BC. By calibrating pH changes in the buffered solution, the amount of lime required to increase pH to the desired level (i.e., the lime requirement) can be calculated.

The Shoemaker, McLean, and Pratt (SMP) single-buffer method for measuring the lime requirement of acidic soils has been widely adopted by U.S. soil-testing laboratories. The buffer solution is a dilute mixture of triethanolamine, paranitrophenol, and K chromate. The SMP method is especially well suited for soils with the following properties: lime requirement > 4 meq/100 g ($>4{,}000$ lb $CaCO_3$/a), pH < 5.8, OM $< 10\%$, and appreciable quantities of soluble (extractable) Al. The SMP method used on soils with low lime requirements will frequently result in overliming.

In very acidic soils, the lime requirement may be based on the $CaCO_3$ needed to lower Al on the CEC. For example, in Colombia it is recommended that 1 ton of $CaCO_3$ be applied for each 1 meq of exchangeable Al^{3+}/100 g soil. In Brazil the amount of lime needed to neutralize exchangeable Al is determined as follows:

$$\text{meq } CaCO_3/100 \text{ g} = 2 \times \text{meq exchangeable Al}/100 \text{ g}$$

Other studies show that the range in $CaCO_3$ recommendation is from 1,000 to 6,000 lb $CaCO_3$/meq Al^{3+}/100 g soil.

Soil pH for Crop Production

For many years the optimum soil pH for crop production was considered to be between 6.5 and 7.0; however, lime sufficient to raise pH to about 5.6 or 5.7 and reduce exchangeable Al^{3+} to less than 10% of the CEC will eliminate pH-related crop production problems (Fig. 3.7). Liming Ultisols and Oxisols to pH 7.0 can impair productivity by

1. Decreasing water percolation.

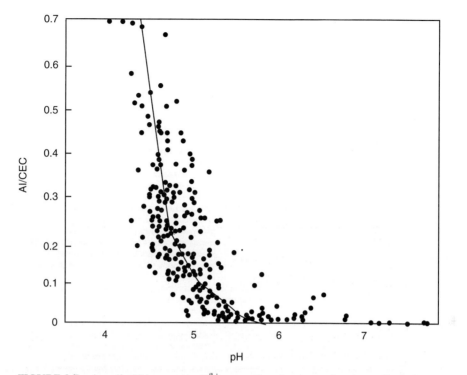

FIGURE 3.7 As soil pH increases, Al^{3+} saturation decreases. In most soils, little or no effect of Al^{3+} toxicity on plant growth is observed above pH 5.0–5.5.

2. Reducing the growth of legumes and nonlegumes.
3. Restricting plant uptake of P and some micronutrients.

Crops vary widely in their tolerance to acidic soils (Fig. 3.8). Reducing soluble Al^{3+} to < 1 ppm is recommended for sensitive crops. On the Alfisols and Mollisols, liming to pH 6.5–6.8 is optimum for most crops except alfalfa, for which liming to pH 6.8–7.0 is recommended.

Liming Materials

The materials commonly used for liming soils are Ca and Mg oxides, hydroxides, carbonates, and silicates (Table 3.4). The accompanying anion *must* lower H^+ activity and hence Al^{3+} in the soil solution. Gypsum ($CaSO_4 \cdot 2H_2O$) and other neutral salts cannot neutralize H^+, as shown in the following reaction:

$$CaSO_4 \cdot 2H_2O + 2H^+ \leftrightarrow Ca^{2+} + 2H^+ + SO_4^{2-} + 2H_2O$$

In fact, the addition of neutral salts actually lowers soil pH. Their addition, especially in a band, results in replacement of adsorbed Al^{3+} in a localized soil zone, sometimes with a significant lowering of the pH in this region. Although gypsum ($CaSO_4$) will not neutralize soil pH, increasing Ca^{2+} in soil solution may enhance

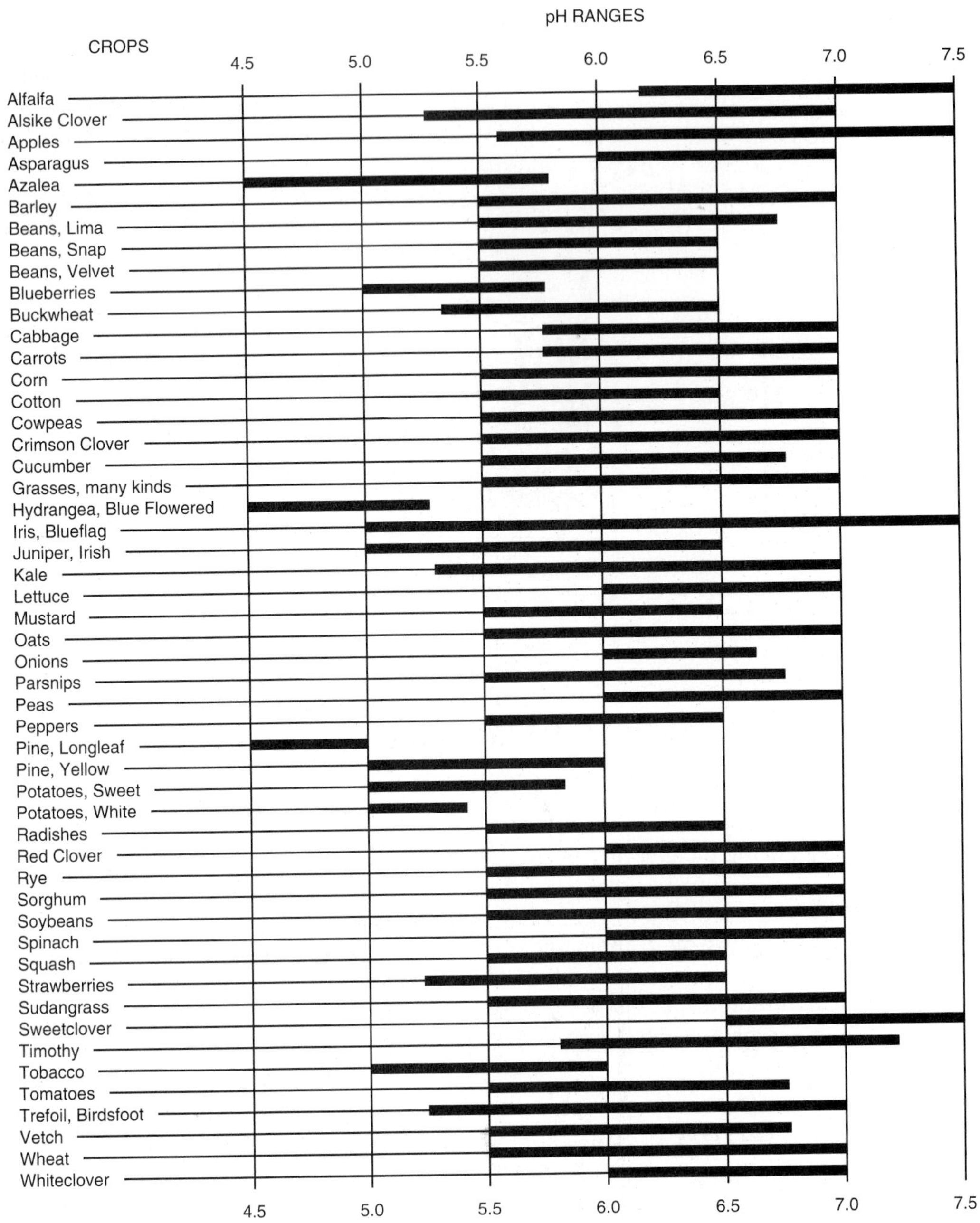

FIGURE 3.8   Range in soil pH for optimum growth of selected crops.   *USDA, 1980.*

growth if Ca^{2+} is marginally deficient or formation of $AlSO_4^+$ ion pairs reduces Al^{3+} concentration in soil solution and subsequent potential Al toxicity.

Liming reactions begin with the neutralization of H^+ in the soil solution by either OH^- or HCO_3^- originating from the liming material. For example, $CaCO_3$ behaves as follows:

$$CaCO_3 + H_2O \rightarrow Ca^{2+} + HCO_3^- + OH^-$$

The rate of the reaction is directly related to the rate at which the OH^- ions are removed from solution. As long as sufficient H^+ ions are in the soil solution, Ca^{2+} and HCO_3^- will continue to go into solution ($CaCO_3$ dissolves). The continued removal of H^+ from the soil solution will ultimately result in the precipitation of Al^{3+} and Fe^{3+} as $Al(OH)_3$ and $Fe(OH)_3$ and their replacement on the CEC with Ca^{2+} and/or Mg^{2+}.

The overall reaction for neutralization of soil acidity can be written as follows:

$$\text{clay} \begin{bmatrix} Al^{3+} \\ Ca^{2+} \\ Mg^{2+} \\ K^+ \\ Al^{3+} \end{bmatrix} + 3CaCO_3 + 3H_2O \rightarrow \text{clay} \begin{bmatrix} K^+ \\ Ca^{2+} \\ Ca^{2+} \\ Mg^{2+} \\ Ca^{2+} \\ Ca^{2+} \end{bmatrix} + 2Al(OH)_3 + 3CO_2$$

Obviously, as pH increases, the BS % also increases, as discussed in Chapter 2.

Calcium Oxide

Calcium oxide (CaO) is the only material to which the term *lime* may be correctly applied. Also known as *unslaked lime, burned lime,* or *quicklime,* CaO is a white powder, shipped in paper bags because of its caustic properties. It is manufactured by roasting $CaCO_3$ in a furnace, driving off CO_2. CaO is the most effective of all liming materials, with a neutralizing value, or calcium carbonate equivalent (CCE), of 179%, compared with pure $CaCO_3$ (Table 3.4). When unusually rapid results are required, either this material or $Ca(OH)_2$ should be selected. Complete mixing of CaO with the soil may be difficult, because immediately after application, absorbed water causes the material to form flakes or granules. These granules may harden due to $CaCO_3$ formation on their surfaces and can thus remain in the soil for long periods of time.

Calcium Hydroxide

Calcium hydroxide [$Ca(OH)_2$], referred to as *slaked lime, hydrated lime,* or *builders' lime,* is a white powder and difficult to handle. Neutralization of acid occurs rapidly. Slaked lime is prepared by hydrating CaO and has a neutralizing value (CCE) of 136 (Table 3.4).

TABLE 3.4   Neutralizing Value (CCE) of Pure Forms of Some Liming Materials

Material	Molecular Wt. (g/mole)	Eq. Wt. (g/eq)	Neutralizing Value (%)
CaO	56	28	179
$Ca(OH)_2$	72	36	136
$CaMg(CO_3)_2$	184	46	109
$CaCO_3$	100	50	100
$CaSiO_3$	116	58	86

Calcium and Calcium-Magnesium Carbonates

Calcium carbonate ($CaCO_3$), or calcite, and calcium-magnesium carbonate [$CaMg(CO_3)_2$], or dolomite, are common liming materials. Limestone is most often mined by open-pit methods. The quality of crystalline limestones depends on the degree of impurities, such as clay. The neutralizing values range from 65 to a little more than 100%. The neutralizing value of pure $CaCO_3$ has been theoretically established at 100%, while pure dolomite has a neutralizing value of 109%. As a general rule, however, the CCE of most agricultural limestones is between 80 and 95% because of impurities.

Marl

Marls are soft, unconsolidated deposits of $CaCO_3$ frequently mixed with earth and usually quite moist. Marl deposits are generally thin, recovered by dragline or power shovel after the overburden has been removed. The fresh material is stockpiled and allowed to dry before being applied to the land. Marls are almost always low in Mg, and their neutralizing value, between 70 and 90%, depends on their clay content.

Slags

Slag ($CaSiO_3$) is a by-product of the manufacture of pig iron. In the reduction of Fe, the $CaCO_3$ in the charge loses its CO_2 and forms CaO, which combines with the molten Si to produce a slag that is either air cooled or quenched with water. The neutralizing value of blast-furnace slags ranges from about 60 to 90%, and they usually contain appreciable amounts of Mg and P, depending on the source of Fe ore and manufacturing process.

Miscellaneous Liming Materials

Other materials used as liming agents in areas close to their source include fly ash from coal-burning power-generating plants, sludge from water treatment plants, lime or flue dust from cement manufacturing, pulp mill lime, carbide lime, acetylene lime, packinghouse lime, and so on. These by-products contain varying amounts of Ca and Mg.

Neutralizing Value or Calcium Carbonate Equivalent of Liming Materials

The value of a liming material depends on the quantity of acid that a unit weight of lime will neutralize, which, in turn, is related to the composition and purity. Pure $CaCO_3$ is the standard against which other liming materials are measured, and its neutralizing value is considered to be 100%. The CCE is defined as the acid-neutralizing capacity of a liming material expressed as a weight percentage of $CaCO_3$.

Consider the following reactions:

$$CaCO_3 + 2H^+ \rightleftarrows Ca^{2+} + H_2O + CO_2$$

$$MgCO_3 + 2H^+ \rightleftarrows Mg^{2+} + H_2O + CO_2$$

In each reaction, 1 mole of CO_3^{2-} will neutralize 2 moles of H^+. The molecular weight of $CaCO_3$ is 100, whereas that of $MgCO_3$ is only 84; thus, 84 g of $MgCO_3$ will neutralize the same amount of acid as 100 g of $CaCO_3$. Therefore, the neutralizing value, or CCE, of equal weights of the two materials is calculated by

$$\frac{84}{100} = \frac{100}{x}$$

$$x = 119$$

Therefore, $MgCO_3$ will neutralize 1.19 times as much acid as the same weight of $CaCO_3$; hence its CCE is 119%. The same procedure is used to calculate the neutralizing value of other liming materials (Table 3.4).

The composition of liming materials is sometimes expressed in terms of the Ca and Mg content of the pure mineral. For example, pure $CaCO_3$ contains 40% Ca and pure $MgCO_3$ contains 28.6% Mg, calculated by the ratio of molecular weights:

$$\frac{24 \text{ g/m Mg}}{84 \text{ g/m MgCO}_3} \times 100 = 28.6\%$$

To convert % Ca to CCE, multiply by 100/40, or 2.5; to convert % Mg to $MgCO_3$, multiply by 84/24, or 3.5.

Fineness of Limestone

The effectiveness of agricultural limestones also depends on the degree of fineness, because the reaction rate depends on the surface area in contact with the soil. CaO and $Ca(OH)_2$ are powders with the smallest particle size, but limestones need to be crushed to reduce particle size. When crushed limestone is thoroughly incorporated into the soil, the reaction rate will increase with increasing fineness. Coarser fractions increase liming effectiveness with increasing exposure time in soil. Decreasing particle size fraction of a liming material decreases the lime rate required to raise soil pH, or decreases the effectiveness of a given lime rate (Fig. 3.9). In this example, a 100-mesh lime material (100% efficient) requires only 1 t/ha to increase soil pH to 7.0, whereas a 50-mesh lime material (40% efficient) requires 2 t/ha to increase pH to 7.0.

Amounts of a particular limestone fraction needed to produce a given rise in pH 2 years after application are shown in Figure 3.10. Much less of the finer fractions than of the coarser fractions is needed to achieve a certain pH, particularly at lower reference pH values.

Because the cost of limestone increases with its fineness, materials that require minimum grinding, yet contain enough fine material to change pH rapidly, are preferred. Agricultural limestones contain both coarse and fine materials. Many states require that 75 to 100% of the limestone pass an 8- to 10-mesh screen and that 25% pass a 60-mesh screen. Thus, coarse and fine particles are fairly well distributed.

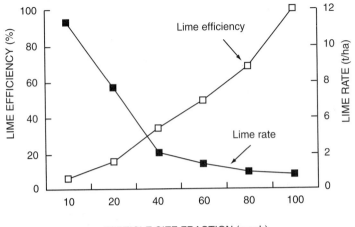

FIGURE 3.9   Relative lime efficiency of different size fractions of limestone in raising soil pH to 7.0. Greater lime rates are needed for coarser lime material to raise soil pH to the same level as a finer material.

Fineness is quantified by measuring the distribution of particle sizes in a given limestone sample (Table 3.5). The fineness factor is the sum of the percentages of the liming agent in each of the three size fractions multiplied by the appropriate effectiveness factor (Table 3.6). The effective calcium carbonate (ECC) rating of a limestone is the product of its CCE (purity) and the fineness factor.

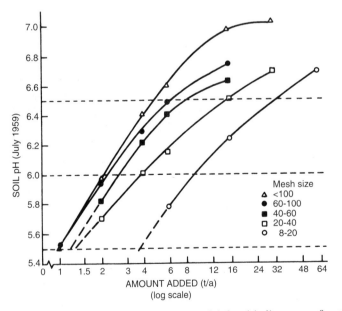

FIGURE 3.10   Effect of various rates of dolomitic limestone fractions on the pH of Withee silt loam after 2 years of equilibration under field conditions.   *Love et al.*, Trans. 7th Int. Congr. Soil Sci., *4:293, 1960.*

TABLE 3.5   Fineness Factors for Agricultural Limestone in the United States and Canada

United States		Canada	
Particle Size	Effectiveness	Particle Size	Effectiveness
(mesh)	(%)	(sieve no.)	(%)
>8	0	>10	0
8–60	50	10–60	40
<60	100	<60	100

TABLE 3.6   Fineness Effects on Effective Calcium Carbonate (ECC) Content of Two Lime Sources

	Solid Agricultural Lime	Suspendable Lime
Percent $CaCO_3$ equivalent	90	98
Sieve analysis		
% on 8-mesh sieve	2	0
% on 60-mesh sieve	21	2
% passing 60-mesh sieve	77	98
Calculated fineness factor		
% on 8-mesh × 0% effectiveness	$2 \times 0 = 0$	$0 \times 0 = 0$
% on 60-mesh × 50 effectiveness	$21 \times 50\% = 11.5$	$2 \times 50\% = 1$
% through 60-mesh × 100% effectiveness	$77 \times 100\% = 77$	$98 \times 100\% = 98$
Fineness factor (%)	88.5	99
Percent ECC = purity × fineness factor	$90 \times 88.5\% = 79.6$	$98 \times 99 = 97.0$

SOURCE: Murphy et al., *Situation 1978, TVA Fert. Conf. Bull. Y-131,* National Fertilizer Development Center–Tennessee Valley Authority (1978).

As a general rule, for the same degree of fineness, the material that costs the least per unit of neutralizing value applied to the land should be used. Assume that a calcitic limestone (CCE = 95%) and a dolomitic limestone (CCE = 105%), both with the same fineness, are available. Assume also that they both cost $12 per ton applied to the land. Based on the neutralizing value, the calcitic limestone will cost 105/95 × 12, or $13.26, per ton, compared with the dolomite at $12 per ton. In addition, the dolomite supplies Mg, which can be deficient in some humid-region soils. Although dolomite has a slightly higher neutralizing value than calcite, dolomite has a lower solubility and thus will dissolve more slowly. For dolomite to be as effective as calcite at the same application rate, dolomite should be ground twice as fine or react twice as long (Fig. 3.11).

Use of Lime in Agriculture

Lime is seldom needed in low-rainfall areas where leaching is minimal, such as parts of the Great Plains states and the arid, irrigated saline, and saline-alkali soils of the southwestern, intermountain, and far western states. The majority of soils in the Prairie Provinces of Canada likewise rarely require lime application.

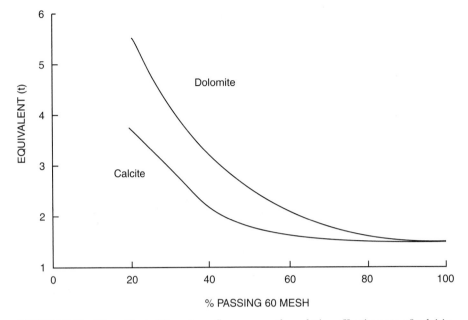

FIGURE 3.11   The effect of limestone fineness on the relative effectiveness of calcitic and dolomitic limestones for increasing crop yield.

Crop responses from the application of lime are usually attributed to decreased toxicity of Al^{3+}, although the plant-nutrient value of the Ca or Mg is also important.

Direct Benefits

Al toxicity is probably the most important growth-limiting factor in many acidic soils, particularly at pH < 5.0–5.5. The toxic effects of excessive Al on root growth can seriously influence plant growth and yield. Figure 3.12 shows how increasing Al^{3+} concentration in the soil solution decreases root length. As we have already seen, decreasing soil pH results in increasing solution and exchangeable Al^{3+}. Excess Al interferes with cell division in plant roots; inhibits nodule initiation; fixes P in less available forms in soils; decreases root respiration; interferes with enzymes governing the deposition of sugars in cell walls; increases cell wall rigidity; and interferes with the uptake, transport, and use of nutrients and water by plants. At pH 4.5 or less, H^+ toxicity damages root membranes and is detrimental to the growth of many beneficial bacteria. The greatest single direct benefit of liming many acidic soils is the reduction in the solubility of Al and Mn.

When lime is added to acidic soils, the activity of Al^{3+} is reduced by precipitation as $Al(OH)_3$ (Table 3.7). Liming raises soil pH while greatly reducing extractable Al and increasing crop yield. Not only is Al^{3+} toxic to plants, but increasing Al^{3+} in the soil solution also restricts the plant uptake of Ca and Mg (Fig. 3.13).

Different crops and varieties of the same crop differ widely in their susceptibility to Al^{3+} toxicity. Different varieties of soybeans, wheat, and barley vary widely in their tolerance to high concentrations of Al^{3+} in the soil solution (Fig. 3.14). Some grasses are quite tolerant of acidic mine spoils.

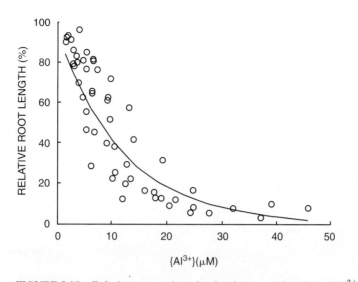

FIGURE 3.12   Relative taproot length of soybean as a function of Al^{3+} in solution. Taproot lengths were measured after 4 days of growth and compared with root elongation in zero Al treatments.

TABLE 3.7   Lime Effects on Wheat Yields, Soil pH, and Al Levels

Lime rate, lb ECC/a	1986–89 avg. wheat yield, bu/a	1989, 0–6-in. depth	
		Soil pH	Al, ppm
0	14	4.6	102
3,000	37	5.1	26
6,000	38	5.9	0
12,000	37	6.4	0

SOURCE: R. Laymond, Kansas State Univ.

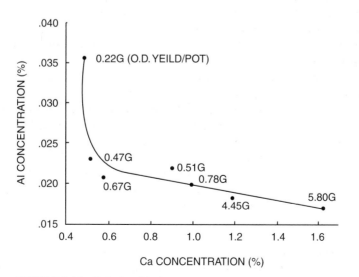

FIGURE 3.13   Relationship between Al and Ca concentration in cotton tops from non-leached subsoil.   *Soileau et al.*, SSSA Proc., *33:919, 1969.*

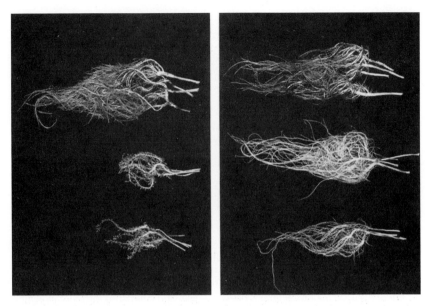

FIGURE 3.14   Differential effects of Al on root growth of Perry (top) and Chief (bottom) soybean varieties grown in solution containing 2 ppm Ca. *Left to right:* 0, 8, 12 ppm Al added.   *Foy et al.,* Agron. J., *61:505, 1969.*

The Al problem is not always economically correctable with conventional liming practices. Differences among genotypes in tolerance to excess Al are in part genetically controlled. A genetic approach has great potential for solving the problem of Al toxicity in acidic soils. Crops also vary in their tolerance to excessive amounts of Mn. For example, rapeseed is very sensitive to Mn toxicity, while barley is more tolerant.

One short-term management strategy for reducing Al^{3+} toxicity to seedlings is band application of fertilizer P (Table 3.8). These data show that band application of P at wheat planting dramatically reduces Al toxicity and increases wheat yield.

Indirect Benefits

EFFECT ON P AVAILABILITY   At low pH values and on soils high in Al and Fe, P precipitates as insoluble Fe/Al-P compounds (see Chapter 5). Liming acidic soils will precipitate Fe and Al as $Fe(OH)_3$ and $Al(OH)_3$, thus increasing plant available P.

Alternatively, liming soils to pH 6.8–7.0 can reduce P availability because of the precipitation of Ca or Mg phosphates. A liming program should be planned so that the pH can be kept between 5.5 and 6.8 if maximum benefit is to be derived from applied P.

MICRONUTRIENT AVAILABILITY   With the exception of Mo, the availability of the micronutrients increases with decreased pH (see Chapter 8). This can be detrimental because of the toxic nature of many micronutrients even at

TABLE 3.8   Lime, P, and Variety Effects on 2-Year Average
Wheat Grain Yields

Variety	Lime rate, lb ECC/a	Method of application and P_2O_5 rate, lb/a		
		None	*40 Broadcast*	*40 Banded*
		— Wheat grain yield, bu/a —		
Karl	0	38	42	54
	3,750	51	51	57
	7,500	49	49	55
2163	0	49	53	56
	3,750	58	57	60
	7,500	58	54	61

SOURCE: R. Laymond, Kansas State Univ.

relatively low solution concentrations. The addition of adequate lime reduces the solution concentration of many micronutrients, and soil pH values of 5.6–6.0 are usually sufficient to minimize toxicity while maintaining adequate availability of micronutrients.

Mo nutrition of crops is improved by liming, and deficiencies are infrequent in soils limed to pH values in excess of 7.0. Because of the effect on other micronutrient availability, liming to this value or above is not normally recommended for most crops in humid areas.

NITRIFICATION   Most of the organisms responsible for the conversion of NH_4^+ to NO_3^- require large amounts of Ca; therefore, nitrification is enhanced by liming to a pH of 5.5–6.5 (see Chapter 4). Decomposition of plant residues and breakdown of soil OM are also faster in this pH range than in more acidic soils. The effect of liming on both mineralization of organic N and nitrification is shown in Table 3.9. Application of lime just before incubation almost doubled the mineralization of organic N. However, lime added 1 or 2 years before sampling had little or no effect on release of mineral N in two of the soils. Although adding lime at the start of the incubation increased nitrification, earlier applications of lime had an even greater effect.

N FIXATION   Symbiotic and nonsymbiotic N_2 fixation is favored by adequate liming (see Chapter 4). Activity of some *Rhizobia* species is greatly restricted by soil pH levels below 6.0; thus, liming will increase the growth of legumes because of increased N_2 fixation. With the nonsymbiotic N_2-fixing organisms, N_2 fixation increases in adequately limed soils, which increases the decomposition of crop residues.

SOIL PHYSICAL CONDITION   The structure of fine-textured soils may be improved by liming, as a result of an increase in OM content and of the flocculation of Ca-saturated clay. Favorable effects of lime on soil structure

TABLE 3.9   Mineralization of Organic N and Nitrification in Three Acidic Soils Incubated for 4 Weeks with and without Lime

Soil	Treatment	Organic N Mineralized (ppm)	Percentage Nitrification
Site 1 (pH 5.5, 0.20% soil N)	No lime	36	8
	Limed at start of incubation	61	66
	Limed 2 yr before in the field	33	94
Site 2 (pH 5.4, 0.13% soil N)	No lime	40	7
	Limed at start of incubation	72	64
	Limed 1 yr before in the field	44	93
Site 3 (pH 5.7, 0.83% soil N)	No lime	90	28
	Limed at start of incubation	177	83
	Limed 1 yr before in the field	134	94

SOURCE: Nyborg and Hoyt, *Can. J. Soil Sci.*, 58:331 (1978).

include reduced soil crusting, better emergence of small-seeded crops, and lower power requirements for tillage operations. However, the overliming of Oxisols and Ultisols can result in the deterioration of soil structure, with a decrease in water percolation. Ca also improves the physical conditions of sodic soils. Increased salt concentration due to $CaCO_3$ dissolution is responsible for preventing clay dispersion and decreases in hydraulic conductivity.

DISEASE   Correction of soil acidity by liming may have a significant role in the control of certain plant pathogens. Clubroot is a disease of cole crops that reduces yields and causes the infected roots to enlarge and become distorted. Lime does not directly affect the clubroot organism, but at soil pH greater than 7.0 germination of clubroot spores is inhibited (Table 3.10).

On the other hand, liming increases the incidence of diseases such as scab in root crops. Severity of take-all infection in wheat, with resultant reductions in yield, is known to be increased by liming soils to near neutral pH.

Application of Liming Materials

Surface applications of lime without some degree of mixing in the soil are not immediately effective in correcting subsoil acidity. In several studies it was observed that 10 to 14 years were required for surface-applied lime without incorporation to raise the soil pH at a depth of 15 cm. For fairly high rates, broadcasting one-half the lime, followed by disking and plowing, and then broadcasting the other half and disking is a satisfactory method of mixing with the plow layer.

Keeping surface soils at the proper pH over a period of years is a practical way of at least partially overcoming the problem of subsoil acidity (Table 3.11). It is evident that maintenance of surface soil at pH 6.0, 6.5, and 7.2 reduced acidity deeper in the root zone.

Neutralization of subsoil acidity through deep incorporation of surface-applied lime is possible with large tillage equipment. The effect of incorporation depth of surface-applied lime on cotton growth showed that the amount and

TABLE 3.10   Effect of Liming on the Harmful Effects of Clubroot Disease in Cauliflower

Lime (t/a)	Lime Applied in 1978			Lime Applied in 1979		
	Yield Percent Marketable	Clubroot Rating*	pH at Harvest	Yield Percent Marketable	Clubroot Rating*	pH at Harvest
0	48	3.3	5.6	28	3.8	5.7
2.5	73	1.8	6.6	39	3.8	6.4
5.0	81	1.1	6.9	63	3.4	6.7
10.0	86	0.2	7.1	74	2.5	7.2

*Clubroot rating $= \dfrac{\text{sum of [number of roots at a rating} \times \text{rating]}}{\text{total number of roots}}$.

Rating: 0, no visible clubroot; 1, fewer than 10 galls on the lateral roots; 2, more than 10 galls on the lateral roots, taproot free of clubroot; 3, galls on taproot; 4, severe clubbing on all roots.

SOURCE: Waring, *Proc. 22nd Annu. Lower Mainland Hort. Improvement Assoc. Growers' Short Course,* pp. 95–96. Lower Mainland Horticultural Improvement Association and British Columbia Ministry of Agriculture (1980).

depth of cotton rooting were increased by mixing lime to a depth of 45 cm (Fig. 3.15). Mixing lime even deeper, to a depth of 60 cm, markedly increased corn yields (Fig. 3.16).

With no-till cropping systems, surface soil pH can decrease substantially in a few years because of the acidity produced by surface-applied N fertilizers and decomposition of crop residues (Table 3.12). Fortunately, the increased acidification is concentrated in the soil surface, where it can be readily corrected by surface liming.

Partial mixing of lime with soil as a means of reducing lime rates may increase soil pH enough to optimize growth (Table 3.13). Maximum yield of wheat dry matter was obtained when lime was confined to 30% of the soil rather than complete mixing.

Equipment

Dry lime applied by the supplier is the most common method. The spinner truck spreader, which throws the lime in a semicircle from the rear of the truck, is often used. Uniform spreading is more difficult with this equipment than with the kind that drops the lime from a covered hopper or conveyor.

Suspending lime in water, frequently referred to as *fluid lime,* is another approach to lime application. Mixing equipment used for conventional suspension fertilizers can be readily utilized to produce fluid lime. Very finely ground lime (100% passing a 100-mesh sieve and 80 to 90% passing a 200-mesh sieve) with

TABLE 3.11   Effect of Maintaining Various Surface pH Levels on Subsoil pH of Wooster Silt Loam Soil

Soil Depth (in.)	pH at Various Depths with Increasing pH in the Surface				
0–7	4.9	5.5	6.0	6.5	7.2
7–14	4.9	5.2	5.9	6.7	7.2
14–21	4.7	4.8	5.2	5.4	6.5

SOURCE: *Ohio Agronomy Guide,* Cooperative Extension Service, Ohio State Univ. (1985).

FIGURE 3.15   Amount and depth of cotton rooting as affected by depth of lime incorporation. From left to right: unlimed; 0–15 cm (0–6 in.) limed; 0–45 cm (0–18 in.) limed.   *Doss et al.*, Agron. J., *71:541, 1979.*

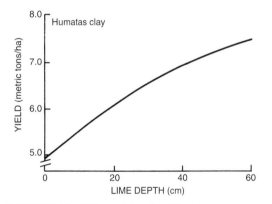

FIGURE 3.16   Effect of depth of lime incorporation on corn yield.   *Bouldin,* Cornell Int. Agr. Bull. 74, *Cornell Univ., 1979.*

TABLE 3.12   Soil pH after 7 Years of Continuous Corn Grown on Maury Silt Loam Soil in Kentucky Affected by Tillage Methods, N Fertilization, and Liming

N Treatment	Soil Depth (cm)	Conventional Tillage		No Tillage	
		Limed	Unlimed	Limed	Unlimed
High N rate	0–5	5.3	4.9	5.5	4.3
(336 kg/ha)	5–15	5.9	5.1	5.3	4.8
	15–30	6.0	5.5	5.8	5.5
Moderate N rate	0–5	5.9	5.2	5.9	4.8
(168 kg/ha)	5–15	6.3	5.6	5.9	5.5
	15–30	6.2	5.7	6.0	5.9

SOURCE: Blevins et al., *Agron. J.,* 40:322 (1978).

TABLE 3.13   Effect of Mixing Lime with Varying Soil Volumes on the Growth of Nugaines Winter Wheat in the Growth Chamber (g/box)

Rate of Lime (t/ha)	Percentage of Soil Limed				
	0	10	30	60	100
0	1.27	—	—	—	—
2.24	—	1.95	2.42	2.03	1.83
6.72	—	1.85	3.10	3.02	2.60
11.20	—	1.72	3.50	2.90	2.60

SOURCE: Kauffman and Gardner, *Agron. J.,* 70:331 (1978).

50% H_2O, along with a suspending agent such as attapulgite clay, may be applied with a fluid fertilizer applicator. Some of the features of fluid lime are as follows:

1. An excellent distribution pattern can be obtained.
2. The finely divided lime reacts rapidly with soil.
3. Only a small amount of liming material is applied at any one time (e.g., 500 to 1,000 lb of material per acre).
4. Very-low-pH soils can be corrected quickly.
5. Regular annual applications help to maintain pH.
6. The cost of fluid lime is usually two to four times higher than that of dry applications.
7. Fluid fertilizer dealers are supplying fluid lime in areas where conventional lime has been difficult to obtain.

Urea-ammonium nitrate (UAN) solutions can be used for suspending the lime component. Incorporation soon after application of the suspension will eliminate potential NH_3 loss following urea hydrolysis. N-K-lime suspensions are also being used successfully. The feasibility of including P sources in lime suspensions is yet to be established. Some lime-herbicide suspensions also may be used.

Regardless of the method employed, care should be taken to ensure uniform application. Nonuniform distribution can result in excesses and deficiencies in different parts of the same field and corresponding nonuniform crop growth.

Factors Determining the Selection of a Liming Program

Intended Crop

Plants differ widely in their sensitivity to soil acidity and thus to added lime (Fig. 3.8). The type of crop to be grown is the most important factor to consider in developing a lime program.

Soil Texture and OM Content

In a coarse-textured, low-OM soil, the lime requirement is less than for a fine-textured or high-OM soil (Fig. 3.5). The overliming of coarse-textured soils is not uncommon, but a knowledge of basic soil chemistry can prevent it.

TABLE 3.14   Adjustment Factor for Lime Rate for Incorporation Depth

Incorporation Depth (in.)	Adjustment Factor
3	0.43
5	0.71
7	1.00
9	1.29
11	1.57

SOURCE: D. Witney, Kansas State Univ.

Time and Frequency of Liming Applications

For rotations that include leguminous crops, lime should be applied 3 to 6 months before the time of seeding; this timing is particularly important on very acidic soils. Lime may not have adequate time to react with the soil if applied just before seeding. If clover is to follow fall-seeded wheat, lime is best applied when the wheat is planted. The caustic forms of lime [CaO and $Ca(OH)_2$] should be spread well before planting to prevent injury to germinating seeds.

The frequency of application generally depends on the texture of the soil, N source and rate, crop removal, precipitation patterns, and lime rate. On sandy soils, frequent light applications are preferable, whereas on fine-textured soils, larger amounts may be applied less often. Finely divided lime reacts more quickly, but its effect is maintained over a shorter period than that of coarse materials.

The most satisfactory means of determining reliming needs is by soil tests. Samples should be taken every 3 years.

Depth of Tillage

Lime recommendations are made on the basis of a 6-in. soil depth. When land is tilled to a depth of 10 in., the lime recommendations should be increased by 50%. In Kansas, for example, recommended lime rates are adjusted according to expected depth of incorporation (Table 3.14). Thus, less lime is applied for incorporation to 3-in. depth compared with 7-in. depth.

Acidulating the Soil

Acidification may be needed for some crops when land is inherently high in $CaCO_3$ or Na_2CO_3, as in semiarid and arid regions (Fig. 3.17). Calcareous soils contain measurable quantities of $CaCO_3$. These soils typically have soil pH > 7.2, although values can be higher depending on salt and/or Na content. Calcareous surface soils occur in areas where annual precipitation is less than 20 in. (Fig. 3.18). As precipitation increases from semiarid to humid regions, depth to $CaCO_3$ increases. Generally, when annual precipitation exceeds 30 to 40 in., no free lime or $CaCO_3$ is present in the rooting zone.

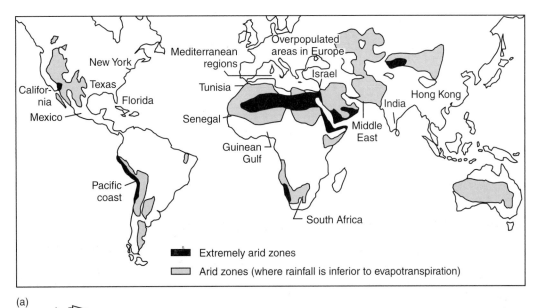

(a)

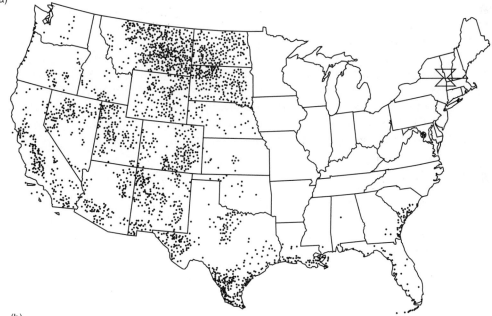

(b)

FIGURE 3.17   Extent of arid and semiarid regions in the world (a) and the areas in the United States (b) where soil salinity limits yield potential ($EC_{se} > 4$ mmho/cm).   *SCS, 1992.*

For soil pH to decrease, all of the measurable $CaCO_3$ would have to be dissolved or neutralized through adding acid or acid-forming material. In most field crop situations, reducing soil pH by neutralizing $CaCO_3$ is not practical. The following example calculation illustrates the inordinately large quantity of elemental S^0 needed to neutralize a soil with only 2% $CaCO_3$ content in the surface 6 in. of soil depth.

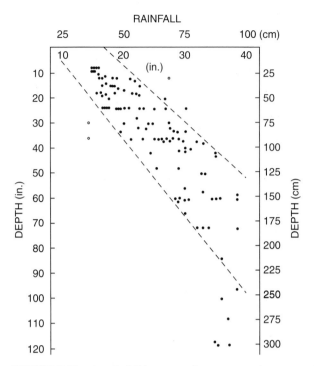

FIGURE 3.18   As rainfall increases from one region to another, the depth to measurable $CaCO_3$ content increases. Thus, in arid and semiarid regions, $CaCO_3$ is present in the surface soil.

2% $CaCO_3$ content in 6-in. surface soil depth
(acre furrow slice [afs] or 2×10^6 lb of soil/a)

$$2\% \; CaCO_3 = \frac{2g \; CaCO_3}{100 \; g \; soil} = \frac{2,000 \; mg \; CaCO_3}{100 \; g \; soil}$$

$$\frac{2,000 \; mg \; CaCO_3}{100 \; g \; soil} \times \frac{1 \; meq}{50 \; mg} = \frac{40 \; meq \; CaCO_3}{100 \; g \; soil}$$

$$\frac{40 \; meq \; CaCO_3}{100 \; g \; soil} = \frac{40 \; meq \; S^0}{100 \; g \; soil}$$

$$\frac{40 \; meq \; S^0}{100 \; g \; soil} \times \frac{16 \; mg}{meq} = \frac{640 \; mg \; S^0}{100 \; g \; soil}$$

$$\frac{0.64 \; g \; S^0}{100 \; g \; soil} \times \frac{2 \times 10^4}{2 \times 10^4} = \frac{6,400 \; g \; S^0}{2 \times 10^6 \, g \; soil} = \frac{12,800 \; lb \; S^0}{2 \times 10^6 \, lb \; soil}$$

$$\frac{12,800 \; lb \; S^0}{2 \times 10^6 \; lb \; soil} = \frac{12,800 \; lb \; S^0}{afs} = \frac{6.4 \; t \; S^0}{afs}$$

Thus, in this example, more than 6 t/a of elemental S^0 would be needed to neutralize all of the free lime content. Once neutralized, soil pH would likely be about the same as before neutralization because the CEC would still be nearly 100% saturated with basic cations (100% BS). To ultimately lower soil pH below 7, additional elemental S^0 would be needed to produce H^+ (or Al^{3+}), which would in turn reduce the BS % necessary to lower soil pH. The additional quantity of elemental S^0 can also be estimated similarly to the pH-BS calculations shown in Chapter 2.

Land leveling to facilitate irrigation and for other purposes often exposes calcareous and high-pH subsoils that are unfavorable for optimum plant growth. Problems of high soil pH are not confined to arid and semiarid areas. Acidifying paddy soil has increased rice yields, which is often related to increased availability of micronutrients. Farmers in humid regions may overlime or dust from limestone-graveled roads may blow onto field borders, causing localized and excessively high pH. In other areas, moderately acidic soils may need further acidification for optimum production of potatoes, blueberries, cranberries, azaleas, rhododendrons, camellias, or conifer seedlings.

The fundamental chemistry of soil acidification is the same as that of liming soils, and several acidic or acid-forming materials can be used.

Elemental S^0

Elemental S^0 is the most effective of the soil acidulents (see Chapter 7). When S^0 is applied to the soil, the following generalized reaction occurs:

$$S^0 + H_2O + 3/2\ O_2 \rightleftarrows 2H^+ + SO_4^{2-}$$

For every mole of S^0 applied and oxidized, 2 moles of H^+ are produced, which decreases soil pH. In calculating the amount to apply to the soil, reference must be made to the buffer curve of that soil (see Chapter 2).

Finely ground S^0 should be broadcast and incorporated several weeks or months before planting because the microbial oxidation reaction may be slow, particularly in cold, alkaline soils. Under some conditions, it may be advisable to acidulate a zone near the plant roots to increase water penetration or P and micronutrient availability. Both of these conditions frequently need to be corrected on saline-alkaline soils. Elemental S^0 can be applied in bands as either granular, dispersible S^0 or S^0 suspensions. When S^0 is applied in a band, much smaller amounts are required than with broadcast applications (see Chapter 7 for a discussion of S^0 management).

Sulfuric Acid

Sulfuric acid (H_2SO_4) has been used for reclaiming Na- or B-affected soils, increasing availability of P and micronutrients, reducing NH_3 volatilization, increasing water penetration, controlling certain weeds and soilborne pathogens, and enhancing the establishment of range grasses. The favorable influence of H_2SO_4 and other treatments on sorghum yield (Table 3.15) and on rice yield (Table 3.16) is partially related to increased nutrient availability.

H_2SO_4 can be added directly to the soil, but it is unpleasant to work with and requires the use of special acid-resistant equipment. It can be dribbled on the surface or applied with a knife applicator, similar to anhydrous NH_3. It can also be

TABLE 3.15   Effects of H_2SO_4 and FeSO$_4$ on Grain Sorghum Yields on a Calcareous Texas Soil

	Fe (kg/ha)		
H_2SO_4	0	112	560
0	434	1,460	2,275
112	605	1,538	2,274
560	2,169	2,429	2,230
5,600	1,885	1,971	1,810

SOURCE: Mather, in Beaton et al. (Eds.), *Fertilizer Technology and Use,* 3d ed., chap. 11. Soil Science Society of America (1985).

TABLE 3.16   Effect of S, Sulfuric Acid, and Gypsum Soil Ameliorants on the Rice Grain Yield of Bluebonnet 50 and IR661

	Mean Yield (t/ha) of	
Soil Ameliorant	*Bluebonnet 50*	*IR661*
Control	2.69	5.83
Gypsum	2.70	6.00
S	3.24	6.72
H_2SO_4	3.69	6.96

SOURCE: Chapman, *Australian J. Exp. Agr. Anim. Husb.,* 20:724 (1980).

applied in ditch irrigation water. H_2SO_4 has the advantage of reacting instantaneously with the soil.

Aluminum Sulfate

Aluminum sulfate [$Al_2(SO_4)_3$] is a popular material among floriculturists for acidulating the soil for production of azaleas, camellias, and similar acid-tolerant ornamentals. When this material is added to water, it hydrolyzes to produce a very acidic solution:

$$Al_2(SO_4)_3 + 6H_2O \rightleftharpoons 2Al(OH)_3 + 6H^+ + 3SO_4{}^{2-}$$

When $Al_2(SO_4)_3$ is added to the soil, in addition to hydrolysis in the soil solution, the Al^{3+} replaces any exchangeable H^+ and other cations on the CEC and reduces the pH even further:

$$Al_2(SO_4)_3 + \text{clay} \left]\begin{matrix}4H^+\\Ca^{2+}\end{matrix} \rightarrow \text{clay} \left]\begin{matrix}Al^{3+}\\Al^{3+}\end{matrix} + Ca^{2+} + 4H^+ + 3SO_4{}^{2-}$$

$Al_2(SO_4)_3$ is not widely used in general agriculture. Iron sulfate (FeSO$_4$) is applied to soils for acidification and behaves similarly to $Al_2(SO_4)_3$.

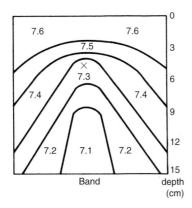

FIGURE 3.19 Application of a P-S fertilizer solution lowers soil pH in the vicinity of fertilizer band ($\times$ denotes point of application). *Leiker, M.Sc. thesis, Kansas State Univ., 1970.*

Ammonium Polysulfide

Liquid ammonium polysulfide (NH_4S_x) is used to lower soil pH and to increase water penetration in irrigated saline-alkaline soils. It can be applied in a band 3 or 4 in. to the side of the seed or metered into the ditch irrigation systems. Band application is more effective in correcting micronutrient deficiencies than application through irrigation water. The polysulfide decomposes into ammonium sulfide and colloidal S^0 when applied. The S^0 and S^{2-} are subsequently oxidized to H_2SO_4. Potassium polysulfide was developed for similar purposes.

Acidification in Fertilizer Bands

Because of the high BC for pH of many calcareous and high-pH soils, it is usually too expensive to use enough acidifying material for complete neutralization of soil alkalinity. It is unnecessary to neutralize the alkalinity of the entire soil mass because soil zones more favorable for root growth and nutrient uptake can be created by confining the acid-forming materials to bands and other localized placement.

Band-applied ammonium thiosulfate and ammonium polyphosphate fertilizer solution acidifies the soil in and near the band, which can increase micronutrient availability (Fig. 3.19).

Saline, Sodic, and Saline-Sodic Soils

In arid and semiarid regions, runoff water collected in depressions evaporates and the salts in the water accumulate. Water also moves upward from artesian sources and shallow water tables. It evaporates, and salts are deposited near or on the soil surface to form saline, sodic, or saline-sodic soils. These soils are widespread in semiarid and arid regions, where the rainfall is not sufficient for adequate leaching, usually less than 20 in./yr (Fig. 3.17). Approximately 10% ($\approx$13 billion acres) of the total land area is affected by salt. Of the cultivated soils, about 20% are saline and 35% are sodic. These types of soils are particularly prevalent in irrigated areas where improper irrigation and drainage methods are used. The salt marshes of the temperate zones, the mangrove swamps of the subtropics and tropics, and the interior salt marshes adjacent to salt lakes are areas where such soils are found.

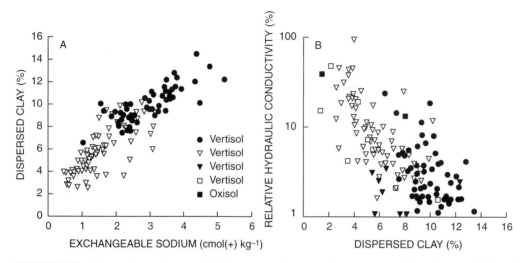

FIGURE 3.20   Increasing exchangeable Na content increases the amount of clay dispersed (A) and decreases the hydraulic conductivity (B). Data represent Vertisols and Oxisols in Australia.   *Sumner, Aust. J. Soil Res. 31:683–750, 1993.*

Rapid extension of irrigated lands over the last four decades has increased the salinization of cultivated lands. Large areas of the Indian subcontinent have been rendered unproductive by salt accumulation and poor water management. Salinity is a major problem in wetland rice.

Salt buildup is an existing or potential danger on almost all of the irrigated land in semiarid and arid regions of the United States, and salinity on nonirrigated cropland and rangeland in these regions is increasing. The accumulated salts contain the cations Na^+, Ca^{2+}, and Mg^{2+} and the anions Cl^-, SO_4^{2-}, HCO_3^-, and CO_3^{2-}. They can be weathered from minerals and accumulate in areas where the precipitation is too low to provide leaching.

Na is particularly detrimental, because of both its toxic effect on plants and effect on soil structure. When a high percentage of the CEC is occupied by Na, the soil aggregates disperse. These soils become impermeable to water, develop hard surface crusts, and may keep a water layer, or "slick spot," on the surface longer than adjacent soils. For example, data from several Vertisols and Oxisols in Australia show that as exchangeable Na increases, the percentage of dispersed clay increases (Fig. 3.20A), resulting in substantial decreases in hydraulic conductivity (or greater impermeability to water) (Fig. 3.20B).

Dispersion problems occur at different exchangeable Na contents. Fine-textured soils with montmorillonitic clays may disperse when about 15% of the exchange complex is saturated with Na. On tropical soils high in Fe and Al oxides and on some kaolinitic soils, 40% Na saturation is required before dispersion is serious. Soils low in clay are also less prone to problems because they are more permeable.

Definitions

SALINE   Saline soils have a *saturated extract conductivity* (EC_{se}) > 4 mmhos/cm, pH < 8.5, and < 15% exchangeable Na% (ESP) (Table 3.17; Fig. 3.21). Saline

TABLE 3.17   Summary of Salt-Affected Soil Classification

Classification	Conductivity (mmhos/cm)	Soil pH	Exchangeable Sodium Percentage	Soil Physical Condition
Saline	>4.0	<8.5	<15	Normal
Sodic	<4.0	>8.5	>15	Poor
Saline-sodic	>4.0	<8.5	>15	Normal

soils were formerly called *white alkali* because of the deposits of salts on the surface following evaporation. The excess salts, mostly Cl^-, SO_4^{2-}, HCO_3^-, and CO_3^{2-}, salts of Na^+, Ca^{2+}, and Mg^{2+}, can be leached out, with no appreciable rise in pH. The concentration of soluble salts is sufficient to interfere with plant growth, although salt tolerance varies with plant species.

SODIC   Sodic soils occur when ESP > 15%, EC_{se} < 4 mmhos/cm, and pH > 8.5 (Table 3.17). They were formerly called *black alkali* because of the dissolved organic matter deposited on the surface along with the salts. In sodic soils, the excess Na disperses the soil colloids and creates nutritional disorders in most plants.

SALINE-SODIC   Saline-sodic soils have both the salt concentration (>4 mmhos/cm) to qualify as saline and high exchangeable Na (>15% ESP) to qualify as sodic; however, soil pH is less than 8.5. In contrast to saline soils, when the salts are leached out, the exchangeable Na hydrolyzes and the pH increases, which results in a sodic soil.

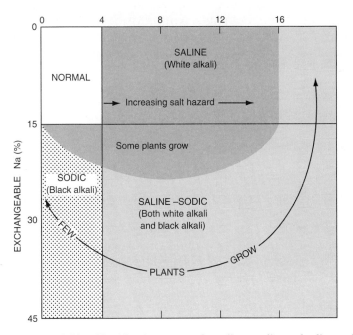

FIGURE 3.21   Classification system for saline, sodic, and saline-sodic soils.   *U.S. Salinity Lab. Handbook 60, 1954.*

Relationships

Several parameters are commonly used to quantify salt- and Na-affected soils, and many of these parameters are interrelated. By measuring the EC_{se} of a soil, the total quantity of salts in the soil solution can be estimated as follows:

$$EC_{se} \times 10 = \text{total soluble cations (meq/l)}$$

If the soluble cations are measured in the saturated extract, the *sodium adsorption ratio (SAR)* can be calculated as follows:

$$SAR = \frac{Na^+}{\sqrt{\dfrac{(Ca^{2+} + Mg^{2+})}{2}}} \text{ (all units in meq/l)}$$

Because of the equilibrium relationships between solution and exchangeable cations in soils, the SAR should be related to the quantity of Na^+ on the CEC, which is expressed as the *exchangeable sodium ratio (ESR)*. The ESR is defined as follows:

$$ESR = \frac{\text{exchangeable } Na^+}{\text{exchangeable } (Ca^{2+} + Mg^{2+})} \text{ (all units in meq/100 g)}$$

Figure 3.22 illustrates the relationship between solution and exchangeable cations in salt-affected soils. This relationship can be used to estimate the ESR if the quantity of exchangeable cations has not been measured. The following equation represents the linear relationship shown in Figure 3.22.

$$ESR = 0.015(SAR)$$

Subsequently, the ESR is related to the *exchangeable sodium percentage (ESP)* previously used to classify Na-affected soils (Table 3.17) and is given by

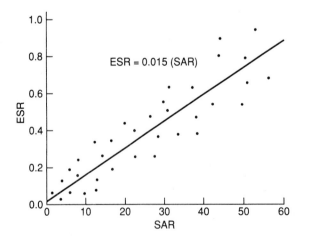

FIGURE 3.22   Relationship between ESR and SAR in salt-affected soils.

$$ESP = \frac{100(ESR)}{1 + ESR}$$

These parameters and the relationships between them are extremely valuable in characterizing the solution and exchange chemistry of salt- and Na-affected soils. The following example illustrates how these relationships can be used. Additional examples are provided in the instructors' manual.

A soil analysis revealed that the saturated extract contained 20 meq Ca^{2+}/l; 10 meq Mg^{2+}/l; and 100 meq Na^+/l. EC_{se} = 2.2 mmho/cm; soil pH = 8.6; CEC = 25 meq/100 g. Evaluate this soil for potential salinity or sodicity problems.

$$SAR = \frac{100}{\sqrt{\dfrac{20 + 10}{2}}} = 25.8$$

$$ESR = 0.015(25.8) = 0.39$$

$$ESP = \frac{100(0.39)}{1.39} = 28\%$$

Since the $EC_{se} <$ 4 mmho/cm and the ESP > 15%, this soil would be classified as *sodic*. Gypsum application would likely be recommended to reduce the ESP (see p. 84 for example calculation).

Effects on Plant Growth

Growth inhibition in salt-sensitive crops, even at low salinity levels, is caused primarily by toxicity from Na^+ and Cl^-. The high osmotic pressure in the soil solution causes a correspondingly low soil water potential, and, when in contact with a plant cell, the solute moves toward the soil solution and the cell collapses (called *plasmolysis*).

Soil salinity limits plant growth by (1) a water imbalance in the plant (physiological drought) and/or (2) ionic imbalances that result in increased energy consumption (carbohydrate respiration) to maintain metabolic processes. Salt-affected plants may exhibit stunted growth and have darker green leaf color. In woody species, excessive soil salinity may cause substantial leaf burn. Saline irrigation water can also result in leaf burn, depending on the crop (Table 3.18). As soil salinity increases above threshold levels (Table 3.19), the plant growth rate decreases. Top growth is usually affected more than root growth.

TABLE 3.18   Sensitivity of Several Crops to Leaf Burn Caused by Chloride Salts Applied by Sprinkler Irrigation

Tolerant	Semitolerant	Sensitive	Very Sensitive
Cotton	Barley	Alfalfa	Potato
Sugar beet	Corn	Sesame	Tomato
Sunflower	Safflower	Soybeans	Fruit crops
	Sorghum		Citrus fruits

SOURCE: Meas et al., *Irrigation Science*, (1982).

TABLE 3.19   Salt Tolerance of Selected Crops

Crop	Threshold EC_{se} (mmho/cm)	% Yield Decrease/ Unit EC_{se} Increase	Salt Tolerance Rating*
Alfalfa	2.0	7.3	MS
Almond	1.5	18	S
Apple	1.0	15	S
Apricot	1.6	23	S
Avocado	1.0	24	S
Barley (Forage)	6.0	7.0	MT
Barley (Grain)	8.0	5.0	T
Bean	1.0	19	S
Beet (Garden)	4.0	9.0	MT
Bent Grass	—	—	MS
Bermuda Grass	6.9	6.4	T
Blackberry	1.5	22	S
Boysenberry	1.5	22	S
Broad Bean	1.6	9.6	MS
Broccoli	2.8	9.1	MT
Bromegrass	—	—	MT
Cabbage	1.8	9.7	MS
Canary Grass (Reed)	—	—	MS
Carrot	1.0	14	S
Clover (Red, Ladino, Alsike)	1.5	12	MS
Clover (Berseem)	1.5	5.8	MT
Corn (Forage)	1.8	7.4	MS
Corn (Grain, Sweet)	1.7	12	MS
Cotton	7.7	5.2	T
Cowpea	1.3	14	MS
Cucumber	2.5	13	MS
Date	4.0	3.6	T
Fescue (Tall)	3.9	5.3	MT
Flax	1.7	12	MS
Grape	1.5	9.5	MS
Grapefruit	1.8	16	S
Hardinggrass	4.6	7.6	MT

continued on page 79.

Although in most situations the decrease in yield is related to total salt concentration in the soil solution, excess soil salinity may induce nutrient imbalances (deficiencies or toxicities). For example, excessive SO_4^{2-} and low Ca and/or Mg can occur in saline soils, causing internal browning in lettuce, blossom-end rot in tomato and pepper, and blackheart in celery.

Plants differ greatly in their tolerance to soil salinity (Table 3.19). For example, old alfalfa is more tolerant than young alfalfa. Barley and cotton have considerable salt tolerance, but high salt content will affect vegetative growth more than grain or bolls of cotton. Cultivar or variety differences also exist. For example, soybean varieties differ in Cl^- exclusion (Table 3.20). Effective excluders of Na^+ and Cl^- may not be very productive because of salt-related water stress. Tolerant crops that do not exclude Na^+ have a capacity to maintain a high K^+/Na^+ ratio in the growing tissue. Conventional breeding and genetic engineering methods are being used to adapt crops to saline environments.

Table 3.19 (*continued*)

Crop	Threshold EC_{se} (mmho/cm)	% Yield Decrease/ Unit EC_{se} Increase	Salt Tolerance Rating*
Lemon	1.0	—	S
Lettuce	1.3	13	MS
Lovegrass	2.0	8.5	MS
Meadow Foxtail	1.5	9.7	MS
Onion	1.2	16	S
Orange	1.7	16	S
Orchard Grass	1.5	6.2	MT
Peach	3.2	19	S
Peanut	3.2	29	MS
Pepper	1.5	14	MS
Plum	1.5	18	S
Potato (White)	1.7	12	MS
Potato (Sweet)	1.5	11	MS
Radish	1.2	13	MS
Rice (Paddy)	3.0	12	MS
Ryegrass (Perennial)	5.6	7.6	MT
Sorghum	4.8	8.0	MT
Soybean	5.0	20	MT
Spinach	2.0	7.6	MS
Strawberry	1.0	33	S
Sudan Grass	2.8	4.3	MT
Sugar Beet	7.0	5.9	T
Sugarcane	1.7	5.9	MS
Tomato	2.5	9.9	MS
Trefoil (Big)	2.3	19	MS
Trefoil (Birdsfoot)	5.0	10	MT
Vetch (Common)	3.0	11	MS
Wheat	6.0	7.1	MT
Wheat Grass (Crested)	3.5	4.0	MT
Wheat Grass (Fairway)	7.5	6.9	T
Wheat Grass (Tall)	7.5	4.2	T
Wild Rye (Beardless)	2.7	6.0	MT

*S, sensitive; MS, moderately sensitive; MT, moderately tolerant; T, tolerant.

Plant tolerance to soil salinity is expressed as the yield decrease with a given amount of soluble salts compared with yield under nonsaline conditions. Threshold salinity levels have been established for most crops and represent the minimum salinity level (EC_{se}) above which salinity limits growth and/or yield (Table 3.19). These values represent general guidelines, since many interactions among plant, soil, water, and environmental factors influence salt tolerance. Above the threshold EC_{se} level, plant growth generally decreases linearly with increasing salinity. Therefore, the relative yield loss (*Y*) at any given EC_{se} level can be calculated for any crop from the values in Table 3.19 for threshold levels (*A*) and the percentage yield decrease per unit increase in EC_{se} level (*B*) above the threshold by the following:

$$Y = 100 - B(EC_{se} - A)$$

For example, alfalfa yield decreases about 7.3% per unit increase in EC_{se} above the 2.0 mmho/cm EC_{se} threshold. Thus, if a soil analysis showed 4.0 mmho/cm EC_{se}, then the estimated relative alfalfa yield would be

$$Y = 100 - 7.3(4.0 - 2.0) = 75.4\%$$

Using these linear relationships, plants can be categorized into groups based on sensitivity or tolerance to soil salinity (Fig. 3.23). These ratings are only relative but can be used to estimate yield depression at specific soil salinity levels.

TABLE 3.20   Leaf Scorch Ratings, Yield, and Cl^- Concentration in Leaves and Seeds of 5 Susceptible and 10 Tolerant Soybean Cultures

| | | | Concentration of Cl^- | |
	Leaf Scorch*	Yield (bu/a)	Leaves (%)	Seed (ppm)
Cl susceptible	3.4	15	1.67	682
Cl tolerant	1.0	24	0.09	111

*1 = none, 5 = severe.
SOURCE: Parker et al. (1986).

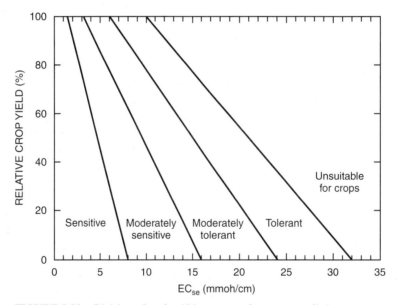

FIGURE 3.23   Divisions for classifying crop tolerance to salinity.

Factors Affecting Salt Tolerance

PLANT FACTORS   For some plants soil salinity influences growth at all growth stages, but for many crops sensitivity varies with the growth stage. For example, several grain crops (e.g., rice, wheat, corn, and barley) are relatively salt tolerant at germination and maturity but are very sensitive during early seedling and, in some cases, vegetative growth stages. In contrast, sugar beet, safflower, soybean, and many bean crops (including soybean) are sensitive during germination. This effect depends on variety, especially with soybean. The amount of growth reduction and/or yield loss often depends on the variety, particularly with many grasses and some legume crops. Differences in salt tolerance have also been observed between different vine and fruit tree rootstocks. Fruit tree and some vine crops are particularly sensitive to Cl toxicity; however, salt-tolerant varieties exhibit reduced Cl accumulation in the roots and/or Cl translocation from roots to above-ground tissues.

SOIL FACTORS   In general, crops grown on nutrient-deficient soils are more salt tolerant than the same crops grown in soils with sufficient levels of nutrients. Lower growth rates and water demand are likely causes for the increased tolerance to soil salinity. In these cases nutrient deficiency is the most limiting factor to maximum yield potential; thus, nutrient additions would increase plant growth and yield potential and subsequently decrease salt tolerance. Because saline and sodic soils have pH values that exceed 7.0, micronutrient deficiencies can be more common (Chapter 8). Overfertilization with N can decrease salt tolerance in some crops because of increased vegetative growth rates and subsequent water demand. At recommended rates, little or no effect on soil salinity or salt tolerance is observed with either inorganic or organic nutrient addition. Continued over-application of manure, as well as N and K fertilizers, can increase soil salinity, especially in poorly drained, irrigated soils. Overapplication with band-applied fertilizers containing relatively high concentrations of N and K can cause salt damage to germinating seeds and seedlings (see Chapter 10).

Proper irrigation management is essential to reducing soil salinity effects on plant growth and yield. Total salt concentration in the soil solution is the highest when the water content has been reduced to its relatively lowest level by evapotranspiration. With irrigation, salts in the soil solution are diluted and EC_{se} decreases. If soil salinity increases above threshold levels (Fig. 3.23) during dry periods, more frequent irrigation will be required to prevent water and salinity stress and thus negative effects on plant growth and yield. Also, the percentage of plant available water decreases with increasing salinity (higher osmotic potential), requiring more frequent irrigation. Excessive irrigation reduces aeration, especially in poorly drained soils, and can reduce salt tolerance in some plants. Under furrow irrigation conditions, excess water leaches into the furrow area; however, soil water movement from the bottom of the furrow to the midrow area deposits salts as water evaporates from the soil surface. Salt-sensitive crops must be planted to the side of the midrow to minimize salt injury to germinating seeds and seedlings (Fig. 3.24). Much lower water rates are used in trickle irrigation compared with furrow irrigation, resulting in greater salt accumulation and potential for yield losses.

ENVIRONMENTAL FACTORS   Under hot, dry conditions, most crops are less salt tolerant than under cool, humid conditions because of greatly increased evapotranspiration demand. These climatic effects of temperature and humidity on salt tolerance are particularly important with the most salt-sensitive crops (Table 3.19).

Managing Saline and Sodic Soils for Crop Production

Saline soils are relatively easy to reclaim if adequate amounts of low-salt irrigation waters are available and internal and surface drainage are feasible.

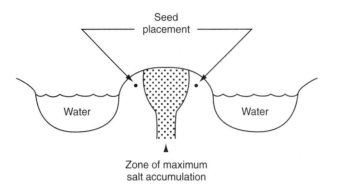

FIGURE 3.24   In furrow-irrigated cropping systems, salts accumulate near the center of the bed. Seeds should be planted on the side of the beds to avoid salinity problems. *Ludwick et al., CSUES, no. 504, 1978.*

Salts must be leached below the root zone and out of contact with subsequent irrigation water.

The quantity of irrigation water needed to leach the salts out of the root zone, or the *leaching requirement (LR),* can be estimated by the following relationship:

$$LR = \frac{EC_{iw}}{EC_{dw}}$$

$$\text{where } LR = \text{leaching requirement}$$
$$EC_{iw} = \text{EC of irrigation water}$$
$$EC_{dw} = \text{EC of drainage water}$$

LR represents the additional water needed to leach out the salts over that needed to saturate the profile. Although this relationship provides an estimate of the water volume needed to reduce the salts in the soil, more sophisticated calculations are generally used to estimate precisely the amount of leaching water needed. The amount of leaching water required depends on (1) the desired EC_{se}, which depends on the salt tolerance of the intended crop; (2) irrigation water quality (EC_{iw}); (3) rooting or leaching depth; and (4) soil water-holding capacity.

As seen in the LR calculation, the quality of the irrigation water used to leach salts below the root zone is an important factor in managing soil salinity. The EC and SAR of the available water must be determined before application. Based on these values, the quality of the water can be evaluated (Fig. 3.25). As the EC and SAR of the irrigation water increase, greater precautions should be taken in using it to leach salts below the root zone.

In soils with a high water table, drain installation may be required before leaching. If there is a dense calcareous or gypsiferous layer or the soil is impervious, deep chiseling or plowing may be used to improve infiltration. When only rainfall or limited irrigation is available, surface organic mulches will reduce evaporation and increase drainage.

Managing the soil to minimize salt accumulation is essential, especially in semi-arid and arid regions. Maintaining the soil near field capacity with frequent wa-

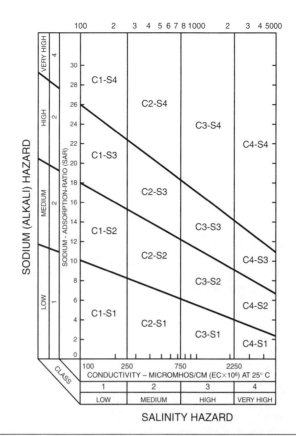

1. Salinity classification

C1 - Low-salinity water can be used for irrigation with most crops on most soils, with little likelihood that a salinity problem will develop. Some leaching is required, but this occurs under normal irrigation practices except in soils of extremely low permeability.

C2 - Medium-salinity water can be used if a moderate amount of leaching occurs. Plants with moderate salt tolerance can be grown in most instances without special practices for salinity control.

C3 - High-salinity water cannot be used on soil with restricted drainage. Even with adequate drainage, special management for salinity control may be required, and plants with good salt tolerance should be selected.

C4 - Very-high-salinity water is not suitable for irrigation under ordinary conditions but may be used occasionally under very special circumstances. The soil must be permeable, drainage must be adequate, irrigation water must be applied in excess to provide considerable leaching, and very-salt-tolerant crops should be selected.

2. Sodium classification

S1 - Low-sodium water can be used for irrigation on almost all soils with little danger of the development of a sodium problem. However, sodium-sensitive crops, such as stone-fruit trees and avocados, may accumulate injurious amounts of sodium in the leaves.

S2 - Medium-sodium water may present a moderate sodium problem in fine-textured (clay) soils unless there is gypsum in the soil. This water can be used on coarse-textured (sandy) or organic soils that take water well.

S3 - High-sodium water may produce troublesome sodium problems in most soils and will require special management, good drainage, high leaching, and additions of organic matter. If there is plenty of gypsum in the soil a serious problem may not develop for some time. If gypsum is not present, it or some similar material may have to be added.

S4 - Very-high-sodium water is generally unsatisfactory for irrigation except at low- or medium-salinity levels, where the use of gypsum or some other amendment makes it possible to use such water.

FIGURE 3.25   Diagram for the classification of irrigation waters.   *U.S. Salinity Laboratory, Handbook 60, 1954.*

tering dilutes salts. Light leaching before planting or light irrigation after planting moves salts below the planting and early rooting zone. If water is available, periodic leaching when crops are not growing will move salts out of the root zone. Much of the salt may precipitate as $CaSO_4 \cdot 2H_2O$ and $CaCO_3$ or $MgCO_3$ during dry periods and will not react as soluble salt, although precipitation of Ca and Mg will increase the proportion of Na^+ present in solution.

Managing soils for improved drainage is essential for controlling soil salinity. When ridge-tillage systems are used, the salt moves upward with capillary H_2O and is deposited on the center of the ridges where the water evaporates. Planting on the shoulders or edge of the ridges helps to avoid problems associated with excess salts (Fig. 3.24).

In sodic and saline-sodic soils, exchangeable Na and/or EC_{se} must be reduced, which can be difficult because the soil clay may be dispersed, preventing infiltration. Exchange of Na^+ is most often accomplished with Ca^{2+} by adding the appropriate rate of gypsum ($CaSO_4 \cdot 2H_2O$). The reaction is

$$
\text{clay}
\begin{bmatrix}
Na^+ \\
Na^+ \\
Ca^{2+} \\
Mg^{2+} \\
Na^+
\end{bmatrix}
+ Ca^{2+} + SO_4^{2-} \rightarrow \quad
\text{clay}
\begin{bmatrix}
Ca^{2+} \\
Ca^{2+} \\
Mg^{2+} \\
Na^+
\end{bmatrix}
+ 2Na^+ + SO_4^{2-}
$$
$$\text{leach below root zone}$$

Estimating the quantity of $CaSO_4 \cdot 2H_2O$ required is similar to the calculation for estimating the $CaCO_3$ required to increase pH. For example, a soil with CEC = 20 meq/100 g contains 15% ESP, and we need to reduce the ESP to 5%; thus, $15\% - 5\% = 10\%$ reduction in ESP.

$$
(0.10) \left(20 \, \frac{\text{meq CEC}}{100 \text{ g}} \right) = 2 \frac{\text{meq Na}^+}{100 \text{ g}} = 2 \frac{\text{meq CaSO}_4 \cdot 2H_2O}{100 \text{ g}}
$$

Thus,

$$
2 \frac{\text{meq CaSO}_4 \cdot 2H_2O}{100 \text{ g}} \left(86 \, \frac{\text{mg CaSO}_4 \cdot 2H_2O}{\text{meq}} \right) (20)
$$

$$
= \frac{3,440 \text{ lb CaSO}_4 \cdot 2H_2O}{a\text{-}6 \text{ in.}}
$$

Selected References

ADAMS, F. (Ed.). 1984. *Soil Acidity and Liming.* Soil Science Society of America, Madison, Wisc.

ALLEY, M. M., and L. W. ZELAZNY. 1987. Soil acidity: soil pH and lime needs. In J. R. BROWN (Ed.), *Soil Testing: Sampling, Correlation, Calibration, and Interpretation.* Special Publication No. 21. Soil Science Society of America, Madison, Wisc.

BRESLER, E., B. L. MCNEAL, and D. L. CARTER. 1982. *Saline and Sodic Soils.* Springer-Verlag, New York.

FOLLETT, R. H., L. S. MURPHY, and R. L. DONAHUE. 1981. *Fertilizers and Soil Amendments.* Prentice-Hall, Englewood Cliffs, N.J.

KAMPRATH, E. J., and C. D. FOX. 1985. Lime-fertilizer-plant interactions in acid soils. In O. P. ENGLESTAD (Ed.), *Fertilizer Technology and Use.* Soil Science Society of America, Madison, Wisc.

MCLEAN, E. O. 1973. Testing soils for pH and lime requirement. In L. M. WALSH and J. D. BEATON (Eds.), *Soil Testing and Plant Analysis.* Soil Science Society of America, Madison, Wisc.

CHAPTER 4

Nitrogen

The N Cycle

N is the most frequently deficient nutrient in crop production; therefore, most nonlegume cropping systems require N inputs. Many N sources are available for use in supplying N to crops. In addition to inorganic fertilizer N, organic N from animal manures and other waste products and from N_2 fixation by leguminous crops can supply sufficient N for optimum crop production. Understanding the behavior of N in the soil is essential for maximizing agricultural productivity and profitability while reducing the impacts of N fertilization on the environment.

The ultimate source of the N used by plants is N_2 gas, which constitutes 78% of the earth's atmosphere. Unfortunately, higher plants cannot metabolize N_2 directly into protein. N_2 must be converted first to plant available N by

1. Microorganisms that live symbiotically on the roots of legumes and certain nonleguminous plants.
2. Free-living or nonsymbiotic soil microorganisms.
3. Atmospheric electrical discharges forming N oxides.
4. Manufacture of synthetic N fertilizers.

The unlimited supply of atmospheric N_2 is in dynamic equilibrium with all fixed forms of N in soil, water, and living and nonliving organisms (Table 4.1). Approximately 78,000 metric tons of N_2 cover every hectare of land (sea level). As N_2 is fixed by these different processes, numerous microbial and chemical processes release N_2 to the atmosphere. The cycling of N in the soil-plant-atmosphere system involves many transformations of N between inorganic and organic forms (Fig. 4.1). The approximate distribution of N throughout the N cycle is shown in Table 4.1. The N cycle can be divided into N inputs or gains, N outputs or losses, and N cycling within the soil, where N is neither gained nor lost (Table 4.2). Except for industrial and combustion fixation, all of these N transformations occur naturally; however, humans influence many of these N processes through soil and crop management activities. The purpose of this chapter is to describe the chemical and microbial cycling of N and how humans influence or manage these transformations to optimize plant productivity.

TABLE 4.1   Approximate Distribution of N throughout the
Soil–Plant/Animal–Atmosphere System

Nitrogen	Metric Tons	Percentage of Total
Atmosphere	3.9×10^{15}	99.3840
Sea (various)	2.4×10^{13}	0.6116
Soil (nonliving)	1.5×10^{11}	0.0038
Plants	1.5×10^{10}	0.00038
Microbes in soil	6×10^9	0.00015
Animals (land)	2×10^8	0.000005
People	1×10^7	0.00000025

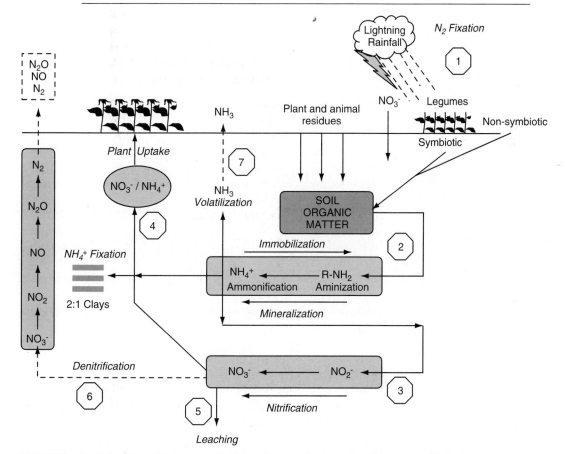

FIGURE 4.1   The N cycle. In step 1, N in plant and animal residues and N derived from the atmosphere through electrical, combustion, and industrial processes (N_2 is combined with H_2 or O_2) is added to the soil. In step 2, organic N in the residues is mineralized to NH_4^+ by soil organisms. Plant roots absorb a portion of the NH_4^+. In step 3, much of the NH_4^+ is converted to NO_3^- by nitrifying bacteria in a process called *nitrification*. In step 4, NO_3^- and NH_4^+ are taken up by plant roots and used to produce the protein in crops that are eaten by humans or fed to livestock. In step 5, some NO_3^- is lost to groundwater or drainage systems as a result of downward movement through the soil in percolating water. In step 6, some NO_3^- is converted by denitrifying bacteria into N_2 and nitrogen oxides (N_2O and NO) that escape into the atmosphere, completing the cycle. In step 7, some NH_4^+ can be converted to NH_3 through a process called volatilization.

TABLE 4.2   N Inputs, Outputs, and Cycling in the Soil-Plant-Atmosphere System*

N Inputs	N Outputs	N Cycling
Fixation	Plant uptake	Immobilization
Biological	Denitrification	Mineralization
Industrial	Volatilization	Nitrification
Electrical	Leaching	
Combustion	Ammonium fixation[†]	
Animal manure		
Crop residue		

[*]Some N output and cycling components can be influenced by management but generally are not managed.

[†]Some fixed ammonium can be released to soil.

Functions and Forms of N in Plants

Forms

Plants normally contain 1 to 5% N by weight and absorb N as both nitrate (NO_3^-) and ammonium (NH_4^+) (Fig. 4.1). In moist, warm, well-aerated soils NO_3^- generally occurs in higher concentrations than NH_4^+. Both ions move to plant roots by mass flow and diffusion.

The rate of NO_3^- uptake is usually high and is favored by low-pH conditions. When plants absorb high levels of NO_3^-, there is an increase in organic anion synthesis within the plant coupled with a corresponding increase in the accumulation of inorganic cations (Ca, Mg, K). The growth medium will become alkaline, and some HCO_3^- can be released from the roots to maintain electroneutrality in the plant and in the soil solution.

NH_4^+ is the preferred N source since energy will be saved when it is used instead of NO_3^- for synthesis of protein. NO_3^- reduction is an energy-requiring process that uses two nitrate reductase (NADH) molecules for each NO_3^- ion reduced in protein synthesis. Plants supplied with NH_4^+ may have increased carbohydrate and protein levels compared with NO_3^-.

Plant uptake of NH_4^+ proceeds best at neutral pH values and is depressed by increasing acidity. Absorption of NH_4^+ by roots reduces Ca^{2+}, Mg^{2+}, and K^+ uptake while increasing absorption of $H_2PO_4^-$, SO_4^{2-}, and Cl^-.

Rhizosphere pH decreases when plants receive NH_4^+, caused by H^+ exuded by the root to maintain electroneutrality or charge balance inside the plant. Differences in 2 pH units have been observed for NH_4^+ versus NO_3^- uptake. This acidification can affect both nutrient availability and biological activity in the vicinity of roots.

NH_4^+ tolerance limits are narrow. High levels of NH_4^+ can retard growth, restrict uptake of K^+, and produce symptoms of K^+ deficiency. In contrast, plants tolerate large excesses of NO_3^- and accumulate it to comparatively high levels in their tissues.

Preference of plants for either NH_4^+ or NO_3^- is determined by the age and type of plant, the environment, and other factors. Cereals, corn, sugar beets, pineapple, rice, and ryegrass use either form of N. Kale, celery, bush beans, and squash

TABLE 4.3   Responses of Corn Hybrids When Supplied with Differing N Sources: Plants Grown under Field Conditions in a Gravel-Hydroponic System

Year	Hybrid	N Source	Grain Yield (g/plant)	Kernel Number (no./plant)	Kernel Weight (mg/kernel)
1986	B73 × LH51	All NO_3	254	688	369
		NO_3/NH_4	275	764	361
	FS 854	All NO_3	277	818	339
		NO_3/NH_4	315	1,000	315
1987	B73 × LH51	All NO_3	154	540	285
		NO_3/NH_4	193	691	279
	B73 × LH38	All NO_3	161	603	267
		NO_3/NH_4	180	742	243
	CB59G × LH38	All NO_3	137	475	288
		NO_3/NH_4	154	545	283
	LH74 × LH51	All NO_3	181	592	306
		NO_3/NH_4	199	607	328

SOURCE: Below and Gentry, *Better Crops*, 72(2): (1988).

grow best when provided with some NO_3^-. Some plants, such as blueberries, *Chenopodium album,* and certain rice cultivars, cannot tolerate NO_3^-. Solanaceous crops, such as tobacco, tomato, and potato, prefer a high NO_3^-/NH_4^+ ratio.

Plant growth is often improved when the plants are nourished with both NO_3^- and NH_4^+ compared with either NO_3^- or NH_4^+ alone. Mixtures of these forms are beneficial at certain growth stages for some genotypes of corn, sorghum, soybeans, wheat, and barley. Increased nonlegume yields with NH_4^+ nutrition are associated with greater tillering. Corn yields increased from 8 to 25% with $NH_4^+ + NO_3^-$, compared with yields with NO_3^- alone, which was related to increased numbers of kernels per plant and not to heavier kernels (Table 4.3). These data also illustrate that corn genotypes differ in their physiological response to NH_4^+.

Recent research results demonstrate that NH_4^+ application postsilking or during grain fill was required to maximize corn yields and that a 50:50 ratio of NH_4^+ to NO_3^- was optimum.

$NH_4^+ + NO_3^-$ nutrition is a major factor influencing the occurrence and severity of plant diseases. Some diseases are more severe when NH_4^+ is the primary form of inorganic N in the root zone; others are more severe when NO_3^- predominates. Two processes may be involved. One is the direct effect of the form of N on pathogenic activity. The other is the influence of NO_3^- or NH_4^+ on the functioning of organisms capable of altering the availability of micronutrient cations. For example, a high NO_3^- supply stimulates certain bacteria, which lowers the availability of Mn to wheat. The effect of N form on rhizosphere soil pH is also partially responsible for the differences observed in the incidence and severity of diseases.

Functions

Before NO_3^- can be used in the plant, it must be reduced to NH_4^+ or NH_3. Nitrate reduction involves two enzyme-catalyzed reactions that occur in roots and/or leaves, depending on the plant species. Both reactions occur in series so that toxic nitrite (NO_2^-) does not accumulate.

	Reduction Reaction	Enzyme	Reaction Site
Step 1	$NO_3 \rightarrow NO_2$	Nitrate reductase	Cytoplasm
Step 2	$NO_2 \rightarrow NH_3$	Nitrite reductase	Chloroplast

The NH_3 produced in these reactions is assimilated into amino acids that are subsequently incorporated into proteins and nucleic acids. Proteins provide the framework for chloroplasts, mitochondria, and other structures in which most biochemical reactions occur. The type of protein formed is controlled by a specific genetic code found in nucleic acids, which determines the quantity and arrangement of amino acids in each protein. One of these nucleic acids, deoxyribonucleic acid (DNA), present in the nucleus and mitochondria of the cell, duplicates the genetic information in the chromosomes of the parent cell to the daughter cell. Ribonucleic acid (RNA), present in the nucleus and cytoplasm of the cell, executes the instructions coded within DNA molecules. Most of the enzymes controlling these metabolic processes are also proteins and are continually metabolized and resynthesized.

In addition to the formation of proteins, N is an integral part of chlorophyll, which is the primary absorber of light energy needed for photosynthesis. The basic chlorophyll structure is the porphyrin ring, composed of four pyrrole rings, each containing one N and four C atoms (Fig. 4.2). A single Mg atom is bonded in the center of each porphyrin ring.

An adequate supply of N is associated with high photosynthetic activity, vigorous vegetative growth, and a dark green color. An excess of N in relation to other nutrients, such as P, K, and S, can delay crop maturity. Stimulation of heavy vegetative growth early in the growing season can be a disadvantage in regions where soil moisture limits plant growth. Early-season depletion of soil moisture without adequate replenishment before the grain-filling period can depress yields.

If N is used properly in conjunction with other needed inputs, it can speed the maturity of crops (Table 4.4). Applications of up to 120 lb N/a lowered the

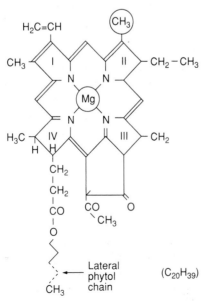

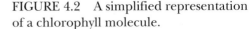

FIGURE 4.2   A simplified representation of a chlorophyll molecule.

TABLE 4.4   Effect of N on the Moisture Content and Yield of Corn Grain (Averages for the Years 1967–77)

N (lb/a)	Yield (bu/a)	Moisture in Grain at Harvest (%)
0	66	36.1
60	101	30.0
120	135	27.9
180	158	26.9
240	167	28.2
300	168	27.2

SOURCE: Ohio State Univ., *17th Annu. Agron. Demonstration, Farm Sci. Rev.* (1979).

moisture content in corn grain at harvest. This favorable influence lessens the energy required to dry grain to 15.5% moisture or permits an earlier harvest.

The supply of N influences carbohydrate utilization. When N supplies are low, carbohydrates will be deposited in vegetative cells, causing them to thicken. When N supplies are adequate and conditions are favorable for growth, proteins are formed from the manufactured carbohydrates. With less carbohydrate deposited in the vegetative portion, more protoplasm is formed, and, because protoplasm is highly hydrated, a more succulent plant results.

Excessive succulence in some crops may have a harmful effect. With cotton, a weakening of the fiber may result, and with grain crops, lodging may occur, particularly with a low K supply or with varieties not adapted to high levels of N. In some cases, excessive succulence may make a plant more susceptible to disease or insect attack. Crops such as wheat and rice have been modified for growth at higher densities and at higher levels of N fertilization. Shorter plant height and improved lodging resistance have been bred into the plants, which respond in yield to much higher rates of N than in the past.

Visual Deficiency Symptoms

When plants are deficient in N, they become stunted and yellow in appearance. The loss of protein N from chloroplasts in older leaves produces the yellowing, or *chlorosis,* indicative of N deficiency. Chlorosis usually appears first on the lower leaves, the upper leaves remaining green; under severe N deficiency, lower leaves turn brown and die. This necrosis begins at the leaf tip and progresses along the midrib until the entire leaf is dead (see color plates inside book cover).

The tendency of the young upper leaves to remain green as the lower leaves yellow or die is an indication of the mobility of N in the plant. When the roots are unable to absorb sufficient N to meet their growing requirement, protein in the older plant parts is converted to soluble N, translocated to the active meristematic tissues, and reused in the synthesis of new protein.

N Addition from the Atmosphere

N compounds in the atmosphere are returned to the earth in rainfall as NH_4^+, NO_3^-, NO_2^-, nitrous oxide (N_2O), and organic N. Because of the small amount of NO_2^- present in the atmosphere, NO_2^- and NO_3^- are combined and reported as NO_3^-. The presence of NO_3^- has been attributed to its formation during atmospheric electrical discharges, but recent studies suggest that only about 10 to 20% of the NO_3^- in rainfall is due to lightning. The remainder probably comes from industrial waste gases or possibly from the soil. Ammonium comes largely from industrial sites where NH_3 is used or manufactured. Ammonia also escapes (volatilization) from the soil surface. The organic N accumulates as finely divided organic residues that are swept into the atmosphere from the earth's surface.

The total amount of N in rainfall ranges between 1 and 50 lb/a/yr, depending on the location. Total N deposition is higher around areas of intense industrial activity and, as a rule, is greater in tropical than in polar or temperate zones. NH_4^+ deposition represents the majority of the total atmospheric N deposited and has increased with time, especially in highly populated and industrial regions (Fig. 4.3). For example, in eastern North Carolina, NH_4^+ deposition has doubled

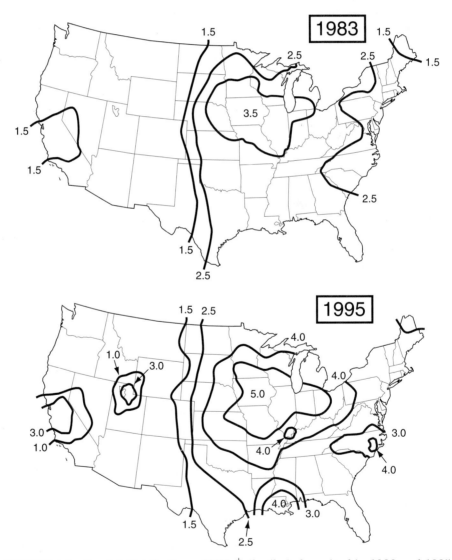

FIGURE 4.3 Spatial distribution of NH_4^+ (kg/ha) deposited in 1983 and 1995 illustrates increasing trend in N deposition. *National Atmospheric Deposition Program, 1996.*

over the last 20 years as a result of human and animal (hogs and poultry) population increases (Fig. 4.4). Figure 4.5 illustrates localized NH_4^+ deposition from a poultry operation that can depress soil pH and reduce crop yield potential if the field is not limed (see Chapter 3). Soil has a pronounced capacity for adsorbing NH_3 gas from the atmosphere. In localized areas where atmospheric NH_3 concentrations are high, 50 to 70 lb NH_3/a/yr may be adsorbed by soils. Sorption is positively related to NH_3 concentration and to temperature.

N Fixation by Rhizobium and Other Symbiotic Bacteria

Biological N_2 Fixation

Many organisms have the unique ability to fix atmospheric N_2 (Table 4.5). Estimates of total annual biological N_2 fixation worldwide range from 130 to 180 × 10^6 metric tons, with about half of this amount fixed by *Rhizobia*. In contrast, world fertilizer N use was about 80×10^6 metric tons in 1995. In the United States, the reliance on biological N_2 fixation for crop production has declined dramatically since the 1950s because of increased production and use of low-cost synthetic N fertilizers (Fig. 4.6). Currently only about 20% of the N applied to crops in the United States is from legumes and crop residues (Fig. 4.7).

N Fixation by Legumes

Symbiotic bacteria (*Rhizobia*) fix N_2 in nodules present on the roots of legumes (Fig. 4.8). This N may be utilized by the host plant, excreted from the nodule into the soil and be used by other nearby plants, or released as nodules or legume residues decompose after the plant dies or is incorporated into the soil.

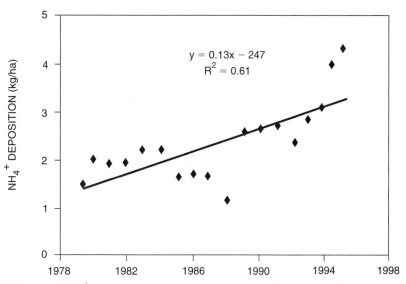

FIGURE 4.4 NH_4^+ deposition in Sampson County, North Carolina, has doubled over the last 20 years. This increase is likely caused by dramatic increases in urban population and confinement hog production.   *National Atmospheric Deposition Program, 1996.*

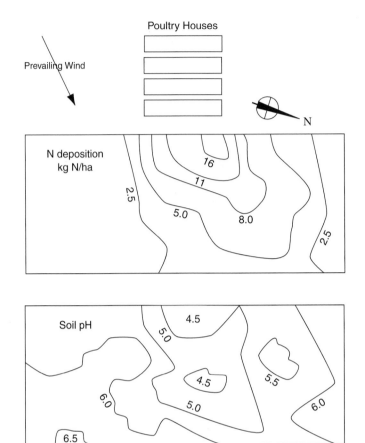

FIGURE 4.5   Ammonia (NH_3) produced from the poultry houses (20,000 animals) deposits ammonium (NH_4^+) downwind and reduces soil pH. NH_4^+ deposition data were collected over one month. Field is approximately 20 acres and poultry houses had been operated for 18 years before air and soil sampling in 1986.   *Speirs and Frost,* Research & Development in Agriculture, *1987.*

TABLE 4.5   Economically Important Microorganisms Involved in Biological N Fixation

Organisms	General Properties	Use in Agriculture
Azotobacter	Aerobic; free fixers; live in soil, water, rhizosphere (area surrounding the roots), leaf surfaces	Proposed benefit to crops has not been confirmed; hormonal effect on root and plant growth
Azospirillum	Microaerobic; free fixers; or found in association with roots of grasses	Potential use in increasing yield of grasses; inoculation benefits crops
	Inside root symbiosis?	Hormonal effect on roots and plant growth
Rhizobium	Fix N in legume-*Rhizobium* symbiosis	Legume crops are benefited by inoculation with proper strains
Actinomycetes, Frankia	Fix N in symbiosis with nonlegume wood trees—alder, *Myrica, Casuarina*	Potentially important in reforestation, wood production
Blue-green algae, *Anabaena*	Contain chlorophyll, as in higher plants; aquatic and terrestrial	Enhance rice in paddy soils; *Azolla* (a water fern)–*Anabaena azolla* symbiosis; used as green manure

SOURCE: Okon, *Phosphorus Agr.,* 82:3 (1982).

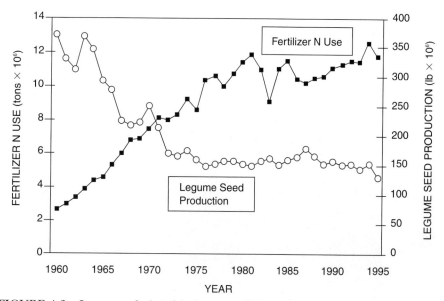

FIGURE 4.6   Inverse relationship between N fertilizer use and legume seed production in the United States.

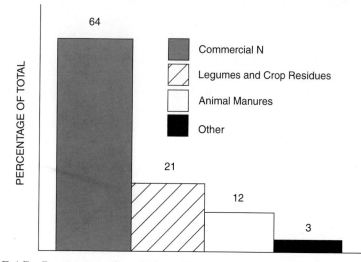

FIGURE 4.7   Percentage of total N applied to U.S. crops by source.   *USDA, 1992.*

Numerous *Rhizobium* species exist, each requiring a specific host legume plant. For example, bacteria that live symbiotically with soybeans will not fix N_2 with alfalfa. Inoculation of the legume seed with the correct inoculum is recommended the first time a field is planted to a new legume species. For example, 40% increases in N fixation by alfalfa have been obtained by carefully matching cultivars and strains of inoculum.

The presence of nodules on legume roots does not necessarily indicate N fixation by active *Rhizobia*. Mature effective alfalfa nodules tend to be large, elongated (2 to 4 by 4 to 8 mm), often clustered on the primary roots, and have pink to red centers. The red color is attributed to the occurrence of leghemoglobin, which

FIGURE 4.8   Example of *Rhizobia* nodules on soybean.   *Courtesy Dr. Dan Israel, NCSU.*

indicates *Rhizobia* are fixing N_2. Ineffective nodules are small (<2 mm in diameter), usually numerous, and scattered over the entire root system. In some cases, they are very large (>8 mm in diameter), few in number, and have white or pale-green centers.

Quantity of N Fixed

Generally, nodule bacteria fix 25 to 80% of the total N in the legume (Table 4.6). N fixation by most perennial legumes ranges from 100 to 200 lb/a/yr, although under optimum conditions N fixation can reach two to three times these values. Short-season annual legumes often fix between 50 and 100 lb N/a/yr. The data in Table 4.7 show that N fixation represented about two-thirds of the total N uptake in the first year of alfalfa production.

The amount of N_2 fixed by *Rhizobia* varies with the yield level; the effectiveness of inoculation; the N obtained from the soil, either from decomposition of OM or from residual N; and environmental conditions. A high-yielding legume crop such as soybeans, alfalfa, or clover contains large amounts of N (Table 4.8).

Soybeans remove about 1.5 lb N/bu from the soil and fix 40% or more of the total N in the plant. However, on lighter-colored soils, soybeans may fix 80% or more. In many environments, the quantity of N removed by soybean grain at harvest exceeds the quantity of N_2 fixed (Table 4.9). For example, when only 40% of the grain N was due to N_2 fixation, soil N exported in the grain exceeded N fixed in the grain, and thus soil N was depleted (−74 lb/a of N). In contrast, when 90% of the N was fixed, soil N was increased (+22 lb/a of N).

TABLE 4.6   N Fixed by Legumes in Temperate Climates

| Legume | N Fixed [lb/a/yr] | |
	Range in Reported Values	Typical
Alfalfa	50–300	200
Beans	20–80	40
Black gram	80–140	100
Chickpeas	20–100	50
Clovers (general)	50–300	150
Cluster beans	30–200	60
Cowpeas	60–120	90
Crimson clover	30–180	125
Fababeans	50–200	130
Green gram	30–60	40
Hairy vetch	50–200	100
Kudzu	20–150	110
Ladino clover	60–240	180
Lentils	40–130	60
Lespedezas (annual)	30–120	85
Peanuts	20–200	60
Peas	30–180	70
Red clover	70–160	115
Soybeans	40–260	100
Sweet clover	20–160	20
Trefoil	30–150	105
Vetch	80–140	80
White clover	30–150	100
Winter peas	10–80	50

TABLE 4.7   Nitrogen Budget for Seeding Year Alfalfa Showing the Allocation of Symbiotically Fixed and Soil-Derived N among Plant Parts

| Nitrogen Budget Component | Seeding Year Harvests | | |
	First (July 12)	Second (Aug. 30)	Third (Oct. 20)
Herbage yield (kg ha^{-1})	3503	3054	1156
Total N yield (herbage, crown, and roots) (kg of N ha^{-1})	118	127	59
Total N$_2$ fixed (kg of N ha^{-1})	57	102	34
Herbage	52	74	22
Roots and crown	5	28	12
Nitrogen from soil (kg of N ha^{-1})	61	25	25
Herbage	54	18	16
Roots and crown	7	7	9

SOURCE: Heichel and Barnes, *Crop Sci,* 21: 330–335 (1981).

Factors Affecting N Fixation

The most important factors influencing the quantity of N fixation by *Rhizobia* are soil pH, mineral nutrient status, photosynthetic activity, climate, and legume management. Any stress in the legume plant by these factors can severely reduce legume yield and N availability to subsequent crops.

TABLE 4.8   Variation of N Fixation Capacity with Legume Species, Legume Productivity, and Initial Soil N Concentration

Species	N from Symbiosis by Harvest %				Dry Matter Yield (lb/a)
	1	2	3	Mean	
Hay and pasture legumes					
Alfalfa*	49	81	58	63	6,809
Red clover*	51	79	65	65	6,230
Birdsfoot trefoil*	27	67	25	40	4,880
Harvest at Grain Maturity					
Grain legumes					
Soybean†		76			2,494
Soybean‡		52			7,837

*3.7% soil OM and 12 ppm soil NO_3-N concentration (0- to 6-in. depth).

†1.8% soil OM and 12 ppm soil NO_3-N concentration (0- to 8-in. depth).

‡4.8% soil OM and 31 ppm soil NO_3-N concentration (0- to 8-in. depth).

SOURCE: Heichel et al., *Crop Sci.*, 21:330–35 (1981).

SOIL pH   Soil acidity can restrict the survival and growth of *Rhizobia* in soil and severely affect nodulation and N fixation processes (Fig. 4.9). *Rhizobia* and roots of the host legume plants can be injured by Al^{3+}, Mn^{2+}, and H^+ toxicity, as well as low levels of available Ca^{2+} and $H_2PO_4^-$.

There are significant differences in the sensitivity of the various *Rhizobial* species to soil acidity. Soil pH values below 6.0 drastically reduce the number of *Rhizobium meliloti* in the root zone of alfalfa, degree of nodulation, and yields of host alfalfa plants, whereas soil pH values between 5.0 and 7.0 have little influence on *R. trifoli* and the host red clover crop.

Liming acidic soils improves conditions for crops such as alfalfa that are dependent on *R. meliloti*. For locations where economical sources of lime may not be available, alternative means of growing alfalfa must be used. Establishing alfalfa under acidic soil conditions where excessive amounts of soluble Mn^{2+} and Al^{3+} are not a problem requires high levels of inoculum and rolling inoculated seeds in a slurry of pulverized lime. Another approach is to select and use acid-tolerant strains of *Rhizobium* (Fig. 4.9).

MINERAL NUTRIENT STATUS   Except in acidic soils, where Ca^{2+} and $H_2PO_4^-$ deficiencies can limit the growth of *Rhizobia,* mineral deficiencies seldom reduce N fixation. N fixation in the nodule requires more Mo than the host plant; thus, Mo deficiency is the most important micronutrient deficiency. Initiation and development of nodules can be affected by Co, B, Fe, and Cu deficiencies. Differences exist in the sensitivity of various *Rhizobia* strains to nutrient stress.

Maximum N fixation occurs only when available soil N is at a minimum. Excess NO_3^- concentration in the soil can reduce nitrogenase activity and thus *Rhizobial* activity and N fixation. Nodules lose their pink color in a high-NO_3^- soil. The reduction in N fixation is related to the competition for photosynthate between NO_3^- reduction and N fixation reactions. The data in Table 4.8

TABLE 4.9   Nitrogen Budget of Soybeans Illustrating the Allocation of Soil and Symbiotic N among Plant Components and the Net Return of N to the Soil with 40 and 90% of Plant Nitrogen from Symbiosis

Crop Component	Dry Matter Content	Total Reduced N Content	Content of Symbiotic N		Soil N Export in Grain		Symbiotic N Return in Residue		Loss (−) or Gain (+) of N	
			40% N Symbiosis	90% N Symbiosis	40% N	90% N	40% N	90% N	40% N	90% N
					lb/a					
Grain	2,100	151	61	136	90	15	—	—	—	—
Residue*	3,424	40	16	37	—	—	16	37	—	—
Total plant	5,524	191	77	173	—	—	—	—	−74	+22

*Pod walls, leaves, stems, roots, and nodules; incomplete grain harvest would increase this value.
SOURCE: Heichel and Barnes, *ASA Spec. Publ. 46,* pp. 46–59 (1984).

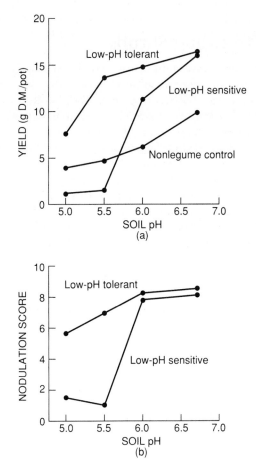

FIGURE 4.9   Forage yields (a) and nodulation scores (b) of alfalfa inoculated with low-pH-tolerant and low-pH-sensitive strains of *Rhizobium meliloti*. Barley was the nonlegume control.   *Rice,* Can. J. Plant Sci. *62:943, 1982.*

also show that legumes grown on soils low in profile NO_3^- obtain more of their N by fixation compared with legumes grown on high-N soils. However, it is sometimes advisable to include a small amount of N fertilizer at planting time to ensure that the young legume seedlings have an adequate supply until the rhizobia can become established on the roots. Early spring N application can be beneficial for legume crops where rhizobial activity is restricted by cold, wet conditions. N fixation by common bean is low and usually unreliable, and N fertilization is recommended.

PHOTOSYNTHESIS AND CLIMATE   A high rate of photosynthate production is strongly related to increased N fixation by *Rhizobia*. Factors that reduce the rate of photosynthesis will reduce N fixation. These factors include reduced light intensity, moisture stress, and low temperature.

LEGUME MANAGEMENT   In general, any management practice that results in reduced legume stands or yield will reduce the quantity of N fixed by legumes. These factors include water and nutrient stress, excessive weed and insect pressure, and improper harvest management. Harvest practices vary greatly with location, but excessive cutting frequency, premature harvest, and delayed harvest, especially in the fall, can reduce legume stands and the quantity of N fixed.

Fixation by Leguminous Trees and Shrubs

N fixation by leguminous trees is important to the ecology of tropical and subtropical forests and to agroforestry systems in developing countries. Numerous leguminous tree species fix appreciable amounts of N. Well-known examples in the United States are *Mimosa, Acacia,* and *black locust.* Three woody leguminous species—*Gliricidia sepium, Leucaena leucocephala,* and *Sesbania biospinosa*—are used as green manure crops in rice-based cropping systems.

Some widely distributed nonleguminous plants also fix N by a mechanism similar to legume and *Rhizobia* symbiosis. Certain members of the following plant families are known to bear root nodules and to fix N: Betulaceae, Elaegnaceae, Myricaceae, Coriariaceae, Rhamnaceae, and Casurinaceae. Alder and *Ceanothus,* two species commonly found in the Douglas fir forest region of the Pacific Northwest, can potentially contribute substantial N to the ecosystem. *Frankia,* an actinomycete, is responsible for N fixation by these nonleguminous woody plants (Table 4.5).

Legume N Availability to Nonlegume Crops

Yields of nonlegume crops are often increased when they are grown following legumes. For example, when corn follows soybeans, the amount of N required for optimum yield is less than that required for corn after corn (Fig. 4.10). Although the difference has been attributed to increased N availability from the previous legume crop, other crop rotation effects are involved. Similar responses have been observed with corn following wheat as with corn after soybeans. These results strongly suggest that rotation benefits are not limited to

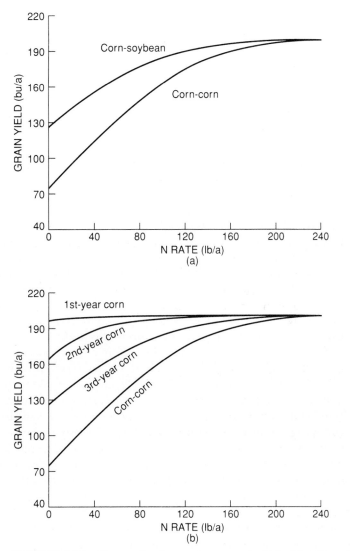

FIGURE 4.10   Generalized corn response to N fertilization when corn follows an annual legume (a) or a perennial legume (b) compared with continuous corn.

legume rotations and that some of the benefits can be related to increased soil N availability.

When a perennial legume such as alfalfa is used in rotation, the response of corn to applied N varies with time (Fig. 4.10). Little or no response of corn to N fertilization is observed in the first year; however, the amount of N required for optimum crop production increases with time as the legume N reserves are depleted. The quantity of fixed N available in the cropping system depends on (1) the quantity of N fixed and (2) the amount and type of legume residue returned. The quantity of legume N incorporated into the soil from first-year alfalfa varies between 35 and 300 kg/ha. Generally, alfalfa can supply all or most

TABLE 4.10   Yield and N Uptake of Barley Grown after Legumes

	Yield of Barley (bu/a)			N Uptake of Barley (lb/a)		
	No Legume	Alfalfa*	Red Clover*	No Legume	Alfalfa	Red Clover
1970	66	41	70	59.4	44	68.2
1971	27	51	51	26.4	63.8	22
1972	26	50	40	26.4	55	41.8
1973	32	52	48	26.4	46.2	33
1974	27	35	37	19.8	28.6	24.2
1975	22	31	26	—	—	—
Total	200	260	272	158.4	237.6	189.2
Mean	34	43	45	—	—	—

*Grown in 1968 and 1969.

SOURCE: Leitch, in *Alfalfa Production in the Peace River Region*, pp. C1–C5. Alberta Agriculture and Agriculture Canada Research Station Beaverlodge, Alberta (1976).

of the N to a nonlegume crop in the first year. Several studies suggest that the N credit commonly attributed to legumes in rotation is overestimated. These contrasting results can probably be explained by soil, climate, and legume management effects.

Crop utilization of N in green manure crops is also highly variable. Legume residue N availability during the first subsequent cropping year ranges between 20 and 50%. Table 4.10 shows that barley grown for a 5-year period after legumes contained a total of 30 to 80 lb of N and yielded 60 to 70 bu more in total than unfertilized (N) barley not following legumes. It can also be seen that the beneficial effects of legumes were most pronounced in the first few years after plowing down the legumes but that some residual effects continued even after 5 years.

The yield benefit of rotations with some legumes may not always be related to the legume N supply. The data in Figure 4.11 illustrate that the corn yield response to fertilizer N was similar in a corn crop following either soybeans or wheat. In this case, the rotation response compared with that of continuous corn is commonly referred to as the *rotation effect* and will be discussed in the next section.

With forage legumes, only part of the N fixed is returned to the soil because most of the forage would be harvested. Forages grown for green manure or as winter cover crops most likely return more fixed N to the soil, depending on species, yield, and management. For example, legume N availability can be greater in a one-cut system compared with a three-cut system because of the increased amount of N incorporated with less frequent harvests (Fig. 4.11). Whether the nonlegume yield response following a legume is due to N or to a rotation effect, the benefit can be observed for several years (Fig. 4.10).

Optimum utilization of legume N by a nonlegume grain crop requires that mineralization of legume N occur over the same time as crop N uptake. Legume N mineralization by soil microbes is controlled by climate, increasing with soil temperature and moisture, as previously noted. The period of N uptake by the crop can vary, depending on the crop. Thus, for maximum utilization of legume N by the nonlegume crop, N uptake must be in synchrony with

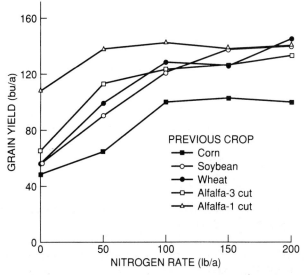

FIGURE 4.11   Corn grain yields as influenced by previous crop and fertilizer N.   *Heichel,* Role of Legumes in Conservation Tillage Systems, *Soil Cons. Serv. Am., p. 33, 1987.*

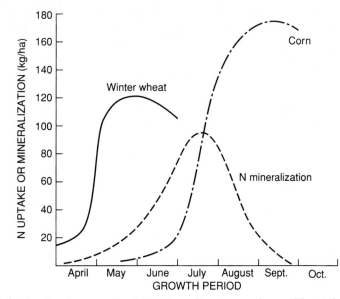

FIGURE 4.12   Synchrony of soil N mineralization and crop N uptake in corn and winter wheat.

N mineralization. For example, the N uptake period for winter wheat is considerably earlier than for corn (Fig. 4.12). The hypothetical distribution of N mineralization shows that corn N uptake is more synchronous with N mineralization than is winter wheat. Therefore, compared with corn, winter wheat may not utilize much legume N and, when mineralization occurs, the inorganic N

TABLE 4.11   Maximum N Transfer from Legumes to Grass; Legume/Grass Ratio Was 2:3 for the Reed Canary–Grass–Alfalfa Community and 3:2 for the Reed Canary–Grass–Trefoil Community

Community	N in Grass from Legume (%)*		
	Harvest 1	Harvest 2	Harvest 3
Grass-alfalfa	64	68	68
Grass-trefoil	68	66	79

*Percentage (proportion × 100, mass basis) of accumulated grass N received from legume via N transfer.

SOURCE: Brophy et al., *Crop Sci.*, 27:372–80 (1987).

is subject to leaching and other losses. Therefore, efficient management of legume N requires careful crop selection.

Legumes in combination with grasses for forage generally supply N for both crops. The data in Table 4.11 show that approximately 70% of the N in the grass originated from N supplied by the legume.

Legume N availability to a companion crop is not well understood. Small amounts of amino acids and other organic N compounds may be excreted by the legume roots. Microbial decomposition of the sloughed-off root and nodule tissue may also contribute N to the crop growing with legumes. Under some conditions, the quantity of fixed N and/or legume N availability is not sufficient, and N fertilization is required for optimum production of both nonlegume and legume crops.

Legume Rotations

One of the reasons for including legumes in a rotation was to supply N, but with the development and availability of inexpensive fertilizer N, agriculture no longer depends on legumes for N (Fig. 4.6). The main purpose of legumes is to supply large amounts of high-quality forage, whether hay or pasture.

In a livestock farming system, legumes serve the dual purpose of providing livestock feed and N for the grain crops. Legumes are generally of superior quality, with higher protein and mineral concentrations compared with N fertilized grasses.

The decision about whether to use legumes or fertilizer N becomes a matter of economics, with selection based on the greatest net return on investment. Like other input costs, the cost of fertilizer N increases with time because of increased manufacturing and transportation costs. As a consequence, interest in legumes to substitute partially for the fertilizer N requirements of nonlegume crops has increased. In some developing countries, commercial N may not be available or is too expensive. Therefore, a cropping system that includes legumes is essential to supply some of the N needed for nonlegumes.

In spite of the advantages of legumes in a rotation, this practice may not always be attractive to growers. Farmers in some areas may not have a ready use or market for forage crops. Thorough extraction of soil moisture reserves might also be a disadvantage in semiarid areas. Examples of various legumes and their use in the United States are shown in Table 4.12.

TABLE 4.12   Examples of Regional Use of Legumes in Cropping or Conservation Tillage Systems

Region	Legume Species	Cropping or Tillage System
Southeast	Crimson clover, hairy vetch	Winter cover crop—no-till corn
	Bigflower vetch, crownvetch, alfalfa, lupine, arrowleaf clover, red clover	Winter cover crops preceding grain sorghum and cotton
Northeast	Alfalfa, birdsfoot trefoil, red clover	Legumes grown for hay or silage in crop rotations that include conventional or no-till corn as feed grain or silage; also used as living mulches
North Central	Soybean, pea	Grown in 1-year rotation with nonlegume, possibly using conservation tillage methods; peas may precede soybeans in a double-cropping system
	Alfalfa, red clover, white clover, alsike clover	Grown for 2 years or more in 3- to 5-year rotations with small grains or corn, possibly by use of conservation tillage methods
	Birdsfoot trefoil, crownvetch, sweetclover	Used for forage, silage, or pasture
Great Plains	Native legumes	Rangeland for grazing
	(Void in genetically adapted material and economically compatible enterprises; water is the limiting factor)	
Pacific Northwest	Dry pea, lentil, chickpea	Rotation or double cropped with grains
	Austrian winter pea	Green manure or alternated with winter wheat
	Alfalfa	Grown in rotation with winter wheat, spring barley, and winter peas
	Faba bean	Grown in rotation for silage
California	Dry bean, lima bean, blackeye pea, chickpea	Grown for grains in various rotations
	Alfalfa	Grown for seed on irrigated land and for erosion control and forage on steeply sloping soils
	Subterranean clover	Rangeland for grazing

SOURCE: Heichal, *Role of Legumes in Conservation Tillage Systems*, Soil Cons. Soc. Am., p. 30 (1987).

N Fixation by Nonsymbiotic Soil Microorganisms

N fixation in soils is also brought about by certain strains of free-living bacteria and blue-green algae (Table 4.5). Blue-green algae are autotropic, requiring only light, water, N_2, CO_2, and the essential mineral elements. Their numbers are normally far greater in flooded than in well-drained soils. Because they need light, they probably make only minor contributions to the N supply in upland agricultural soils after closure of crop canopies. In desert or semiarid regions, blue-green algae or lichens containing them become active following occasional rains and fix considerable quantities of N during their short-lived activity. N fixation by blue-green algae is of economic significance in tropical rice soils. N availability to other organisms by blue-green algae is of considerable importance during the early stages of soil formation.

There is a symbiotic relationship between *Anabaena azolla* (a blue-green alga) and *Azolla* (a water fern) in temperate and tropical waters. The blue-green alga located in cavities in leaves of the water fern is protected from external adverse conditions and is capable of supplying all of the N needs of the host plant. An important feature of this association is the water fern's very large light-harvesting surface, a property that limits the N_2-fixing capacity of free-living blue-green algae. The organism *Beijerinckia*, found almost exclusively in the tropics, inhabits the leaf surfaces of many tropical plants and fixes N on these leaves rather than in the soil.

In southeast Asia, *Azolla* has been used for centuries as a green manure in wetland rice culture, as a fodder for livestock, as a compost for production of other crops, and as a weed suppressor.

In California the *Azolla-Anabaena* N-fixing association has supplied 105 kg N/ha per season, or about 75% of the N requirements of rice. When used as a green manure, it provided 50 to 60 kg N/ha and substantially increased yields over unfertilized rice.

Certain N-fixing bacteria can grow on root surfaces and to some extent within root tissues of corn, grasses, millet, rice, sorghum, wheat, and many other higher plants. *Azospirillum brasilense* is the dominant N-fixing bacterium that has been identified. Inoculation of cereal crops with *A. brasilense* has been reported to improve growth and N nutrition, although the response to inoculation has been highly variable. In most of the studies in which inoculation was beneficial, the response was related to factors other than increased N fixation. Some of the possibilities are increased scavenging of plant nutrients, altered root permeability, hormonal action, and enhanced NO_3^- reduction in the roots.

Azotobacter- and *Clostridium-*inoculated seed may provide a maximum of only 5 kg N/ha; therefore, these nonsymbiotic organisms are of little value to N availability in intensive agriculture.

Industrial Fixation of N

From the standpoint of commercial agriculture and world food security, the industrial fixation of N is by far the most important source of N as a plant nutrient. The production of N by industrial fixation is based on the Haber-Bosch process, in which H_2 and N_2 gases react to form NH_3:

$$3H_2 + N_2 \xrightarrow[\text{1,200°C, 500 atm}]{\text{Catalyst}} 2NH_3$$

The NH_3 produced can be used directly as a fertilizer (anhydrous NH_3), although numerous other fertilizer N products are manufactured from NH_3 (see pp. 139–141).

Forms of Soil N

Total N content of soils ranges from less than 0.02% in subsoils to more than 2.5% in peats. The N concentration in the top 1 ft of most cultivated soils in the United

States varies between 0.03 and 0.4%. Soil N occurs as inorganic or organic N, with 95% or more of total N in surface soils present as organic N.

Inorganic N Compounds

The inorganic forms of soil N include ammonium (NH_4^+), nitrite (NO_2^-), nitrate (NO_3^-), nitrous oxide (N_2O), nitric oxide (NO), and elemental N (N_2), which is inert except for its utilization by *Rhizobia* and other N-fixing microorganisms.

From the standpoint of plant nutrition, NH_4^+, NO_2^-, and NO_3^- are the most important and are produced from aerobic decomposition of soil organic matter or from the addition of N fertilizers. These three forms usually represent 2 to 5% of the total soil N. N_2O and NO are important forms of N lost through denitrification.

Organic N Compounds

Organic soil N occurs as proteins, amino acids, amino sugars, and other complex N compounds. The proportion of total soil N in these various fractions is as follows: bound amino acids, 20 to 40%; amino sugars such as the hexosamines, 5 to 10%; and purine and pyrimidine derivatives., 1% or less. Very little is known about the chemical nature of the 50% or so of the organic N not found in these fractions.

Proteins are commonly found in combination with clays, lignin, and other materials resistant to decomposition. The biological oxidation of free amino acids is an important source of NH_4^+. Relative to other forms, the quantities of free amino acids in soils are low.

N Transformations in Soils

The quantities of NH_4^+ and NO_3^- available to plants depend largely on the amounts applied as N fertilizers and mineralized from organic N in soil. The amount of plant available N released from organic N depends on many factors affecting N mineralization, immobilization, and losses of NH_4^+ and NO_3^- from the soil.

Soil organic matter (OM) comprises organic materials in all stages of decomposition. It is composed of relatively stable material, termed *humus,* that is somewhat resistant to further rapid decomposition (Fig. 4.13). Organic materials that are subject to fairly rapid decomposition range from fresh crop residues to relatively stable humus. The primary microbial processes involved in fresh residue and humus turnover or cycling in soils are mineralization and immobilization. These reactions, combined with other physical, chemical, and environmental factors, are important in organic matter stability in soils and in inorganic N availability to plants.

The conceptual relationship between crop residues and their degradation through autotrophic and heterotrophic microbial processes that ultimately form relatively stable soil humus is shown in Figure 4.13. The size of these components depends on climate, soil type, and soil and crop management, all of which influence the quantity of crop residues produced and returned to the soil. The relatively small heterotrophic biomass (1 to 8% of total soil OM) represents soil mi-

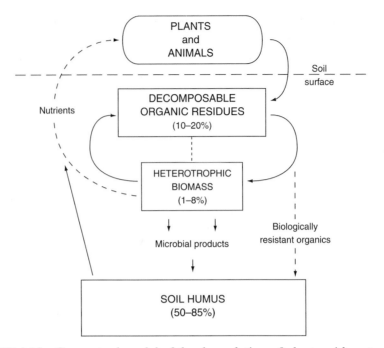

FIGURE 4.13   Conceptual model of the degradation of plant residues to stable soil humus. Relative sizes of the microbial and organic biomass components are shown.   *Doran and Smith,* SSSA Spec. Publ. *19, p. 55, 1987.*

croorganisms and fauna and is responsible for the majority of mineralization and immobilization reactions that influence availability of N and other nutrients. Soil humus, the largest component of soil OM, is relatively resistant to microbial degradation; however, it is essential for establishing and maintaining optimum soil physical conditions important for plant growth.

N Mineralization

N mineralization is the conversion of organic N to NH_4^+ (Fig. 4.1). Mineralization of organic N involves two reactions, *aminization* and *ammonification,* which occur through the activity of heterotrophic microorganisms. Heterotrophs require organic C compounds for their source of energy.

Mineralization increases with a rise in temperature and is enhanced by adequate, although not excessive, soil moisture and a good supply of O_2. Decomposition proceeds under waterlogged conditions, although at a slower rate, and is incomplete. Aerobic, and to a lesser extent anaerobic, respiration releases NH_4^+.

Soil moisture content regulates the proportions of aerobic and anaerobic microbial activity (Fig. 4.14). Maximum aerobic activity and N mineralization occur between 50 and 70% water-filled pore space. Soil temperature strongly influences microbial activity and N mineralization (Fig. 4.14). Optimum soil temperature for microbial activity ranges between 25 and 35°C. The influence of soil moisture and temperature fluctuations on N mineralization throughout a growing season is shown in Figure 4.12.

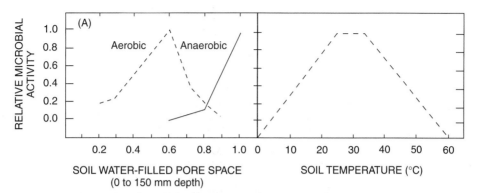

FIGURE 4.14 Influence of soil moisture (water-filled pore space) and temperature on relative microbial activity in soil. *Doran and Smith*, SSSA Spec. Publ. *19, p. 60 & 65, 1987.*

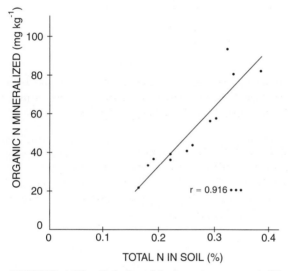

FIGURE 4.15 Relationship between organic N mineralized on incubation of nonlimed samples and total N in soils.

Soil OM contains about 5% N, and during a single growing season 1 to 4% of organic N is mineralized to inorganic N. As total soil N content increases, the quantity of N mineralized from soil organic N increases (Fig. 4.15). Therefore, soil and crop management strategies that conserve or increase soil OM will result in a greater contribution of mineralizable N to N availability to crops. The quantity of N mineralized during the growing season can be estimated. For example, if a soil contained 4% OM and 2% mineralization occurred, then

$$4\% \text{ OM} \times (2 \times 10^6 \text{ lb soil/a} - 6 \text{ in.}) \times (5\% \text{ N}) \times (2\% \text{ N mineralized}) = 80 \text{ lb N/a}$$

Thus, each year, 80 lb N/a as NH_4^+ are mineralized, which enter the soil solution and can be utilized by plants or other soil N processes (Fig. 4.1).

AMINIZATION   Heterotrophic bacteria and fungi are responsible for one or more reactions in OM decomposition. Bacteria dominate the breakdown of proteins in neutral and alkaline environments, with some involvement of fungi, while fungi predominate under acidic conditions. The end products of the activities of one group furnish the substrate for the next and so on until the material is decomposed. The final stage is the decomposition of proteins and the release of amines, amino acids, and urea. This step is termed *aminization* and is represented by

$$\text{Proteins} \xrightarrow[\substack{\text{Bacteria}\\\text{Fungi}}]{\text{H}_2\text{O}} \underset{\substack{\text{Amino acids}}}{\text{R}-\overset{\displaystyle\text{NH}_2}{\underset{\displaystyle\text{H}}{\text{C}}}-\text{COOH}} + \underset{\text{Amines}}{\text{R}-\text{NH}_2} + \underset{\text{Urea}}{\overset{\displaystyle\text{NH}_2}{\underset{\displaystyle\text{NH}_2}{\text{C}}}=\text{O}} + \text{CO}_2 + \text{energy}$$

AMMONIFICATION   Amines, amino acids, and urea produced by aminization of organic N are decomposed by heterotrophs and release NH_4^+. This step is termed *ammonification* and is represented by

$$R\text{—}NH_2 + H_2O \rightarrow NH_3 + R\text{—}OH + \text{energy}$$
$$\xrightarrow{+\,H_2O} NH_4^+ + OH^-$$

A diverse population of aerobic and anaerobic bacteria, fungi, and actinomycetes is capable of liberating NH_4^+, which is subject to several fates (Fig. 4.1). It can be

1. Converted to NO_2^- and NO_3^- by *nitrification.*
2. Absorbed directly by higher plants.
3. Utilized by heterotrophic organisms in further decomposing organic C residues.
4. Fixed in a biologically unavailable form in the lattice of certain clay minerals.
5. Converted to N_2 and slowly released back to the atmosphere.

N Immobilization

N immobilization is the conversion of inorganic N (NH_4^+ or NO_3^-) to organic N and is basically the reverse of N mineralization (Fig. 4.1). If decomposing OM contains low N relative to C, the microorganisms will immobilize NH_4^+ or NO_3^- in the soil solution. The microbes need N in a C:N ratio of about 8:1; therefore, inorganic N in the soil is utilized by the rapidly growing population. N immobilization during crop residue decomposition can reduce NH_4^+ or NO_3^- concentrations in the soil to very low levels. Soil microorganisms compete very effectively with plants for NH_4^+ or NO_3^- during immobilization, and plants can readily become N deficient. Fortunately, in most cropping systems, sufficient fertilizer N is applied to compensate for immobilization and crop requirements. After decomposition of the low N residue, microbial activity subsides and the immobilized N, which occurs as proteins in the microbes, can be mineralized back to NH_4^+ (Fig. 4.1).

If added organic material contains high N relative to C, N immobilization will not proceed because the residue contains sufficient N to meet the microbial demand during decomposition. Inorganic N in solution will actually increase from mineralization of some of the organic N in the residue material.

C/N Ratio Effects on N Mineralization and Immobilization

The ratio of % C to % N (C/N ratio) defines the relative quantities of these two elements in crop residues and other fresh organic materials, soil OM, and soil microorganisms (Table 4.13). The N content of humus or stable soil OM ranges from 5.0 to 5.5%, whereas C ranges from 50 to 58%, giving a C/N ratio ranging between 9 and 12.

Whether N is mineralized or immobilized depends on the C/N ratio of the OM being decomposed by soil microorganisms. For example, a typical soil mineralized 0.294 mg N, as measured by plant uptake (Table 4.14). When residues of

TABLE 4.13   C/N Ratios in a Selection of Organic Materials

Organic Substances	C/N Ratio	Organic Substances	C/N Ratio
Soil microorganisms	8:1	Bitumens and asphalts	94:1
Soil organic matter	10:1	Coal liquids and shale oils	124:1
Sweet clover (young)	12:1	Oak	200:1
Barnyard manure (rotted)	20:1	Pine	286:1
Clover residues	23:1	Crude oil	388:1
Green rye	36:1	Sawdust (generally)	400:1
Corn/sorghum stover	60:1	Spruce	1,000:1
Grain straw	80:1	Fir	1,257:1
Timothy	80:1		

TABLE 4.14   N Mineralized from Various Residues as Measured by Plant Uptake

Plant Residue*	C/N Ratio	N Uptake (mg)
Check soil	1.8	0.294
Tomato stems	45.3	0.051
Corn roots	48.1	0.007
Corn stalks	33.4	0.038
Corn leaves	31.9	0.020
Tomato roots	27.2	0.029
Collard roots	19.6	0.311
Bean stems	17.3	0.823
Tomato leaves	15.6	0.835
Bean stems	12.1	1.209
Collard stems	11.2	2.254
Collard leaves	9.7	1.781

*Residues above the dashed line have a C/N ratio >20:1.

Residues below the dashed line have a C/N ratio <20:1.

SOURCE: Iritani and Arnold, *Soil Sci.*, 89:74 (1960).

variable C/N ratio are added to the soil, N mineralization or immobilization would be indicated if plant uptake was greater or less than 0.294 mg N, respectively. In this study, a C/N ratio of approximately 20:1 was the dividing line between immobilization and mineralization.

The progress of N mineralization and immobilization following residue addition to soils can be illustrated in a generalized diagram (Fig. 4.16). During the initial stages of the decomposition of fresh organic material there is a rapid increase in the number of heterotrophic organisms, accompanied by the evolution of large amounts of CO_2. If the C/N ratio of the fresh material is greater than 30:1, N immobilization occurs, as shown in the shaded area under the top curve. As decay proceeds, the residue C/N ratio narrows and the energy supply diminishes. Some of the microbial population dies because of the decreased food supply, and ultimately a new equilibrium is reached, accompanied by mineralization of N (indicated by the crosshatched area under the top curve). The result is that the final soil level of inorganic N may be higher than the original level.

Generally, when organic substances with C/N ratios greater than 30:1 are added to soil, soil N is immobilized during the initial decomposition process. For ratios between 20 and 30, there may be neither immobilization nor release of mineral N. If the organic materials have a C/N ratio of less than 20, there is usually a release of mineral N early in the decomposition process. There may also be an increase in OM or humus, depending on the quantity and type of fresh organic material added. The time required for this decomposition cycle to run its

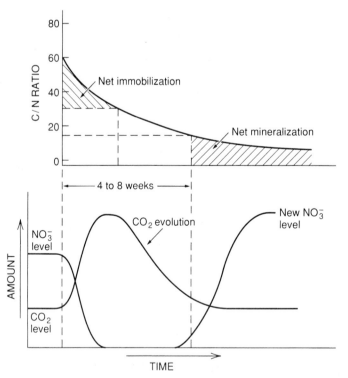

FIGURE 4.16   Changes in NO_3^- levels of soil during the decomposition of low-N crop residues.   *B. R. Sabey, Univ. of Illinois.*

course depends on the quantity of OM added, the supply of inorganic N, the resistance of the material to microbial attack (a function of the amount of lignins, waxes, and fats present), the temperature, and the soil moisture level.

The N content of the residue being added to soil also can be used to predict whether N is immobilized or mineralized (Fig. 4.17). Concentrations of 2.0% N are usually sufficient to minimize immobilization of soil N under aerobic conditions. Under anaerobic conditions in submerged soils, the N requirement for decomposition of crop residues may be only about 0.5%.

When high C/N ratio residues are added to soil, N in the residue and inorganic soil N are used by the microorganisms during residue decomposition. The quantity of inorganic soil N immobilized by the microbes can be estimated. For example, incorporating 2,000 lb/a residue containing 45% C and 0.75% N (C/N ratio of 60:1) into the soil represents 900 lb C:

$$2,000 \text{ lb residue} \times 45\% \text{ C} = 900 \text{ lb C in the residue}$$

Increasing microbial activity will utilize 35% of the residue C, while the remaining 65% is respired as CO_2 (Fig. 4.16). Thus, the microbes will use 315 lb C in the residue:

$$900 \text{ lb C} \times 35\% = 315 \text{ lb C used by microbes}$$

The increasing microbial population will require N governed by the microbe C/N ratio of 8:1 (Table 4.13):

$$\frac{315 \text{ lb C}}{X \text{ lb N}} = \frac{8}{1}$$

$$X = 39 \text{ lb/a N need by microbes}$$

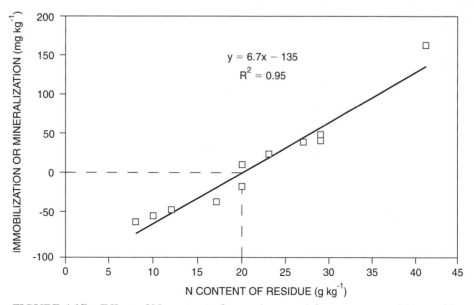

FIGURE 4.17   Effect of N content of organic materials on apparent N immobilization or mineralization.   *Goos,* J. Nat. Resources Life Sci. Educ. *25:126, 1995.*

The microbes will readily use the 15 lb/a N in the residue during decomposition.

$$\frac{900 \text{ lb C}}{X \text{ lb N}} = \frac{60}{1}$$

$$X = 15 \text{ lb/a N in residue}$$

The residue N content can also be calculated as follows: 2,000 lb residue $\times$ 0.75% N = 15 lb N. Thus, the quantity of N immobilized is

$$39 \text{ lb N needed} - 15 \text{ lb N in residue} = 24 \text{ lb/a N immobilized}$$

Therefore, at least 24 lb N/a will be needed to compensate for immobilization of inorganic N. Routine fertilizer N recommendations usually account for N immobilization requirements.

N Mineralization and Immobilization Effects on Soil OM

In virgin (uncultivated) soil the humus content is determined by soil texture, topography, and climatic conditions. Generally, OM content is higher in cooler than in warmer climates and, with similar annual temperature and vegetation, increases with an increase in effective precipitation (Fig. 4.18). These differences are related to reduced potential for OM oxidation with cooler temperatures and increased biomass production with increased rainfall. Humus content is greater in fine-textured than in coarse-textured soils and is related to increased biomass production in finer-textured soils because of improved soil water storage and reduced humus oxidation potential. OM contents are higher under grassland vegetation than under forest cover. These relations are generally true for well-drained soil conditions. Under conditions of poor drainage, aerobic decomposition is impeded and organic residues build up to high levels, regardless of temperature or soil texture.

The C/N ratio of the undisturbed *topsoil* in equilibrium with its environment is about 10 or 12. Generally, it narrows in the subsoil because of the lower amounts of C. An uncultivated soil has a relatively stable soil microbial population, a relatively constant amount of plant residue returned to the soil, and usually a low rate of N mineralization. If the soil is disturbed with tillage, N mineralization immediately and rapidly increases. Continued cultivation without the return of adequate crop residues ultimately leads to a decline in the humus content of soils. The influence of soil and crop management on soil OM and its relationship to soil and crop productivity is discussed in Chapter 13.

Any change in OM content dramatically reduces the quantity of N mineralized and thus soil N availability to crops. The differences in N mineralization can be readily calculated. For example, suppose that a virgin soil has a 5% OM content, and as the soil is cultivated (conventional tillage) the rate of OM loss is 4% per year. The quantity of N mineralized in the first year is thus

$$5\% \text{ OM} \times (2 \times 10^6 \text{ lb/a} - 6 \text{ in.}) \times 4\% \text{ OM loss/yr} = 4,000 \text{ lb/a OM loss/yr}$$

Since OM contains about 5% N, the amount of N mineralized is

$$4,000 \text{ lb OM/a} \times 5\% \text{ N} = 200 \text{ lb N/a/yr}$$

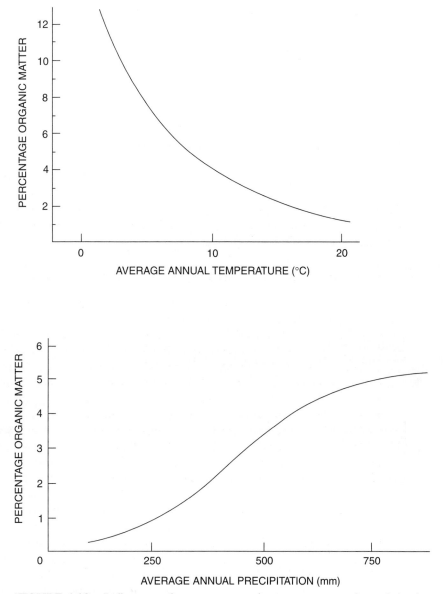

FIGURE 4.18   Influence of average annual temperature and precipitation on soil OM content.

Assume that after 50 years of cultivation, the OM declined to 2.5% or one-half the original level, as depicted in Fig. 4.19. Assume that 2% of the OM oxidizes per year; thus, the quantity of N mineralized is

$$2.5\% \text{ OM} \times (2 \times 10^6 \text{ lb/a} - 6 \text{ in.}) \times 2\% \text{ OM loss/yr} \times 5\% \text{ N} = 50 \text{ lb N/a/yr}$$

These estimates of the change in N mineralized illustrate that cultivation of virgin soils mineralized sufficient N to optimize yields of most crops, especially at the low yield levels experienced 50 years ago. The excess N not utilized by the crop was subject to several losses, which include leaching and denitrification. However,

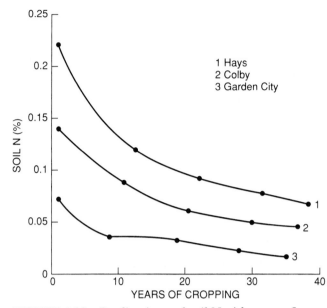

FIGURE 4.19   Decline in total soil N with years of cropping at three locations in Kansas. Each site was in wheat-fallow-wheat, with all residues incorporated with tillage. Total soil N represents soil organic matter, since 95% of total N is organic N.   *Haas and Evans,* USDA Tech. Bull. no. 1164, *1957.*

at present yield levels, mineralization of 50 lb N/a is insufficient and fertilizer N is needed to optimize yields.

Nitrification

Some of the NH_4^+ in soil is converted to NO_3^-, a process called *nitrification* (Fig. 4.1). Nitrification is a two-step process in which NH_4^+ is converted first to NO_2^- and then to NO_3^-. Biological oxidation of NH_4^+ to NO_2^- is represented by

$$2NH_4^+ + 3O_2 \xrightarrow{\text{\textit{Nitrosomonas}}} 2\,NO_2^- + 2H_2O + 4H^+$$
$$(-3)\text{-----------------}\rightarrow(+3) \text{ increasing oxidation state of N}$$

Nitrosomonas are autotrophic bacteria that obtain their energy from the oxidation of N and their C from CO_2. Other autotrophic bacteria (*nitrosolobus, nitrospira,* and *nitrosovibrio*), and to some extent heterotrophic bacteria, can also oxidize NH_4^+ and other reduced N compounds (i.e., amines) to NO_2^-.

In the second reaction, NO_2^- is oxidized to NO_3^- by

$$2NO_2^- + O_2 \xrightarrow{\text{\textit{Nitrobacter}}} 2NO_3^-$$
$$(+3)\text{-------------}\rightarrow(+5) \text{ increasing oxidation state of N}$$

NO_2^- oxidation occurs with autotrophic bacteria called *nitrobacter,* although some heterotrophs are also involved.

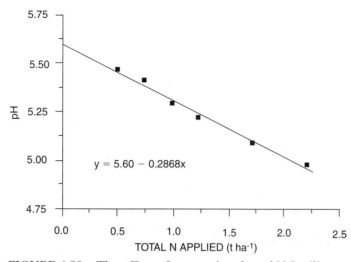

FIGURE 4.20   The effect of ammonium-based N fertilizer on soil pH (0- to 22-cm depth).   *Rasmussen and Rohde,* Soil Sci. Soc. Am. J., *53:119, 1989.*

The source of NH_4^+ can be mineralization of organic N or N fertilizers containing or forming NH_4^+. The reaction rates associated with nitrification in most well-drained soils are NO_2^- to $NO_3^- > NH_4^+$ to NO_2^-. As a result, NO_2^- generally does not accumulate in soils, which is fortunate since NO_2^- is toxic to plant roots. Both reactions require molecular O_2; thus, nitrification readily takes place in well-aerated soils. The reactions also show that nitrification of 1 mole of NH_4^+ produces 2 moles of H^+. Increasing soil acidity with nitrification is a natural process, although soil acidification is accelerated with continued application of NH_4^+-containing or NH_4^+-forming fertilizers (Fig. 4.20). Since NO_3^- is readily produced, and is very mobile and subject to leaching losses, understanding the factors affecting nitrification in soils will provide insight into management practices that minimize NO_3^- loss by leaching.

FACTORS AFFECTING NITRIFICATION   Because nitrification is a microbial process, soil environmental conditions influence nitrification rate. Generally the environmental factors favoring the growth of most agricultural plants are those that also favor the activity of nitrifying bacteria. Factors affecting nitrification in soils are (1) supply of NH_4^+, (2) population of nitrifying organisms, (3) soil pH, (4) soil aeration, (5) soil moisture, and (6) temperature.

Supply of NH_4^+. A supply of NH_4^+ is the first requirement for nitrification. If conditions do not favor mineralization of NH_4^+ (or if NH_4^+-containing or NH_4^+-forming fertilizers are not added to the soils), nitrification does not occur. Temperature and moisture levels that enhance nitrification are also favorable to mineralization.

Large amounts of small grain straw, corn stalks, or similar materials with a wide C/N ratio incorporated into soils with low inorganic N content will result in N immobilization by microorganisms as residues are decomposed. If crops are planted immediately after residue incorporation, they may become N deficient. Deficiencies can be prevented by adding sufficient N to supply the needs of the microorganisms *and* the growing crop.

Population of Nitrifying Organisms. Soils differ in their ability to nitrify NH_4^+ even under similar conditions of temperature, moisture, and NH_4^+ content. One factor that may be responsible is the variation in the numbers of nitrifying organisms present in different soils.

Variation in populations of nitrifiers results in differences in the lag time between the addition of the NH_4^+ and the buildup of NO_3^-. Because of the tendency of microbial populations to multiply rapidly in the presence of an adequate supply of C, the total amount of nitrification is not affected by the number of organisms initially present, provided that temperature and moisture conditions are favorable for sustained nitrification.

Soil pH. Nitrification takes place over a wide range in pH (4.5 to 10), although the optimum pH is 8.5. Nitrifying bacteria need an adequate supply of Ca^{2+}, $H_2PO_4^-$, and a proper balance of micronutrients. The influence of both soil pH and available Ca^{2+} on the activity of the nitrifying organisms suggests the importance of liming in crop production.

Soil Aeration. Aerobic nitrifying bacteria will not produce NO_3^- in the absence of O_2 (Fig. 4.21). Maximum nitrification occurs at the same O_2 concentration in the above-ground atmosphere.

Soil conditions that permit rapid diffusion of gases into and out of the soil are important for maintaining optimum soil aeration. Soils that are coarse textured or possess good structure facilitate rapid exchange of gases and ensure an adequate supply of O_2 for nitrifying bacteria. Return of crop residues and other organic amendments will help maintain or improve soil aeration.

Soil Moisture. Bacteria are sensitive to soil moisture. Nitrification rates are generally highest at soil water contents at field capacity or 1/3 bar matric suction (80% of total pore space). N mineralization and nitrification are reduced

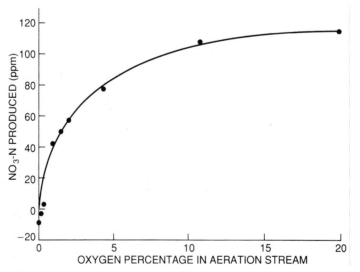

FIGURE 4.21   Production of NO_3^- incubated with added $(NH_4)_2 SO_4$ and aerated with air-N mixtures with varying O_2 percentages.   *Black*, Soil–Plant Relationships, *1957. Reprinted with permission of John Wiley & Sons, Inc., New York.*

when soil moisture exceeds field capacity. Between 15 bars and air dryness, mineralization and nitrification continue to decline gradually. For example, in a soil incubated at the wilting point (15 bars), more than half of the NH_4^+ is nitrified in 28 days (Fig. 4.22). At 7 bars, 100% of the NH_4^+ is converted to NO_3^- at the end of 21 days. Apparently, the *Nitrobacter* are able to function well even in dry soils. Obviously, soil moisture and soil aeration are closely related in their effects on nitrification.

Temperature. Most biological reactions are influenced by temperature. The temperature coefficient, Q_{10}, is 2 over the range 5 to 35°C. Thus, a twofold change in the mineralization or nitrification rate is associated with a shift of 10°C within this temperature range (Fig. 4.23).

Optimum soil temperature for nitrification of NH_4^+ to NO_3^- is 25 to 35°C, although some nitrification occurs over a wide temperature range (Fig. 4.24). For off-season application of NH_3 or NH_4^+ containing or NH_4^+ forming fertilizers winter soil temperatures should be low enough to retard formation of NO_3^-, thereby reducing the risk of leaching and denitrification losses. Fall applications of NH_4^+-containing or NH_4^+-forming fertilizers are most efficient when minimum air temperatures are below 40°F (4.4°C) or when soil temperatures are below 50°F (10°C).

Even if temperatures are occasionally high enough to permit nitrification of fall-applied NH_4^+, this is not detrimental if leaching does not occur. In many areas moisture movement through the soil profile during the winter months is insufficient to remove any NO_3^- that may accumulate.

For example, NH_4^+ sources may be applied in late summer or early fall in the Great Plains before winter wheat planting. The same is true for spring cereal

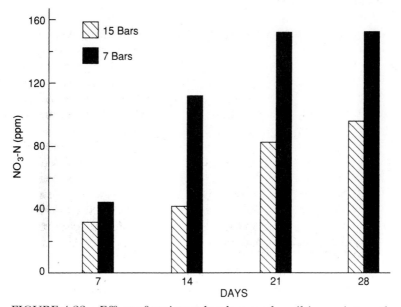

FIGURE 4.22 Effect of moisture levels near the wilting point on the nitrification of 150 ppm of N applied as $(NH_4)SO_4$ to a Millville loam and incubated at 25°C. *Justice et al., SSSA Proc., 26:246, 1962.*

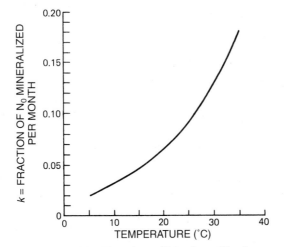

FIGURE 4.23   Fraction of N mineralized per month, *k*, in relation to temperature (*k* was estimated graphically for observed average monthly air temperatures).   *Stanford et al.*, Agron. J., *69:303, 1977.*

crops in the northern Plains. Improved positioning and distribution of N will often result from its overwinter movement in dry regions. In humid areas water movement through the soil profile is excessive, and NO_3^- losses occur. Whether NH_4^+ can be applied in the fall without significant NO_3^- loss depends on local soil and weather conditions.

It is possible to apply NH_4^+ sources in the fall in cool and/or dry climates to soils of fine texture without appreciable loss by leaching, provided that temperatures remain below 40°F. The presence of NH_4^+ does not ensure its loss against leaching. It is necessary that the soil have a sufficiently high CEC to retain the

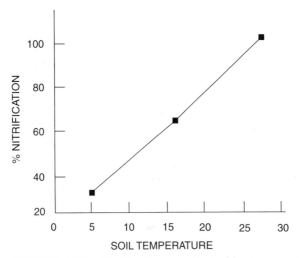

FIGURE 4.24   Nitrification as affected by temperature.   *Chandra*, Can. J. Soil Sci., *42:314, 1962.*

added NH_4^+ and prevent its loss in percolating water. Sandy soils with low CEC permit appreciable movement of NH_4^+ into the subsoil.

Nitrate Leaching

NO_3^- is very soluble in water and is not influenced by soil colloids. Consequently, it is highly mobile and subject to leaching losses when both soil NO_3^- content and water movement are high (Fig. 4.25). NO_3^- leaching is generally a major N loss mechanism from field soils in humid climates and under irrigated cropping systems.

NO_3^- leaching from field soils must be carefully controlled because of the serious impact that it can have on the environment. High NO_3^- levels in surface runoff and water percolating through the soil can pollute drinking water sources and stimulate unwanted plant and algae growth in lakes and reservoirs.

Some of the factors that influence the magnitude of NO_3^- leaching losses are (1) rate, time, source, and method of N fertilization; (2) use of nitrification inhibitors; (3) intensity of cropping and crop uptake of N; (4) soil characteristics that affect percolation; and (5) quantity, pattern, and time of precipitation and/or supplemental irrigation.

Examples of NO_3^- leaching from Corn Belt soils into water draining from tile lines located several feet below the soil surface are given in Figure 4.25. NO_3^- leaching losses occurred at all of these locations. It is important to match crop needs for N with total supplies of soil and applied N so that the quantities of leachable NO_3^- are minimized.

Nitrogen leaching is a natural process and even under natural or uncultivated systems some N transport below the root zone occurs. In agricultural

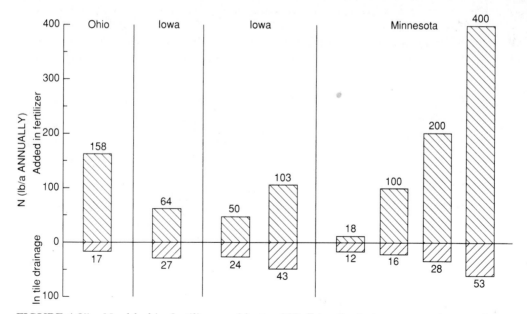

FIGURE 4.25   N added in fertilizer and lost as NO_3^- in tile drainage water in experiments in Ohio, Iowa, and Minnesota.   *Council for Agricultural Science and Technology,* Agriculture and Groundwater Quality, *Report no. 103, p. 23, Ames, Iowa, 1985.*

systems, proper soil, crop, and nutrient management present little or no risk of NO_3^- contamination of surface and groundwater. However, poor management practices may result in substantial leaching losses of applied N. Chapter 10 and 13 discuss N management strategies for maximizing crop productivity and minimizing the contribution of N leaching on water quality.

Ammonium Fixation

Certain clay minerals, particularly vermiculite and illite, are capable of fixing NH_4^+ by a replacement of NH_4^+ for interlayer cations in the expanded lattices of clay minerals (Fig. 4.26). The fixed NH_4^+ can be replaced by cations that expand the lattice (Ca^{2+}, Mg^{2+}, Na^+, H^+) but not by those that contract it (K^+). Fixation of freshly applied NH_4^+ can occur in clay-, silt-, and sand-size particles if they contain substantial amounts of vermiculite. Coarse clay (0.2 to 2 μm) and fine silt (2 to 5 μm) are the most important fractions in fixing added NH_4^+.

The moisture content and temperature of the soil affect NH_4^+ fixation (Table 4.15). Alternate cycles of wetting-drying and freezing-thawing are believed to contribute to the stability of recently fixed NH_4^+.

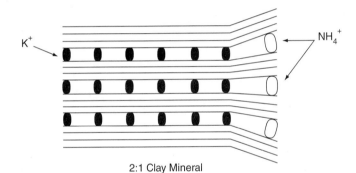

2:1 Clay Mineral

FIGURE 4.26   Diagram of an expanding clay mineral capable of fixing native or applied NH_4^+.

TABLE 4.15   Average Amounts of Native Fixed NH_4^+ and Added NH_4^+ Fixed under Moist, Frozen, and Oven-Dry Conditions in Several Wisconsin Soils

Horizon Groupings	Average Native Fixed NH_4^+ (meq/100 g)	Average Fixation of Applied NH_4^+ (meq/100 g) under Three Conditions		
		Moist	Frozen	Dried
Gray-brown podzolic soils				
Ap + A$_1$	0.54	0.08	0.14	0.68
A$_2$ + A$_3$	0.41	0.06	0.06	0.35
B$_1$ + B$_2$	0.60	0.15	0.25	0.82
Brunizem soils				
Ap + A$_1$	0.64	0.07	0.10	0.56
A$_3$	0.65	0.07	0.11	0.72
B$_1$ + B$_2$	0.60	0.15	0.16	0.67

SOURCE: Walsh et al., *Soil Sci.*, 89:183 (1960).

The presence of K^+ often restricts NH_4^+ fixation since K^+ can also fill fixation sites (see "K Fixation" in Chapter 6). Consequently, it has been suggested that K fertilization before NH_4^+ application is a practical way of reducing NH_4^+ fixation where it is a problem in the field.

The availability of fixed NH_4^+ ranges from negligible to relatively high. Clay fixation of NH_4^+ provides some degree of protection against rapid nitrification and subsequent leaching. Some fixed NH_4^+ is exchangeable and nitrified to NO_3^-.

Although the agricultural significance of NH_4^+ fixation is not great, it is important in certain soils. For example, a group of soils from Oregon and Washington were found to fix 1 to 30% of the applied anhydrous NH_3. In certain soils of eastern Canada, relatively large portions of fertilizer NH_4^+ can be fixed, often ranging from 14 to 60% in surface soil and as high as 70% in subsurface soil. Native fixed NH_4^+ is significant in many of these soils and can amount to about 10 to 31% of the total fixation capacity.

Gaseous Losses of N

The major losses of N from the soil are due to crop removal and leaching; however, under certain conditions, inorganic N can be converted to gases and lost to the atmosphere (Fig. 4.1). The primary pathways of gaseous N losses are by denitrification and NH_3 volatilization (Table 4.16).

DENITRIFICATION   When soils become waterlogged, O_2 is excluded and anaerobic conditions occur. Some anaerobic organisms obtain their O_2 from NO_2^- and NO_3^-, with the accompanying release of N_2 and N_2O. The most probable biochemical pathway leading to these losses is

$$2\,NO_3^- \longrightarrow 2\,NO_2^- \longrightarrow 2NO \longrightarrow N_2O\uparrow \longrightarrow N_2\uparrow$$
$$+5 \qquad\qquad +3 \qquad\qquad +2 \qquad +1 \qquad\qquad 0\ \text{N oxidation state}$$

Examples of the loss of NO_2^- and NO_3^- and the formation of N_2 and N_2O by denitrification are shown in Figure 4.27. The bacteria responsible for denitrification belong to the genera *Pseudomonas, Bacillus,* and *Paracoccus.* Several autotrophs also involved in denitrification include *Thiobacillus denitrificans* and *T. thioparus.*

Large populations of denitrifying organisms exist in arable soils, and are most numerous in the vicinity of plant roots. Carbonaceous exudates from active roots support the growth of denitrifying bacteria in the rhizosphere. The potential for denitrification is high in most field soils, but conditions must arise that cause these organisms to shift from aerobic respiration to a denitrifying metabolism involving NO_3^- as an electron acceptor in the absence of O_2.

Amounts of gaseous N lost by denitrification are variable because of the fluctuations in environmental conditions between seasons and years. The proportions of the two major products of denitrification, N_2 and N_2O, also vary; however, N_2 predominates, sometimes accounting for about 90% of the total N lost. The occurrence of N_2O becomes greater as soil O_2 supplies improve.

Factors Affecting Denitrification. The magnitude and rate of denitrification are strongly influenced by soil and environmental factors, the most important of

TABLE 4.16   Gaseous Losses of N from Soils

Form of N Lost	Source of N	General Reaction
N and NO_2 gases	A. Denitrification of NO_3^-	$NO_3^- \rightarrow NO_2^- \rightarrow NO \rightarrow N_2O\uparrow \rightarrow N_2\uparrow$
	B. Nitrification of NH_4^+	$NH_4^+ \rightarrow NH_2OH \rightarrow$ (e.g., $H_2N_2O_2$) $\rightarrow NO_2^- \rightarrow NO_3^-$ $\qquad\qquad\qquad\qquad\qquad\qquad\downarrow$ $\qquad\qquad\qquad\qquad\qquad N_2O$
	C. Chemical reactions of nitrites with	
	Ammonium	$NH_4^+ + NO_2^- \rightarrow N_2\uparrow + 2H_2O$
	α-Amino acids	$NO_2^- + NH_2R \rightarrow N_2\uparrow + ROH + OH^-$ (Van Slyke reaction)
	D. Lignin	$NO_2^- + \text{lignin} \rightarrow N_2\uparrow + N_2O\uparrow + CH_3ONO$
	E. Phenol	(aromatic ring reaction) $\quad$ phenol $\xrightarrow[\text{pH}>5]{NO_2} \xrightarrow{NO_2} N_2\uparrow + N_2O\uparrow + \text{organic residue}$
	Decomposition of nitrite with transition metal cation	$3NO_2^- + 2H^+ \rightarrow 2NO + NO_3^- + H_2O$ $Mn^{2+} + NO_2^- + 2H^+ \rightarrow Mn^{3+} + NO + H_2O$ $Fe^{2+} + NO_2^- + 2H^+ \rightarrow Fe^{3+} + NO + H_2O$
NH_3	A. Fertilizers	
	anhydrous NH_3	$NH_3\text{(liquid)} \rightarrow NH_3\uparrow \text{(gas)}$
	urea	$(NH_2)_2CO + H_2O \rightarrow 2NH_3\uparrow + CO_2$
	NH_4^+ salts	$NH_4^+ + OH^- \rightarrow NH_3\uparrow + H_2O \text{ (pH} > 7)$
	B. Decomposition of residues and manures	Release and volatilization of NH_3

SOURCE: Modified from Kurtz, *ASA Spec. Publ. 38*, p. 5 (1980).

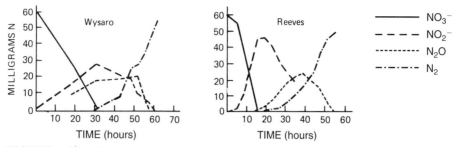

FIGURE 4.27 Sequence and magnitude of N products formed and utilized during anaerobic denitrification of Reeves loam (pH 7.8) and Wysaro clay (pH 6.1) at 30°C. *Cooper and Smith*, Soil Sci. Soc. Am. J., *27:659, 1963.*

which are the amount and nature of OM present; soil moisture; aeration; pH; temperature; and level and form of inorganic N (i.e., NO_3^- versus NH_4^+).

DECOMPOSABLE OM. The amount of readily decomposable soil OM or C strongly influences denitrification in soil (Fig. 4.28). The following equations illustrate the amount of available C required for microbial reduction of NO_3^- to N_2O or N_2:

$$4(CH_2O) + 4NO_3^- + 4H^+ = 4CO_2 + 2N_2O + 6H_2O$$

$$5(CH_2O) + 4NO_3^- + 4H^+ = 5CO_2 + 2N_2 + 7H_2O$$

Under field conditions, freshly added crop residues can stimulate denitrification.

SOIL WATER CONTENT. Soil water content is one of the most important factors in determining denitrification losses. Waterlogging of soil results in rapid denitrification by impeding the diffusion of O_2 to sites of microbiological activity. The

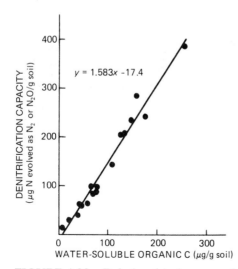

FIGURE 4.28 Relationship between denitrification capacity and water-soluble organic C. *Burford and Bremner*, Soil Biol. Biochem., *7:389, 1975.*

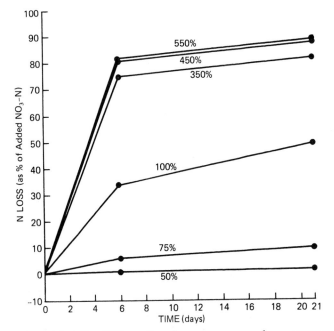

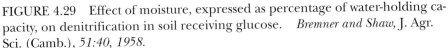

FIGURE 4.29   Effect of moisture, expressed as percentage of water-holding capacity, on denitrification in soil receiving glucose.   *Bremner and Shaw,* J. Agr.
Sci. (Camb.), *51:40, 1958.*

effect of increasing the degree of waterlogging on denitrification is clearly shown
in Figure 4.29.

Rapid conversion of NO_3^- to N_2O and N_2 occurs when rain saturates a warm,
biologically active soil. Potential denitrification losses of 10 to 30 lb N/a following saturation have been measured.

Saturation of soil with water during snow melt in the spring is also suspected of
causing major denitrification losses of N. The duration of snow cover on fields
and the time when the melting takes place are two factors that seem to affect denitrification associated with spring thawing.

In flooded rice soils, NO_3^--containing fertilizers are ineffective because of N lost
by denitrification. Some NO_3^- is always present in such soils, however, since a portion
of the NH_4^+ in the aerobic zone of the plant-soil-water system is converted to NO_3^-.
When this NO_3^- diffuses into anaerobic parts of the soil, it is rapidly denitrified.

AERATION. Formation of NO_3^- and NO_2^- depends on an ample supply of
O_2. Denitrification proceeds only when the O_2 supply is too low to meet microbiological requirements. The denitrification process can operate in seemingly
well-aerated soil, presumably in anaerobic microsites where the biological O_2 demand exceeds the supply (Fig. 4.30). Large losses of N by denitrification are possible with the simultaneous occurrence of a low rate of O_2 diffusion into the soil
and a high respiratory demand within it.

Decreased partial pressure of O_2 will increase denitrification losses. These
losses do not become appreciable, however, until the O_2 level is drastically reduced to concentrations of 10% or less (Fig. 4.31).

SOIL PH. Soil acidity can influence denitrification since the bacteria responsible for denitrification are sensitive to low pH. As a result, many acidic soils con-

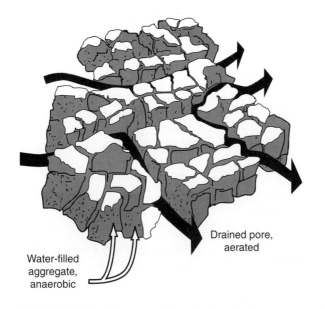

Drained pore, aerated

Water-filled aggregate, anaerobic

FIGURE 4.30   Diagram of microsites within an aerated soil that represent anaerobic, water-saturated aggregates in which native or applied N can be denitrified.

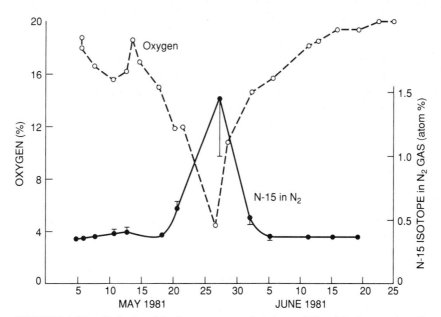

FIGURE 4.31   Relationship between production of N_2 with decreasing O_2 concentration (15-cm soil depth). 100 kg N/ha were applied on May 5. Release of N_2 peaked when soil O_2 dropped to 5% after 60 mm of rain. Total N loss averaged about 20% of the N applied.   *Colburn et al.*, J. Soil Sci., *35:542–43, 1984.*

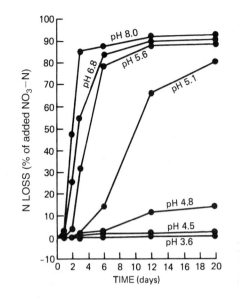

FIGURE 4.32   Effect of soil pH on denitrification. *Bremner and Shaw,* J. Agr. Sci. (Camb.), *51:40, 1958.*

tain small populations of denitrifiers. Denitrification is negligible in soils of pH below 5.0 but very rapid in high-pH soils (Fig. 4.32).

Acidity also regulates the sequence and relative abundance of the various N gases formed during denitrification. At pH values below 6.0 to 6.5, N_2O predominates and represents more than half of the N gases released in acidic environments. Formation of NO is usually confined to pH < 5.5. NO_2 may be the first gas detectable in neutral or slightly acidic soil, but it is reduced microbiologically so that N_2 is the principal product above pH 6. The occurrence of N_2O under acidic conditions is believed to be due to its resistance to further reduction to N_2.

TEMPERATURE. Denitrification is very sensitive to soil temperature, increasing rapidly in the 2 to 25°C range. Denitrification will proceed at slightly higher rates when the temperature is increased in the 25 to 60°C range. It is inhibited by temperatures above 60°C. The increase in denitrification at elevated soil temperatures suggests that thermophilic microorganisms play a major role in denitrification. Thus, denitrification losses coinciding with spring thawing are related to accelerated denitrification rates when soils are quickly warmed from 2 to 12°C or higher.

NO_3^- LEVELS. A supply of NO_3^- in soil is a prerequisite for denitrification. High NO_3^- concentrations increase the rate of denitrification and exert a strong influence on the ratio of N_2O to N_2 released by denitrification (Table 4.17). Although NO_2^- inhibited reduction of N_2O to N_2, the data showed that NO_3^- had a much greater depressive action.

PRESENCE OF PLANTS. Under field conditions, denitrification rates are increased by plants because of their release of readily available C in root exudates and sloughed-off root tissues. Denitrification in most fertilized soil-cropping systems is controlled by the supply of organic C. Plants may also increase denitrification by (1) consuming O_2 through root activity and (2) stimulating high microbial populations in the rhizosphere. On the other hand, they may restrict denitrification by (1) uptake of NO_3^- or NH_4^+; (2) reducing soil water content with resultant increase in O_2 supply; and (3) directly increasing O_2 levels in the rhizospheres of certain plants that transport O_2 (e.g., paddy rice).

TABLE 4.17 Effects of Different Amounts of NO_3^- and NO_2^- on N_2 Produced

Soil	pH	NO_3^- Added	NO_2^- Added	N_2 Produced in 4 Hours
			$\mu g/g \ soil$	
Clarion	7.2	0	0	43
		20	0	17
		10	10	20
		5	15	21
		0	20	27
Tama	6.6	0	0	33
		20	0	5
		10	10	8
		5	15	12

SOURCE: Gaskell et al., *SSSA J,* 45:1124–27 (1981).

Agricultural and Environmental Significance of Denitrification. Under reducing conditions, inorganic soil N is subject to continuous denitrification losses to the atmosphere. Since the earth's atmosphere is largely N_2, while its oceans are virtually NO_3^- free, denitrification is responsible for returning N_2 to the atmosphere, thus offsetting gains from biological N_2 fixation.

There appear to be two categories of N loss by denitrification: (1) rapid and extensive flushes associated with heavy rains, irrigation, and snow melt and (2) continuous small losses over extended periods in anaerobic microsites. Losses may account for 0 to 70% of applied N, with 10 to 30% being more typical. The rate and extent of denitrification losses of N under field conditions are still approximate despite much research. In spite of the fall-applied N, when heavy winter snows persist into late spring, N deficiencies can be observed. Losses of 25 to 50% in efficiency of N fertilizers are attributed mainly to denitrification under these conditions.

Contrasted with data of substantial N losses by denitrification are the findings by other scientists that total volatile loss of $N_2O + N_2$ from moderately well-drained, irrigated clay loam soil was only 1 to 3% of the N applied to corn. Generally, annual losses by denitrification in Britain range from 1 to 30 kg/ha N for arable land.

There is concern that increased use of N fertilizers may substantially increase emissions of N_2O from soils and thereby lead to partial destruction of the stratospheric ozone layer protecting the biosphere from biologically harmful ultraviolet radiation from the sun. Although there is evidence that denitrification of fertilizer-derived NO_3^- is responsible for emission of N_2O, contributions from NO_3^- produced by the natural transformations of soil OM and fresh crop residues have been largely ignored or discounted.

Denitrification can be useful in removing excessive amounts of NO_3^- from irrigation water and from various wastewaters (Chapter 13). For direct treatment of water, it may be necessary to inoculate with denitrifying organisms and provide sufficient readily mineralizable C in forms such as methanol. Where treatment systems involve disposal of contaminated wastewaters on soil, measures must also be taken to ensure that levels of mineralizable C are adequate in the soil areas being treated.

Chemical Reactions Involving NO_2^-. In addition to microbial denitrification, there are certain conditions in which losses of soil and fertilizer N can occur through chemical reactions involving NO_2^- (Table 4.16). Although NO_2^- does not usually accumulate in soil, detectable amounts occur in calcareous soils and in localized soil zones influenced by additions of NH_4^+-containing or NH_4^+-forming fertilizers.

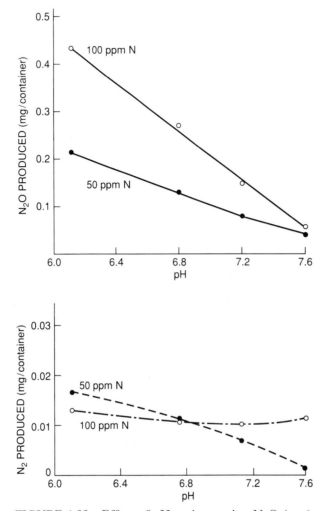

FIGURE 4.33   Effect of pH on increasing N_2O (top) and N_2 (bottom) production with increasing application of NO_2^- as $NaNO_2$.   *Christianson et al.,* Can. J. Soil Sci., *59:147, 1979.*

Factors Favoring NO_2^- Accumulation. NO_2^- does not usually accumulate in soil, but when it does it can adversely affect plants and microorganisms. Toxic levels of NO_2^- are generally caused by reduced *Nitrobacter* activity related to high pH and NH_4^+ levels. At 7.5 to 8.0 pH the potential for converting NH_4^+ to NO_2^- exceeds that for converting NO_2^- to NO_3^-, but at neutral pH the reverse is true. Although buildup of NO_2^- in soil is favored by high pH, its breakdown into N_2O and N_2 is restricted by high soil pH (Fig. 4.33).

NO_2^- formed in the fall may undergo chemical denitrification, even if soils freeze (Fig. 4.34). The rise in the chemodenitrification rate in frozen soil could be the result of forcing dissolved salts, including NO_2^-, into a narrow unfrozen water layer near the surface of soil colloids. This effectively increases NO_2^- concentration, which in turn enhances chemical denitrification.

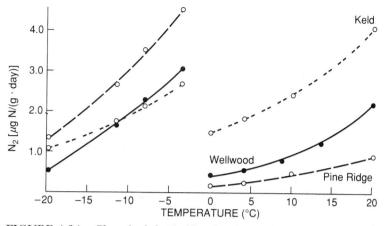

FIGURE 4.34   Chemical denitrification intensity of three soils under frozen and unfrozen conditions: 100 ppm NO_2-N.   *Christianson and Cho,* 23rd Annu. Manitoba Soil Sci. Meet., *p. 109, Univ. of Manitoba, 1979.*

Influence of Fertilizers on NO_2^- Accumulation. High rates of band-applied urea, anhydrous NH_3, aqua NH_3, and $(NH_4)_2HPO_4$ fertilizers cause temporary elevation of NH_4^+ and pH, which encourages NO_2^- accumulation in the band, regardless of initial soil pH. Diffusion and/or dilution of the NH_4^+ in fertilizer bands will restore conditions suitable for conversion of NO_2^- to NO_3^-. NO_2^- can diffuse beyond the microsite region high in pH and NH_4^+ to reach a soil environment, where the normal functioning of *Nitrobacter* will quickly convert it to NO_3^-. Small quantities of N_2O can be generated by autotrophic nitrification of NH_4^+ from fertilizers. Research results indicate that about 0.15% of the N applied is lost as N_2O. However, anhydrous NH_3 produces substantially more N_2O than other NH_4^+ sources.

VOLATILIZATION OF NH_3 Volatilization of NH_3 is a mechanism of N loss that occurs naturally in all soils (Fig. 4.1). However, compared with NH_3 volatilization from N fertilizers, NH_3 loss from N mineralized from organic N is relatively small. Thus, NH_3 volatilization will be discussed relative to surface application of N fertilizers. Numerous soil, environmental, and N fertilizer management factors influence the quantity of NH_3 volatilized from fertilizers. Understanding how these factors interact requires an understanding of the chemical reactions of N fertilizers with soil.

Volatilization of NH_3 ultimately depends on the quantity of NH_3 and NH_4^+ in the soil solution, which is highly dependent on pH (Fig. 4.35). The relationship is described as follows:

$$NH_4^+ \rightarrow NH_3 + H^+ \qquad (pK_a\ 9.3) \qquad (1)$$

Appreciable quantities of NH_3 appear only when soil solution pH exceeds 7.5. For example, at pH 8 and 9.3, NH_3 represents 10 and 50% of the total $NH_3 + NH_4^+$ in solution, respectively. Therefore, NH_3 loss is favored by high soil pH or by reactions that temporarily raise the pH. When NH_4^+-containing fertilizers are added to acidic or neutral soils, little or no NH_3 volatilization occurs because soil solution pH is not increased. Recall that soil pH decreases slightly when the NH_4^+ is nitrified to NO_3^-. When NH_4^+-forming fertilizers (e.g., urea) are added to

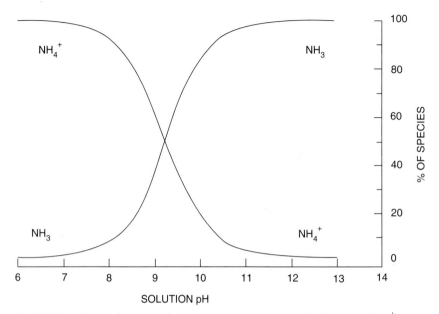

FIGURE 4.35   Influence of pH on the proportion of NH_3 and NH_4^+ in solution.

acidic or neutral soils, solution pH around the urea granule increases during hydrolysis, as shown by the following equation:

$$CO(NH_2)_2 + H^+ + 2H_2O \rightarrow 2NH_4^+ + HCO_3^- \tag{2}$$

Solution pH increases above 7 because H^+ is consumed in the reaction; thus, the NH_4^+–NH_3 equilibrium shifts to the right (eq. 1) to favor NH_3 volatilization loss. Therefore, in neutral and acidic soils, NH_4^+-containing fertilizers are less subject to NH_3 loss than urea and urea-containing fertilizers.

In calcareous soils, solution pH is buffered at about 7.5; thus, NH_4^+-containing fertilizers may be subject to NH_3 volatilization losses. For example, when $(NH_4)_2SO_4$ is applied to a calcareous soil, it reacts according to the following equations:

$$(NH_4)_2SO_4 + 2CaCO_3 + 2H_2O \rightarrow \tag{3}$$
$$2NH_4^+ + 2HCO_3^- + Ca^{2+} + 2OH^- + CaSO_4$$

$$NH_4^+ + HCO_3^- \rightarrow NH_3 + CO_2 + H_2O \tag{4}$$

The solution pH is increased because of the OH^- produced. The Ca^{2+} and OH^- may further combine with $(NH_4)_2SO_4$ as follows:

$$(NH_4)_2SO_4 + Ca^{2+} + 2OH^- \rightarrow 2NH_3 + H_2O + CaSO_4 \tag{5}$$

When all three equations (eq. 3, 4, and 5) are combined, the overall reaction can be represented as follows:

$$(NH_4)_2SO_4 + CaCO_3 \rightarrow 2NH_3 + CO_2 + H_2O + CaSO_4 \tag{6}$$

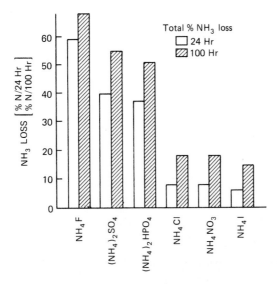

FIGURE 4.36   Total NH_3 loss at the end of 24 and 100 hours, as influenced by the anions of several NH_4^+ salts. NH_4-N was applied on the soil surface at 550 kg N/ha.   *Fenn and Kissel,* Soil Sci. Soc. Am. J., *37:855, 1973.*

Since the $CaSO_4$ produced is only slightly soluble, the reaction proceeds to the right and NH_3 volatilization is favored. Similar reactions occur with other NH_4^+-containing fertilizers that produce insoluble Ca precipitates (e.g., $[NH_4]_2HPO_4$). In comparison, volatilization losses are reduced with NH_4^+-containing fertilizers that produce soluble Ca reaction products (e.g., NH_4NO_3, NH_4Cl). Figure 4.36 demonstrates the impact that anions of various NH_4^+ salts can have on NH_3 volatilization. A rise in soil pH accompanies the formation of insoluble precipitates.

Generally NH_3 volatilization losses in calcareous soils are greater with urea fertilizers than with NH_4^+ salts, except those forming insoluble Ca precipitates. NH_3 losses also increase with increasing fertilizer rate and with liquid compared with dry N sources.

Volatilization of NH_3 is much greater with broadcast applications compared with subsurface or surface band methods (Table 4.18). These data show increased crop response to fertilizer N when urea-ammonium nitrate (UAN) is band applied compared with surface broadcasting. Immediate incorporation of broadcast N greatly reduces the NH_3 volatilization potential.

TABLE 4.18   Mean (1979–80) No-Till Corn Grain Yield as Affected by N Rate and Method of Application for UAN Solution

N Rate (kg/ha)	Broadcast Spray	Surface Band	Incorporate Band
		Yield (t/ha)	
90	5.61	7.40	7.87
180	6.77	8.34	8.84
270	7.18	8.69	9.66
Mean	6.52	8.14	8.46

SOURCE: Touchton and Hargrove, *Agron. J.,* 74:825 (1982).

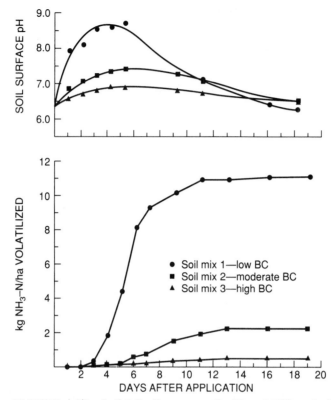

FIGURE 4.37   Soil BC effects on soil pH and NH_3 volatilization after N fertilizer application.   *Ferguson et al.,* SSSAJ, *48:578, 1984.*

The Buffer Capacity of the soil greatly influences NH_3 volatilization loss (Fig. 4.37). Soil pH and subsequent NH_3 loss will be much less in a soil with high buffering compared with one with low buffering. BC works in two ways: (1) to resist the increase in pH with fertilizer addition and (2) to remove part of the NH_4^+ and NH_3 from solution. Soil BC will increase with increasing CEC and OM content.

NH_3 losses are also influenced by environmental conditions during the reaction period of urea and NH_4^+ salts with soil. In general, volatilization increases with increasing temperature up to about 45°C, which is related to higher reaction rates and urease activity. If the surface soil is dry, the microbial and chemical reactions involved in NH_3 volatilization do not readily take place. Maximum NH_3 loss occurs when the soil surface is at or near field capacity moisture content and when slow drying conditions exist for several days. Water evaporation from the soil surface encourages NH_3 volatilization.

The presence of surface crop residues can greatly increase the potential for NH_3 volatilization. Crop residues increase NH_3 losses by maintaining wet, humid conditions at the soil surface and by reducing the quantity of urea diffusing into the soil. Crop residues also have a high urease activity (Fig. 4.38). Partial incorporation of the residue can significantly reduce NH_3 losses from surface-applied urea fertilizer.

Although substantial losses of NH_3 have been measured in laboratory studies, their validity should be closely examined. It should be recognized that experi-

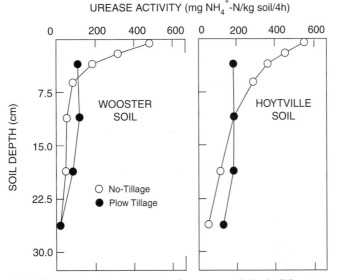

FIGURE 4.38 Distribution of urease activity in Wooster and Hoytville soil profiles as affected by tillage. *Dick, SSSAJ, 48:569, 1984.*

mental systems will impose artificial conditions of air movement, temperature, and relative humidities quite different from those occurring naturally.

For example, NH_3 volatilization losses as high as 70% of the N fertilizer applied have been reported from laboratory studies. Field studies conducted under a wide range of conditions show that volatilization losses with $(NH_4)_2SO_4$ broadcast on a calcareous soil can be about 50% of the fertilizer N applied, while NH_3 volatilization losses can be as high as 25% with urea. In an acidic soil, NH_3 losses are greater for urea than for $(NH_4)_2SO_4$. The quantity of NH_3 loss depends on the interaction of many soil, environmental, and N fertilizer factors. NH_3 volatilization losses are greatest in coarse-textured, calcareous soils with a surface residue cover.

NH_3 EXCHANGE BY PLANTS   Field crops exposed to air containing normal atmospheric concentrations of NH_3 may obtain as much as 10% of their N requirement by direct absorption of NH_3. Researchers have demonstrated that corn seedlings are a natural sink for atmospheric NH_3, absorbing up to 43% of the NH_3 from air containing 1 ppm NH_3. NH_3 produced near the ground surface of grass-clover pasture can be completely absorbed by the plant cover.

The opposite reaction, one of NH_3 volatilization from plant foliage, has also been observed from a number of crops, including alfalfa pasture, corn, Rhodes grass, and winter wheat. NH_3 release was related to the stage of plant growth, with losses occurring during ripening and senescence. Others have suggested that as much as one-third of the N in a wheat crop is volatilized as NH_3 after anthesis. Losses have also been reported for rice and soybeans.

Research results indicate that both absorption and loss of NH_3 can occur in field crops. The quantity depends on the wetness of the soil surface and the extent of evaporation, which both influence the amount of NH_3 released into the air coming into contact with plant canopies.

N Sources for Crop Production

Both organic and inorganic N sources are available to supply the N required for optimum crop productivity. From a management standpoint, it is important to understand that the processes and reactions of N in the soil (nitrification, volatilization, denitrification, leaching, and so forth) occur regardless of the N source used. Therefore, management practices that minimize N loss mechanisms and that increase the quantity of applied N recovered by the crop will increase production efficiency and reduce potential impacts of N use on the environment. Nutrient management technologies are discussed in greater detail in Chapter 10, but the commonly available N sources used in agricultural production systems and their reactions with soil are presented here.

Organic N Forms

Before 1850 virtually all of the fertilizer N consumed in the United States was in the form of natural organic materials, primarily animal manure and legume N. Presently these materials account for 40% of the total N use in the United States (Table 4.19). However, depending on the rate of manure applied, considerable quantities of N and other nutrients are added with manure. A complete discussion of fertilization with manure is found in Chapter 10. The average N concentration in organic materials is typically between 1 and 13%.

MANURE   The total quantity of manure produced annually in the United States is nearly 200 million tons (dry weight), with about 60% produced and deposited by grazing animals. The remaining 40% is produced in confinement animal production systems, where disposal through land application has led to serious environmental concerns. The average annual manure production ranges between 4 and 16 tons of manure, depending on the animal (Table 4.20).

The quantity of N in manure (Table 4.20) and the availability to plants varies greatly and depends on (1) nutrient content of the animal feed, (2) method of manure handling and storage, (3) quantity of added materials (i.e., bedding, water, and so on), (4) method and time of application, (5) soil properties, and (6) intended crop. Most wastes exiting the animal contain 75 to 90% water. Storage and handling usually reduce water content in solid storage systems and increase water content in liquid systems, such as the lagoon storage common with swine production. With total N contents ranging between <1 and 6%, 50 to 75% of the total N is organic N, while the remaining 25 to 50% of total N is NH_4^+

TABLE 4.19   Estimated Percentage of Total N Added to U.S. Cropland by Various Sources

Nitrogen Source	Total Amount (million tons)	Percentage of Total
Commercial N	8.55	57
Legumes, crop residues	3.74	25
Animal manures	2.14	14
Other sources	0.52	4

SOURCE: *USDA, 1992.*

TABLE 4.20   Annual Manure Production by Selected Animals and N Content

Animal Species	Manure* Production (t/yr)	Percentage of Total		N Form (% of Total)				
		Feces	Urine	Amino Acid	Urea	NH_4^+	Uric Acid	Other
				%				
Poultry	4.5	25	75	27	4	8	61	1
Sheep	6.0	50	50	21	34	<1.5	—	43
Horse	8.0	60	40	24	25	<1.0	—	49
Beef	8.5	50	50	20	35	0.5	—	44
Dairy	12	60	40	23	28	<0.5	—	49
Swine	16	33	67	27	51	<0.5	—	22

*Based on 1,000 lb animal weight.

(Table 4.20). Thus, manure N availability to plants predominantly depends on mineralization of the organic N in the manure. The mineralization process is the same as described previously for soil OM.

The organic N fraction in manure is composed of stable and unstable forms (Figure 4.39). Urea and uric acid (Table 4.20) are the main components of unstable organic N and are readily mineralized to plant available NH_4^+. Since NH_4^+ can be converted to NH_3 under optimum soil and environmental conditions, significant volatilization losses of manure N are possible, ranging from 15 to 40% of total N. In lagoon systems 60 to 90% of total manure N can be lost through

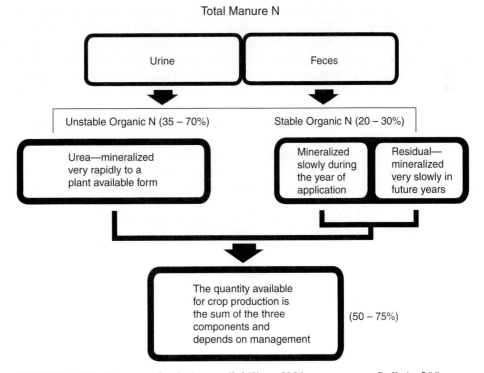

FIGURE 4.39   Form and relative availability of N in manure.   *Bulletin 308, Univ. of Maryland, 1986.*

TABLE 4.21   Mineralization Factors for Selected Animal Wastes and Storage/Handling Methods[*]

Animal	Mineralization Factor[†]	
	Solid Storage[‡]	Liquid Storage[‡]
Swine	0.50–0.60	0.30–0.35
Beef cattle	0.25–0.35	0.25–0.30
Dairy cattle	0.25–0.35	0.25–0.30
Sheep	0.25	—
Poultry	0.50–0.60	0.50–0.70
Horses	0.20	—

[*]Factors represent the proportion of organic N mineralized in the first year of application.

[†]Mineralization factors are reduced for surface application compared with subsurface injection.

[‡]Higher factors in a storage column represent *anaerobic* liquid storage and solid manure *without* bedding or litter added. Lower factors in a storage column represents *aerobic* liquid storage and solid manure *with* bedding or litter added.

volatilization during storage and land application. The remaining stable organic N will mineralize in the first and subsequent years after application. The less resistant stable organic N will generally mineralize in the year of application. This fraction represents 30 to 60% of total N in the applied manure, depending on manure type and placement. Subsurface injection results in greater N mineralization (and N availability) than surface-applied manure (Table 4.21). The more resistant stable organic N mineralizes slowly over the next several years, where about 50, 25, and 12.5% of the N mineralized in the first year is mineralized in the second, third, and fourth years, respectively.

Examples of the differences in N mineralization rate for selected liquid manures applied to soil are shown in Figure 4.40. The kinetics of N mineralization from manure, legume, or native soil OM can often be described by a first-order rate equation:

$$N_{min} = N_0 \, (1 - e^{-kt})$$

where N_{min} = the amount of N mineralized
N_0 = the mineralizable N pool
k = the rate constant
t = time

This equation indicates larger quantities of N mineralized initially, with decreasing N mineralized with time (Fig. 4.40).

LEGUME   Legumes can provide substantial proportions of plant available N to crop production (Table 4.6). Please refer to the material on N fixation in this chapter.

SEWAGE SLUDGE   About 75% of the sewage handled by municipal sewage treatment plants is of human origin, and the remaining 25% is from industrial sources. The end products of all sewage treatment processes are sewage sludge and sewage effluent. Sewage sludge refers to the solids produced during sewage

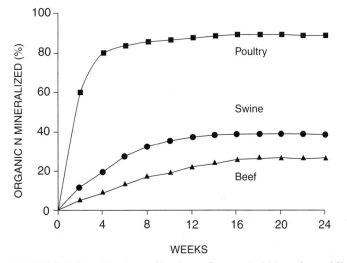

FIGURE 4.40   N mineralization of organic N in selected liquid animal manures applied to a sandy soil.   *Van Faassen, 1987, Netherlands. In v.d. Meer (ed). Animal Manure on Grassland & Fodder Crops. Martinus Nijhoff Publ. Dordrecht, Netherlands. Van Faassen and Van Dijk pp. 28–45, 1987.*

treatment. Sewage effluent is essentially clear water containing low concentrations of plant nutrients and traces of OM, which may be chlorinated and discharged into a stream or lake.

Sludge is a heterogeneous material, varying in composition from one city to another and even from one day to the next in the same city. Sewage sludge contains <1 to 3% N (Table 10.21). Approximately 25% of the municipal sludge produced is land applied, although very little is applied to land used for crop production.

Before developing plans for land application of sludge, it is essential to obtain representative samples of the sludge over a period of time and determine its typical chemical analysis (Table 10.21). Sewage sludge is disposed of by (1) application on cropland by approved methods; (2) incineration, with loss of OM and N; and (3) burial in landfill sites, where it will produce methane for many years. See Chapter 10 for management information.

Synthetic Fertilizer N Sources

Synthetic or chemical fertilizers are the most important sources of N. Over the last 20 years, world N consumption has increased from 22 to 80 million metric tons. In the United States, anhydrous NH_3, urea, and N solutions represented about 70% of total N use in 1995 (Fig. 4.41). Anhydrous NH_3 is the basic building block for almost all chemically derived N fertilizer materials. Most of the NH_3 in the world is produced synthetically by reacting N_2 and H_2 gases (Haber-Bosch process; see p. 125). From NH_3, many different fertilizer N compounds are manufactured. A few materials do not originate from synthetic NH_3, but they constitute only a small percentage of N fertilizers. For convenience, the various N compounds are grouped into three categories: ammoniacal, nitrate, and slowly available. The composition of some common N fertilizers is shown in Table 4.22.

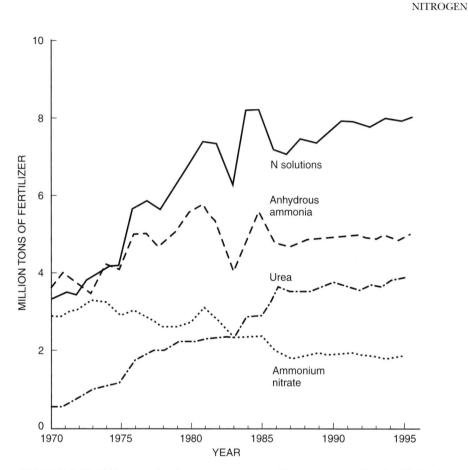

FIGURE 4.41   Changes in the most common N sources used in the United
States.   *USDA, 1994.*

AMMONIACAL SOURCES

Anhydrous NH₃. Anhydrous NH_3 contains 82% N, the highest amount of any
N fertilizer (Table 4.22). In some respects NH_3 behaves like water, since they both
have solid, liquid, and gaseous states. The great affinity of anhydrous NH_3 for water is apparent from its solubility (Table 4.23). As a result, NH_3 is rapidly absorbed
by water in human tissue. Because NH_3 is very irritating to the eyes, lungs, and skin,
safety precautions must always be taken with anhydrous NH_3 use. Safety goggles,
rubber gloves, and an NH_3 gas mask are required safety equipment. A large con-
tainer of water attached to the NH_3 tank is also required for washing skin and eyes
exposed to NH_3. Current regulations require certification of anyone applying NH_3.

Under normal atmospheric conditions, anhydrous NH_3 in an open vessel boils
and escapes into the atmosphere. To prevent escape, it is stored under pressure
and/or refrigeration (−28°F), as is often done at large, modern bulk-storage facilities. When liquid NH_3 is released from a pressurized vessel, it expands rapidly,
vaporizes, and produces a white cloud of water vapor. This cloud is formed by the
condensation of water in the air surrounding the liquid NH_3 as it vaporizes.

Because anhydrous NH_3 is a gas at atmospheric pressure, some may be lost to
the above-ground atmosphere during and after application. If the soil is hard or

TABLE 4.22   Typical Composition of Some Common Chemical Sources of Fertilizer N

Source	N	P_2O_3	K_2O	CaO	MgO	S	Cl	Physical State
								Nutrient Content (%)
Anhydrous ammonia	82.0	—	—	—	—	—	—	Gas
Aqua ammonia	20–25	—	—	—	—	—	—	Liquid
Ammonium chloride	25.0–26.0	—	—	—	—	—	66	Solid
Ammonium nitrate	33.0–34.0	—	—	—	—	—	—	Solid
Ammonium sulfate	21.0	—	—	—	—	24.0	—	Solid
Monoammonium phosphate	11.0	48.0–55.0	—	2.0	0.5	1.0–3.0	—	Solid
Diammonium phosphate	18.0–21.0	46.0–54.0	—	—	—	—	—	Solid
Ammonium phosphate-sulfate	13.0–16.0	20.0–39.0	—	—	—	3.0–14.0	—	Solid
Ammonium polyphosphate	10.0–11.0	34.0–37.0	—	—	—	—	—	Liquid
Ammonium thiosulfate	12.0	—	—	—	—	26.0	—	Liquid
Calcium nitrate	15.0	—	—	34.0	—	—	—	Solid
Potassium nitrate	13.0	—	44.0	0.5	0.5	0.2	1.2	Solid
Sodium nitrate	16.0	—	—	—	—	—	0.6	Solid
Urea	45.0–46.0	—	—	—	—	—	—	Solid
Urea-sulfate	30.0–40.0	—	—	—	—	6.0–11.0	—	Solid
Urea-ammonium nitrate	28.0–32.0	—	—	—	—	—	—	Liquid
Urea-ammonium phosphate	21.0–38.0	13.0–42.0	—	—	—	—	—	Solid
Urea phosphate	17.0	43.0–44.0	—	—	—	—	—	Solid

full of clods during application, the slit behind the applicator blade will not close or fill, and some NH_3 will escape to the atmosphere.

Anhydrous NH_3 convertors are often used to reduce the need for deep injection and preapplication tillage. The convertors serve as depressurization chambers for compressed anhydrous NH_3 stored in the applicator or nurse tank. Anhydrous NH_3 freezes as it expands in the convertors, separating the liquid NH_3 from the vapor and greatly reducing the pressure. The temperature of liquid NH_3 is about $-32°C$ ($-26°F$). Approximately 85% of the anhydrous NH_3 turns to liquid; the remainder stays in vapor form. The liquid flows by gravity through regular application equipment into the soil. Vapor collected at the top of the convertor is injected into the soil in the usual manner.

NH₃ RETENTION ZONES. Immediately after injection of NH_3 into soil, a localized zone high in both NH_3 and NH_4^+ is created. The horizontal, roughly circular- to oval-shaped zone is about 1-½ to 5 in. in diameter, depending on the method and rate of application, spacing, soil texture, and soil moisture content. Vertical movement is normally about 2 in., with most of it directed toward the soil surface.

A number of temporary yet dramatic changes occur in NH_3 retention zones that markedly influence the soil chemical, biological, and physical conditions in the retention zone. Some of the conditions that develop include

1. Increased concentrations of NH_3 and NH_4^+ (1,000 to 3,000 ppm).
2. pH increases to 9 or above.
3. NO_2^- increases to 100 ppm or more.
4. Osmotic suction of soil solution that exceeds 10 bar.
5. Lower populations of soil microorganisms.
6. Solubilization of OM.

TABLE 4.23   Properties of Anhydrous NH_3

Color	Colorless
Odor	Pungent, sharp
Chemical formula	NH_3
Molecular weight	17.03
Weight per gallon of liquid at 60°F	5.15 lb
Specific gravity of the gas (air = 1)	0.588
Specific gravity of the liquid (water = 1)	0.617
Boiling point	$-28°F$
Vapor pressure at 0, 68, and 100°F	16, 110, and 198 psig, respectively
One gallon of liquid at 60°F expands to	113 standard ft^3 of vapor
One pound of liquid at 60°F expands to	22 standard ft^3 of vapor
One cubic foot of liquid at 60°F expands to	850 standard ft^3 of vapor
Solubility in water at 60.8°F	0.578 lb/lb of water

	ppm N
Slight detectable odor	1
Detectable odor but no adverse effects on unprotected workers for exposure periods of up to 8 hours	25
Noticeable irritation of the eyes and nasal passages within a few minutes	100
Irritation to eyes and throat; no direct adverse effects, but exposure should be avoided	400–700
May be fatal after short exposure	2,000
Convulsive coughing, respiratory spasms, strangulation, and asphyxiation	5,000+

Free NH_3 is extremely toxic to microorganisms, higher plants, and animals. It can readily penetrate cell membranes, which are relatively impermeable to NH_4^+. There is a very close relationship between pH and concentration of free or nonionized NH_3 and NH_4^+. Between pH 6.0 and 9.0, there is a 500-fold increase in NH_3 concentration (Fig. 4.35).

Figure 4.42 summarizes schematically the effects of pH, osmotic suction, and/or NH_4^+ concentration on the formation of NO_2^- and NO_3^-. The influence of high osmotic suction or NH_4^+ in the soil solution is primarily on *Nitrosomonas* bacteria. Activity is retarded by pH values above 8.0, especially in the presence of large amounts of NH_3. NO_2^- accumulates at pH values between 7 and 8, whereas below pH 7, NO_3^- becomes abundant.

NH_3 is lost to the atmosphere if it does not react rapidly with water and various organic and inorganic soil components. Possible NH_3 retention mechanisms are as follows:

1. Chemical
   a. $NH_3 + H^+ \rightarrow NH_4^+$.
   b. $NH_3 + H_2O \rightarrow NH_4^+ + OH^-$.
   c. Reaction of NH_3 with OH^- groups and tightly bound water of clay minerals.
   d. Reaction with water of hydration around the exchangeable cations on the exchange complex.
   e. Reaction with OM.

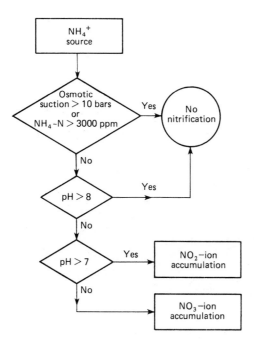

FIGURE 4.42   Schematic diagram showing the effects of osmotic suction and pH on nitrification. *Wetselaar et al.,* Plant Soil, *36:168, 1972.*

2. Physical
   a. NH_4^+ fixation by expanding clay minerals.
   b. Adsorption by clay minerals and organic components through H bonding.

The relative importance of these mechanisms varies from soil to soil and is also influenced by environmental conditions.

The capacity of soils to retain NH_3 increases with soil moisture content, with maximum NH_3 retention occurring at or near field capacity. As soils become drier or wetter than field capacity, they lose their ability to hold NH_3 (Fig. 4.43). The size of the initial NH_3 retention zone decreases with increasing soil moisture. Diffusion of NH_3 from the injection zone is impeded by high soil moisture, because of the strong affinity of NH_3 for water.

The NH_3-holding capacity of soils increases with the clay content. NH_3 movement is greater in sandy soils than in clay soils since NH_3 can diffuse more freely in the larger pores in coarse-textured soils. Soil textural differences in NH_3 retention are often obscured by other properties, such as OM and moisture content.

As might be expected, NH_3 retention increases with increasing depth of injection and varies considerably, depending on soil properties and conditions. Studies have shown that an injection depth of 5 cm was effective for a silt loam soil, but placement at 10 cm was necessary in a fine, sandy loam soil. In dry soil, NH_3 loss declines with increasing placement depth (Fig. 4.43).

At a given rate, the NH_3 applied per unit volume of soil decreases with decreasing injection spacing. With the greater retention achieved with narrow spacings, there is less chance of NH_3 loss, particularly in sandy soils with limited capacity for holding NH_3.

The OM component of soils contributes significantly to NH_3 retention. At least 50% of the NH_3-holding capacity of soils is attributed to OM. The nature and extent

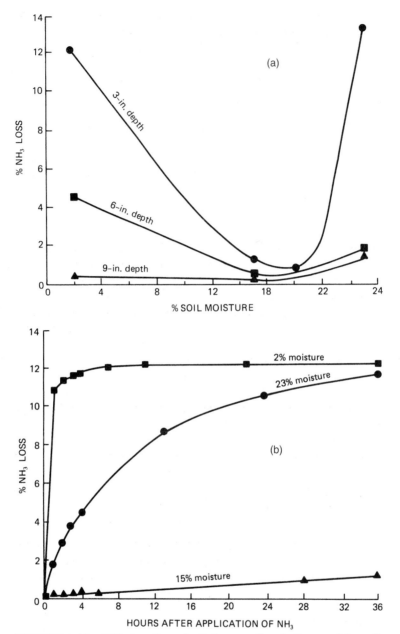

FIGURE 4.43   NH$_3$ loss from a silt loam soil as influenced by soil moisture content and depth of application (a) and time (b). Anhydrous NH$_3$ applied at 100 lb N/a on 40-in. row spacing.   *Stanley and Smith*, SSSAJ, *20:557, 1956.*

of changes in soil properties with NH$_3$ applications can have an important bearing on crop responses to N fertilizers. The high concentration of NH$_3$ and NH$_4^+$, which produces high soil pH and high osmotic potential, results in a partial and temporary sterilization of soil within the retention zone (Table 4.24). Bacterial activity is probably affected most by free NH$_3$, while fungi are depressed by high pH. Partially sterilized conditions at the center of the retention zone are known to persist for as long

TABLE 4.24   Numbers of Fungi, Bacteria, and Actinomycetes in Loamy Fine Sand in the NH_3 Injector Row Compared with Untreated Areas

Day after Treatment	Bacteria ($\times 10^6$/g)		Actinomycetes ($\times 10^6$/g)		Fungi ($\times 10^3$/g)	
	Check	NH_3	Check	NH_3	Check	NH_3
0	2.3	0.3	1.5	0.4	20.1	5.1
3	1.3	6.3	0.9	1.0	20.2	10.4
10	3.1	9.2	0.9	2.0	15.0	9.3
24	1.3	4.2	0.5	1.3	22.7	9.2
31	4.5	3.4	0.4	1.0	20.0	13.3
38	0.9	0.9	0.3	0.7	24.0	4.0

SOURCE: Eno and Blue, *Soil Sci. Soc. Am. J.,* 18:178 (1954).

as several weeks. A rapid recovery in the activity of bacteria and actinomycetes generally occurs. As a consequence of reduced microbial activity, nitrification of NH_4^+ to NO_2^- and NO_3^- will be reduced until conditions return to normal.

High concentrations of NH_3, NH_4^+, and NO_2^- can severely damage germinating seedlings (Fig. 4.44). Concentrations in excess of 1,000 ppm of NH_3 near the seed were associated with substantial reductions in corn plants. Deeper injection offsets the harmful effects of high rates of NH_3 more than extending the time for the NH_3 effects to dissipate. Closer spacing of the NH_3 injection would also reduce the injurious effect of large amounts of NH_3.

The OH^- produced by the reaction of anhydrous NH_3 in soil will dissolve or solubilize soil OM. Most of these effects on OM are only temporary. Solubilization of OM may temporarily increase the availability of nutrients associated with OM.

Contrasting beneficial and harmful effects on soil structure have been reported following the use of anhydrous NH_3. Several long-term studies have shown no difference among N sources on soil physical properties. Impairment of soil structure is not expected to be serious or lasting except in situations involving low-OM soils, in which any alteration or loss of OM would likely be harmful.

Aqua NH_3 (20 to 25% N). The simplest N solution is aqua NH_3, which is made by forcing compressed NH_3 gas into water. It has a pressure of less than 10 lb/in.^{2} and usually is composed of 25 to 29% NH_3 by weight.

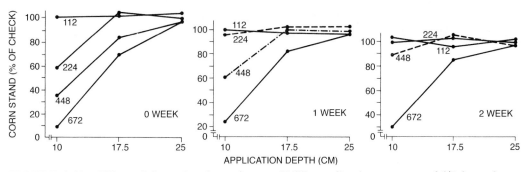

FIGURE 4.44   Effect of time, depth, and rate of NH_3 application on a stand 27 days after planting (numbers by lines are kg/ha of N).   *Colliver and Welch,* Agron. J., *62:341, 1970.*

Transportation and delivery costs limit aqua NH_3 production to small, local fluid fertilizer plants. Aqua NH_3 is used for direct soil applications or to produce other liquid fertilizers.

The NH_3 will volatilize quickly at temperatures above 50°F; thus, aqua NH_3 is usually injected into soil to depths of 2 to 4 in. At temperatures over 50°F, surface applications of aqua NH_3 should be immediately incorporated into the soil.

Nonpressure N Solutions. Of the liquid N fertilizers used for direct application in the United States, nonpressure N solutions are next to anhydrous NH_3 in popularity (Fig. 4.41). In 1995 usage of N solutions was more than 9.5 million tons, equivalent to approximately 24% of the total N consumed. Some of the reasons for the rapid growth in use of N solutions are:

1. Easier and safer to handle and apply than other N fertilizers (especially NH_3).
2. Applied more uniformly and accurately than solid N sources.
3. Many pesticides are compatible with N solutions and both can be applied simultaneously, eliminating one pass across the field.
4. Applied through various types of irrigation systems.
5. Safely transported in pipelines, barges, and railcars, which are less expensive and hazardous than the containers required for anhydrous NH_3.
6. Low-cost storage facilities can be used to store them more economically than most other N products.
7. Excellent sources of N for use in formulation of fluid N, P, K, and S fertilizers.
8. Lower cost of production than most solid N sources.

N solutions are usually produced from urea, NH_4NO_3, and water and are referred to as *UAN solutions* (Table 4.25). Each UAN solution has a specific *salting-out* temperature, which is the temperature below which dissolved salts begin to precipitate out of solution. The salting-out temperature determines the extent to which outside winter storage may be practiced and the time of year at which these solutions may be field applied. Salting-out temperatures vary directly with the concentration of plant nutrients in solution (Table 4.25).

The most positive feature of nonpressurized solutions is their ease of handling and application. Direct application, either broadcast or band applied, is common.

N solutions are often added directly to grasses and small grains. When grasslands are not dormant, spray applications of UAN can cause foliage scorching. A temporary leaf burn, usually lasting for less than a week, will sometimes occur when broadleaf herbicides and N combinations are sprayed on small grains.

TABLE 4.25   Physical and Chemical Characteristics of Urea-Ammonium Nitrate (UAN)

	Grade (% N)		
	28	30	32
Composition by weight (%)			
Ammonium nitrate	40.1	42.2	43.3
Urea	30.0	32.7	35.4
Water	29.9	25.1	20.3
Specific gravity at 15.6°C (60°F)	1.283	1.303	1.32
Salting-out temperature, °C (°F)	−18 (+1)	−10 (+14)	−2 (+28)

Ammonium Nitrate (NH_4NO_3). Fertilizer-grade NH_4NO_3 contains between 33 and 34% N and is a more popular fertilizer in Europe than in North America. In 1995 about 1.7 million metric tons of NH_4NO_3 were used in the United States.

The NO_3^- and NH_4^+ components of NH_4NO_3 are readily available to crops, and thus this N fertilizer is widely used in cropping situations in which growing crops are topdressed with N.

NH_4NO_3 has some disadvantages, which include

1. Very hygroscopic; care must be taken to prevent caking and physical deterioration in storage and handling.
2. High risk of fire or even explosions unless suitable precautions are taken; intimate contact with oxidizable forms of C such as fuel oil produces an explosive mixture that is widely used as a blasting agent.
3. Less effective for flooded rice than urea or NH_4^+ fertilizers.
4. More prone to leaching and denitrification than NH_4^+ products.

Ammonium Sulfate [$(NH_4)_2SO_4$]. Ammonium sulfate represents only about 2% of total N fertilizer use in the United States. The main advantages of $(NH_4)_2SO_4$ are low hygroscopicity and chemical stability. It is a good source of both N and S. The strongly acid-forming reaction of $(NH_4)_2SO_4$ in soil can be advantageous in high-pH soils and for acid-requiring crops. Its use can be undesirable in acidic soils already in need of liming.

The main disadvantage of $(NH_4)_2SO_4$ is its relatively low N content (21% N); it is generally too expensive to use as a N source. It can, however, be an economical source of N when transportation costs are low, when it is a relatively inexpensive by-product, and when it is used with crops requiring S.

Ammonium Phosphates. Monoammonium ($NH_4H_2PO_4$) and diammonium [$(NH_4)_2HPO_4$] phosphates are considered to be more important as sources of P than N. Therefore, their properties and reactions in the soil are covered in Chapter 5.

Ammonium Chloride (NH_4Cl). Fertilizer-grade NH_4Cl usually contains 25% N. About two-thirds of the world capacity for manufacture of this material is located in Japan, with the remaining one-third situated in India and China. Most of it is produced by the dual-salt process, in which NH_4Cl and $(Na)_2CO_3$ are formed simultaneously. Another production method is the direct neutralization of NH_3 with HCl.

Some of its advantages include a higher N concentration than $(NH_4)_2SO_4$ and superiority over $(NH_4)_2SO_4$ for rice. Ammonium chloride is an excellent source of both N and Cl^- for coconut, oil palm, and kiwifruit, which are Cl^--responsive crops.

Ammonium chloride is as acid forming as $(NH_4)_2SO_4$ per unit of N, and this effect is undesirable in acidic soil, especially if liming costs are excessive. Other shortcomings are its low N analysis in comparison to urea or NH_4NO_3, and its high Cl^- content limits its use to tolerant crops.

Urea [$CO(NH_2)_2$]. Favorable economics of manufacturing, handling, storage, and transportation have made urea a very competitive source of fertilizer N. Worldwide urea use is almost five times that of NH_4NO_3. Urea is the principal form of dry fertilizer N in the United States, approaching 16% of total N use (Fig. 4.41).

Granular urea has noteworthy characteristics, including (1) less tendency to stick and cake than NH_4NO_3, (2) lack of sensitivity to fire and explosion, and (3) less corrosiveness to handling and application equipment. Substantial savings in handling, storage, transportation, and application costs are possible because of urea's high N content.

BIURET LEVELS. The concentration of biuret (NH_2-CO-NH-CO-NH_2) in urea is of special concern because of its phytotoxicity. Biuret levels of 2% can be tolerated in most fertilizer programs. Because citrus, pineapple, and other crops are sensitive to biuret in urea applied as a foliar spray, less than 0.25% biuret is recommended. Solutions made from urea containing 1.5% biuret are acceptable for foliar application on corn and soybeans. Urea high in biuret should not be placed near or in the seed row.

BEHAVIOR OF UREA IN SOILS. When applied to soil, urea is hydrolyzed by the enzyme urease to NH_4^+. Depending on soil pH, the NH_4^+ may form NH_3, which can be volatilized at the soil surface, as represented in the following reactions:

$$CO(NH_2)_2 + H^+ + 2H_2O \xrightarrow{\text{Urease}} 2NH_4^+ + HCO_3^-$$
$$NH_4^+ \longrightarrow NH_3 + H^+$$

Urea hydrolysis proceeds rapidly in warm, moist soils, with most of the urea transformed to NH_4^+ in several days.

Urease, an enzyme that catalyzes the hydrolysis of urea, is abundant in soils. Large numbers of bacteria, fungi, and actinomycetes in soils possess urease. Urease activity increases with the size of the soil microbial population and with OM content. The presence of fresh plant residues often results in abundant supplies of urease (Fig. 4.38).

Urease activity is greatest in the rhizosphere, where microbial activity is high and where it can accumulate from plant roots. Rhizosphere urease activity varies depending on the plant species and the season of the year. Although temperatures up to 37°C favor urease activity, hydrolysis of urea occurs at temperatures down to 2°C and lower. This evidence of urease functioning at low temperatures, combined with urea's ability to melt ice at temperatures down to 11°F (−12°C), suggests that a portion of fall- or early-winter-applied urea may be converted to NH_3 or NH_4^+ before the spring.

The effects of soil moisture on urease activity are generally small in comparison to the influence of temperature and pH. Hydrolysis rates are highest at soil moisture contents optimum for plants.

Free NH_3 inhibits the enzymatic action of urease. Since significant concentrations of free NH_3 can occur at pH values above 7, some temporary inhibition of urease by free NH_3 occurs after the addition of urea because soil pH in the immediate vicinity of the urea source may reach values of up to 9.0. High rates of urea fertilization in localized placement could create conditions restrictive to the action of urease.

MANAGEMENT OF UREA FERTILIZERS. Careful management of urea and urea-based fertilizers will reduce the potential for NH_3 volatilization losses and increase the effectiveness of urea fertilizers.

Surface applications of urea are most efficient when they are washed into the soil or applied to soils with low potential for volatilization. Conditions for best performance of surface-applied urea are cold, dry soils at the time of application and/or the occurrence of significant precipitation, probably more than 0.25 cm (0.1 in.), within the first 3 to 6 days following application. Movement of soil moisture containing dissolved NH_3 and diffusion of moisture vapor to the soil surface during the drying process probably contribute to NH_3 volatilization at or near the soil surface.

Incorporation of broadcast urea into soil minimizes NH_3 losses by increasing the volume of soil to retain NH_3. Also, NH_3 not converted in the soil must diffuse over much greater distances before reaching the atmosphere. If soil and other environmental conditions appear favorable for NH_3 volatilization, deep incorporation is preferred over shallow surface tillage.

Band placement of urea results in soil changes comparable to those produced by applications of anhydrous NH_3. Diffusion of urea from banded applications can be 2.5 cm (1 in.) within 2 days of its addition, while appreciable amounts of NH_4^+ can be observed at distances of 3.8 cm (1.5 in.) from the band. After dilution or dispersion of the band by moisture movement, hydrolysis begins within 3 to 4 days or less under favorable temperature conditions.

Placement of urea with the seed at planting should be carefully controlled because of the toxic effects of free NH_3 on germinating seedlings (see reactions, p. 170). The harmful effects of urea placed in the seed row can be eliminated or greatly reduced by banding at least 2.5 cm (1 in.) directly below and/or to the side of the seed row of most crops. Seed placed urea should not exceed 5 to 10 lbs N/a.

The effect on germination of urea placed near seeds is influenced by available soil moisture. With adequate soil moisture in medium-textured loam soils at seeding time, urea at 30 lb N/a can be used without reducing germination and crop emergence. However, in low-moisture, coarse-textured (sandy loam) soils, urea at 10 to 20 lb N/a often reduces both germination and crop yields. Seedbed moisture is less critical in fine-textured (clay and clay loam) soils, and urea can usually be drilled in at rates of up to 30 lb N/a.

To summarize, the effectiveness of urea depends on the interaction of many factors, which cause some variability in the crop response to urea. However, if managed properly, urea will be about as effective as the other N sources.

UREA-BASED FERTILIZERS. Urea phosphate $[CO(NH_2)_2H_3PO_4]$ is a crystalline product formed by the reaction of urea with orthophosphoric acid. The common grade is 17-44-0, which is primarily used to produce other grades of lower analysis. Urea phosphates with lower purity standards may be adequate for production of suspension fertilizers and for fertigation. Urea has also been combined with $(NH_4)_2HPO_4$ into a solid 28-28-0.

Granular urea sulfate with grades ranging from 40-0-0-4 to 30-0-0-13 have been produced. The N/S ratio in this product may vary from 3:1 to 7:1, thus providing enough scope to correct N and S deficiencies in most soils. Although numerous urea-based fertilizers have been produced in pilot plants, they are not commonly used in North America.

NO_3^- SOURCES

In addition to NH_4NO_3, several other NO_3^--containing fertilizers, including sodium nitrate ($NaNO_3$), potassium nitrate (KNO_3), and calcium nitrate $Ca(NO_3)_2$, should be mentioned because of their importance in certain regions. These NO_3^- sources are quite soluble and thus very mobile in the soil solution. They are quickly available to crops and are susceptible to leaching under conditions of high rainfall. They may be immobilized by soil microorganisms in the decomposition of organic residues. Like all other NO_3^- sources, they are subject to denitrification.

Unlike NH_4^+ fertilizers, NO_3^- salts of Na^+, K^+, and Ca^{2+} are not acid forming. Because the NO_3^- is often absorbed by crops more rapidly than the accompanying cation, HCO_3^- and organic anions are exuded from roots, resulting in a slightly higher soil solution pH. Prolonged use of $NaNO_3$, for example, will maintain or even raise the original soil pH.

At one time sodium nitrate [$NaNO_3$ (16% N)] was the major source of N fertilizer in many countries. Most of it originated in a large ore body on the Chilean coastal range, and NO_3^- production continues to be a major industry in Chile.

Potassium nitrate (KNO_3, 13% N) contains two essential nutrients and is manufactured by reaction of concentrated HNO_3 with KCl. KNO_3 finds its greatest use in fertilizers for intensively grown crops such as tomatoes, potatoes, tobacco, leafy vegetables, citrus fruits, peaches, and other crops. The properties of KNO_3 that make it attractive for these crops include moderate salt index, rapid NO_3^- uptake, favorable N/K_2O ratio, negligible Cl^- content, and alkaline reaction in soil.

Calcium nitrate [$Ca(NO_3)_2$, 15% N] is produced by treating $CaCO_3$ with HNO_3. It is extremely hygroscopic, which detracts from its utility as a fertilizer. Except in very dry climates, $Ca(NO_3)_2$ is prone to liquefication, and storage in moisture-proof bags is usually mandatory.

Because of its fast-acting NO_3^- component, $Ca(NO_3)_2$ is a useful fertilizer for winter-season vegetable production. It is sometimes used in foliar sprays for celery, tomatoes, and apples. On sodium-affected soils, $Ca(NO_3)_2$ can be used as a Ca^{2+} source to displace Na^+ on the CEC.

SLOWLY AVAILABLE N COMPOUNDS   In addition to N uptake by crops, commercial fertilizers are subject to many different fates in soil; thus, crop recovery of applied N seldom exceeds 60 to 70%. Development of N fertilizers with greater efficiency is desirable because their production is energy intensive and because of environmental concerns over excessive movement of N into surface waters and groundwaters, as well as the effects of gaseous N losses on the upper atmosphere.

It would be desirable to have sources capable of releasing N over an extended period, thus avoiding the need for repeated applications of conventional water-soluble products. These materials also would reduce hazards of injury to germinating crops when used at high rates with or near the seed.

The ideal product would be one that releases N in accordance with crop needs throughout the growing season. Most of the materials that have been developed for controlled N availability can be grouped as follows:

1. Substances of low water solubility that must undergo chemical and/or microbial decomposition to release plant available N.
2. Nitrification and urease inhibitors.

   Substances of Low Water Solubility Requiring Decomposition. This group is composed of chemical compounds that are only slightly soluble in water or in the soil solution. The rate of N release is related to their water solubility and to the rate of microbiological action and chemical hydrolysis.

The best-known products in this category are urea-formaldehydes, or urea-forms. They are white, odorless solids containing about 38% N that are made by reacting urea with formaldehyde in the presence of a catalyst.

A typical ureaform may contain 30% of its N in forms that are soluble in cold water (25°C). N in the cold-water fraction nitrifies almost as quickly as urea. Solubility in boiling water is a measure of the quality of the remaining 70% of its N. At least 40% of the N insoluble in cold water should be soluble in hot water for an acceptable agronomic response; typical values are 50 to 70%.

Consumption of ureaform in the United States is approximately 50,000 tons annually. Most of it is used in nonfarm markets for turfgrass, landscaping, ornamental use, horticulture, and greenhouse crops and as an aid in overcoming planting shock of transplanted coniferous seedlings.

Sulfur-coated urea (SCU) is a controlled-release N fertilizer consisting of an S shell around each urea particle. The N concentration is between 36 and 38%, and all of it is supplied as urea. The release rate of SCU can be adjusted by changing the quantity of S used for coating. The S coating must be oxidized by soil microorganisms before the urea is exposed and subsequently hydrolyzed.

SCU has the greatest potential for use in situations in which multiple applications of soluble N sources are needed during the growing season, particularly on sandy soils under conditions of high rainfall or irrigation. It is advantageous for use on sugarcane, pineapple, grass forages, turf, ornamentals, fruits such as cranberries and strawberries, and rice under intermittent or delayed flooding. SCU might also find general use under conditions in which decomposition losses are significant.

Another advantage of SCU is its S content. Although S in the coating may not be sufficiently available to correct deficiencies during the first year after application, it can be an important source of plant available S in succeeding years.

Water Soluble Materials Requiring Decomposition. Triazone is a controlled release N compound containing between 28 and 30% N. Because of the closed-ring structure and strong C-N bonds, the N is released slowly. Triazone is predominately used as a foliar applied N source, exhibiting excellent absorption properties with no toxicity to plant tissues.

Nitrification and Urease Inhibitors. Certain substances are toxic to the nitrifying bacteria and will, when added to the soil, temporarily inhibit nitrification. Many chemicals have been tested in recent years for their ability to inhibit nitrification in soils and thereby manage additions of fertilizer N more effectively. A nitrification inhibitor should ideally (1) be nontoxic to plants, other soil organisms, fish, and mammals; (2) block the conversion of NH_4^+ to NO_3^- by specifically inhibiting *Nitrosomonas* growth or activity; (3) not interfere with the transformation of NO_2 by *Nitrobacter;* (4) be able to move with the fertilizer so that it will be distributed uniformly throughout the soil; (5) be able to maintain inhibitory action for periods ranging from several weeks to months; and (6) be relatively inexpensive.

The most effective compound is N-Serve [2-chloro-6(trichloromethyl)pyridine] frequently referred to as *nitrapyrin*. About 5 million acres of U.S. cropland are being treated annually with nitrapyrin. Dicyandiamide, or DCD, has been tested as both a nitrification inhibitor and a slow-release N source. DCD is readily soluble and stable in anhydrous NH_3, and nitrification is effectively inhibited for up to 3 months by the addition of 15 kg/ha of DCD. Urea containing 1.4% DCD (by weight) and UAN solution with 0.8% DCD are currently available in North America for increasing the effectiveness of fertilizer N.

Nitrification inhibitors prevent N losses only when conditions suitable for unwanted transformations to NO_3^- coincide with the effective period of the inhibitor. If soil and environmental conditions are favorable for NO_3^- losses, treatment with an inhibitor often increases fertilizer N efficiency. Generally, coarse-textured, low-OM soils are responsive to nitrification inhibitors added to N fertilizers (Fig. 4.45). Although the circumstances favoring loss of NO_3^- N are generally known, it is very difficult to predict accurately when and how much N will be lost. Also, protective action is unlikely when the situations for NO_3^- loss develop after the effects of the inhibitor have dissipated. When conditions are optimum, yield responses on corn can be 2 to 20 bu/a. The higher yield responses would be observed on sandy, low-OM soils in which NO_3 leaching potential is high. Little or no responses are observed on fine-textured, high-OM soils or with high N application rates.

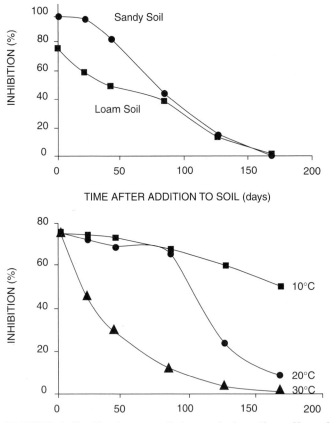

FIGURE 4.45   Persistence of nitrapyrin in soil as affected by soil texture (a) and soil temperature (b).   *McCarty and Bremner,* Communications in Soil Sci. & Plant Anal. *21:639–648, 1990.*

A large number of urease inhibitors have been evaluated for their ability to control urea hydrolysis in soils. Compounds that are effective urease inhibitors should be (1) effective at low concentrations, (2) relatively nontoxic to higher forms of life, (3) inexpensive, (4) compatible with urea, and (5) as mobile in soil as urea. Prospects for improving the effectiveness of urea through urease inhibition do not appear as promising as the more direct alternative of concentrating urea in the soil by using band placement.

The most common urease inhibitor commercially available is Agrotain or n-butyl-thiophosphoric triamide (NBPT). NBPT is used with surface-broadcast applications of urea granules or UAN solution. Yield responses up to 20 bu/a of corn have been observed, although 3 to 6 bu/a is more common. The higher yield responses occur under conditions of high N volatilization potential (see pp. 131–135), optimum or lower N rates, and where urea is broadcast in heavy surface residue environments.

Recently, ammonium thiosulfate (ATS) has been evaluated as a urease inhibitor. Laboratory results show that ATS mixed with UAN (10% by volume) will inhibit urease for about a month, depending on soil properties and environmental conditions. Results from field studies have been inconsistent in demonstrating increased N fertilizer efficiency with ATS.

Selected References

BARBER, S. A. 1984. *Soil Nutrient Bioavailability: A Mechanistic Approach*. John Wiley & Sons, New York.

BOSWELL, F. C., J. J. MEISINGER, and N. L. CASE. 1985. Production, marketing, and use of nitrogen fertilizers, pp. 229–92. In O. P. ENGELSTAD (Ed.), *Fertilizer Technology and Use*. Soil Science Society of America, Madison, Wisc.

FOLLETT, R. F. (Ed.). 1989. Nitrogen management and groundwater protection. In *Developments in Agricultural Managed-Forest Ecology #21*. Elsevier, New York.

FOLLETT, R. H., L. S. MURPHY, and R. L. DONAHUE. 1981. *Fertilizers and Soil Amendments*. Prentice-Hall, Englewood Cliffs, N.J.

HAUCK, R. D. 1984. *Nitrogen in Crop Production*. American Society of Agronomy, Madison, Wisc.

HAUCK, R. D. 1985. Slow-release and bioinhibitor-amended nitrogen fertilizers, pp. 293–322. In O. P. ENGELSTAD (Ed.), *Fertilizer Technology and Use*. Soil Science Society of America, Madison, Wisc.

POWER, J. F., and R. I. PAPENDICK. 1985. Organic sources of nitrogen, pp. 503–20. In O. P. ENGELSTAD (Ed.), *Fertilizer Technology and Use*. Soil Science Society of America, Madison, Wisc.

STEVENSON, F. J. (Ed.). 1982. *Nitrogen in Agricultural Soils*. American Society of Agronomy, Madison, Wisc.

STEVENSON, F. J. 1986. *Cycles of Soil*. John Wiley & Sons, New York.

CHAPTER 5

Phosphorus

Phosphorus (P) does not occur as abundantly in soils as N and K. Total P in surface soils varies between about 0.005 and 0.15%. The average total P content of soils is lower in the humid Southeast than in the Prairie and Western states. Unfortunately, the quantity of total P in soils has little or no relationship to the availability of P to plants. Although prairie soils are often high in total P, many of them are characteristically low in plant available P. Therefore, understanding the relationships and interactions of the various forms of P in soils and the numerous factors that influence P availability is essential to efficient P management.

The P Cycle

Figure 5.1 illustrates the interrelationships between the various forms of P in soils. The decrease in soil solution P concentration with absorption by plant roots is buffered by both inorganic and organic P fractions in soils. Primary and secondary P minerals dissolve to resupply $H_2PO_4^-$/HPO_4^{2-} in solution. Inorganic P adsorbed on mineral and clay surfaces as $H_2PO_4^-$ or HPO_4^{2-} (labile inorganic P) can also desorb to buffer decreases in solution P. Numerous soil microorganisms digest plant residues containing P and produce many organic P compounds in soil that are mineralized through microbial activity to supply solution P.

Water-soluble fertilizer P applied to soil readily dissolves and increases the concentration of soil solution P. Again, the inorganic and organic P fractions can buffer the increase in solution P. In addition to P uptake by roots, solution P can be adsorbed on mineral surfaces and precipitated as secondary P minerals. Soil microbes immobilize solution P as microbial P, eventually producing readily mineralizable P compounds (labile organic P) and organic P compounds more resistant to microbial degradation. Maintenance of solution P concentration (intensity) for adequate P nutrition in the plant depends on the ability of labile P (quantity) to replace soil solution P taken up by the plant. The ratio of quantity to intensity factors is called the *buffer capacity,* which expresses the relative ability of the soil to buffer changes in soil solution P. The larger the buffer capacity, the greater the ability to buffer solution P.

The P cycle can be simplified to the following relationship:

$$\text{Soil solution} \leftrightarrow \text{labile P} \leftrightarrow \text{nonlabile P}$$

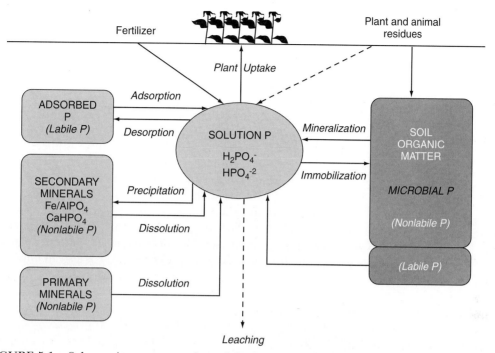

FIGURE 5.1   Schematic representation of the P cycle in soil.

where labile and nonlabile P represent both inorganic and organic fractions (Fig. 5.1). Labile P is the readily available portion of the quantity factor that exhibits a high dissociation rate and rapidly replenishes solution P. Depletion of labile P causes some nonlabile P to become labile, but at a slow rate. Thus, the quantity factor comprises both labile and nonlabile P fractions.

The interrelationships among the various P fractions are complex; however, understanding the dynamics of P transformations in soils will provide the basis for sound management of soil and fertilizer P to ensure adequate P availability to plants. Before we detail the various soil components of the P cycle and their interrelationships, we discuss plant P and its function in plant growth and metabolism.

Forms and Functions of P in Plants

Forms

Phosphorus concentration in plants ranges between 0.1 and 0.5%, considerably lower than N and K. Plants absorb either $H_2PO_4^-$ or HPO_4^{2-} orthophosphate ions. Absorption of $H_2PO_4^-$ is greatest at low pH values, whereas uptake of HPO_4^{2-} is greatest at higher values of soil pH (see pg. 159).

Plants may also absorb certain soluble organic phosphates. Nucleic acid and phytin occur as degradation products of the decomposition of soil OM and can be taken up by growing plants. Because of the instability of many organic P compounds in the presence of an active microbial population, their importance as sources of P for higher plants is limited.

Functions

The most essential function of P in plants is in energy storage and transfer. Adenosine di- and triphosphates (ADP and ATP) act as "energy currency" within plants (Fig. 5.2). When the terminal phosphate molecule from either ADP or ATP is split off, a relatively large amount of energy (12,000 cal/mol) is liberated. Energy obtained from photosynthesis and metabolism of carbohydrates is stored in phosphate compounds for subsequent use in growth and reproductive processes.

Donation or transfer of the energy-rich phosphate molecules from ATP to energy-requiring substances in the plant is known as *phosphorylation*. In this reaction ATP is converted back to ADP. The compounds ADP and ATP are formed and regenerated in the presence of sufficient P. Almost every metabolic reaction of any significance proceeds via phosphate derivatives (Table 5.1). P is also an important

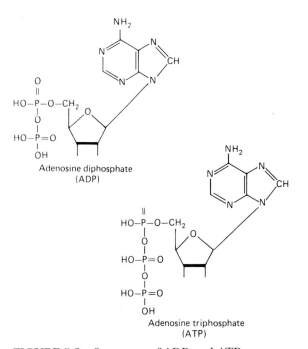

FIGURE 5.2   Structure of ADP and ATP.

TABLE 5.1   Processes or Pathways Involving ADP and ATP

Membrane transport	Generation of membrane electrical potentials
Cytoplasmic streaming	Respiration
Photosynthesis	Biosynthesis of cellulose, pectins, hemicellulose, and lignin
Protein biosynthesis	
Phospholipid biosynthesis	Lipid biosynthesis
Nucleic acid synthesis	Isoprenoid biosynthesis → steroids and gibberellins

structural component of nucleic acids, coenzymes, nucleotides, phosphoproteins, phospholipids, and sugar phosphates.

An adequate supply of P early in the life of a plant is important in the development of its reproductive parts. Large quantities of P are found in seed and fruit, and it is considered essential for seed formation.

A good supply of P is associated with increased root growth. When soluble phosphate compounds are applied in a band, plant roots proliferate extensively in that area of treated soil. Similar observations are made with both NO_3^- and NH_4^+ applied in a band near roots (Fig. 5.3). The greatly increased root proliferation should encourage extensive exploitation of the treated soil areas for nutrients and moisture. P is also associated with early maturity of crops, particularly grain crops. Ample P nutrition reduces the time required for grain ripening (Fig. 5.4).

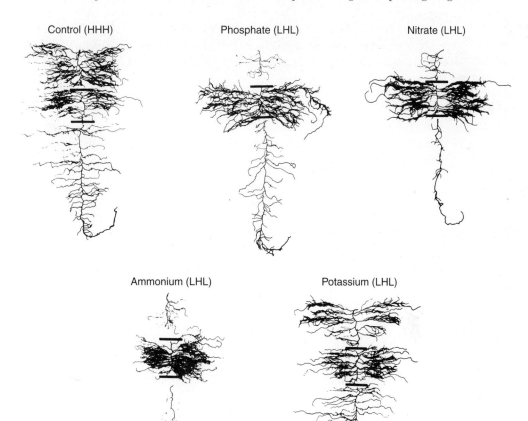

FIGURE 5.3   Effect of a localized supply of phosphate, nitrate, ammonium, and potassium on root form. Control plants (HHH) received the complete nutrient solution to all parts of the root system. The other roots (LHL) received the complete nutrient solution only in the middle zone, the top and bottom being supplied with a solution deficient in the specified nutrient.   *Drew,* New Phytol., *75:486, 1975.*

FIGURE 5.4   Effect of P fertilization on the maturity of small grains. Notice the more advanced maturity of the small grains receiving the P (*left*) in contrast to those that received no P (*right*).   *Courtesy of O. H. Long, Univ. of Tennessee.*

An adequate supply of P is associated with greater straw strength in cereals. The quality of certain fruit, forage, vegetable, and grain crops is improved and disease resistance increased when these crops have satisfactory P nutrition. The effect of P on raising the tolerance of small grains to root-rot diseases is particularly noteworthy. Also, the risk of winter damage to small grains can be substantially lowered by applying P, particularly on low-P soils and with unfavorable growing conditions.

Visual Deficiency Symptoms

P is mobile in plants, and when a deficiency occurs, it is translocated from older tissues to the active meristematic regions. Because of the marked effect of P deficiency on retarding overall growth, the striking foliar symptoms that are evident with N or K deficiency are seldom observed. In corn and some other grass species, P deficiency symptoms are also expressed by purple discoloration of the leaves or leaf edges (see color plates inside book cover).

Forms of Soil P

Soil Solution P

The amount of $H_2PO_4^-$ and HPO_4^{2-} present in the soil solution depends on soil solution pH (Fig. 5.5). At pH 7.2 there are approximately equal amounts of

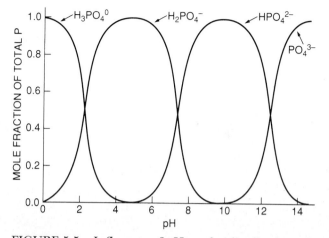

FIGURE 5.5 Influence of pH on the distribution of orthophosphate species in solution.

$H_2PO_4^-$ and HPO_4^{2-}. Below this pH, $H_2PO_4^-$ is the major form in soil solution, whereas HPO_4^{2-} is the predominant form above pH 7.2. Plant uptake of HPO_4^{2-} is much slower than with $H_2PO_4^-$.

The average soil solution P concentration is about 0.05 ppm and varies widely among soils. The solution P concentration required by most plants varies from 0.003 to 0.3 ppm and depends on the crop species and level of production (Table 5.2). Maximum corn grain yields may be obtained with 0.01 ppm P if the yield potential is low, but 0.05 ppm P is needed under conditions of high yield potential (Fig. 5.6).

The actively absorbing surface of plant roots is the young tissue near the root tips. Relatively high concentrations of P accumulate in root tips, followed by a zone of lesser accumulation, where cells are elongating, and then by a second region of higher concentration, where the root hairs are developed. Rapid replenishment of soil solution P is important where roots are actively absorbing P.

TABLE 5.2 Estimated Concentration of P in Soil Solution Associated with 75 and 95% of Maximum Yield of Selected Crops

Crop	Approximate P in Soil Solution for Yield Indicated (ppm)	
	75% of Max.	*95% of Max.*
Cassava	0.003	0.005
Peanuts	0.003	0.01
Corn	0.008	0.025
Wheat	0.009	0.028
Cabbage	0.012	0.04
Potatoes	0.02	0.18
Soybeans	0.025	0.20
Tomatoes	0.05	0.20
Head lettuce	0.10	0.30

SOURCE: Fox, *Better Crops Plant Food,* 66:24 (1982).

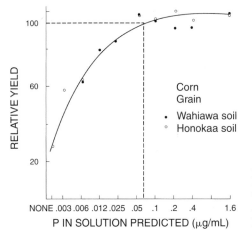

FIGURE 5.6   Influence of inorganic P in soil solution on corn grain yield. *Fox,* Chemistry in the Soil Environment, *p. 232, ASA, 1981.*

As roots absorb P from soil solution, diffusion and mass flow transport additional P to the root surface (see Chapter 2). Mass flow in low-P soils provides only a small portion of the requirement. For example, assume a transpiration ratio[1] of 400 and 0.2% P concentration in the crop. If the average solution concentration is 0.05 ppm P, then the quantity of P moving to the plant by mass flow is estimated by

$$\frac{400 \text{ g H}_2\text{O}}{\text{g plant}} \times \frac{100 \text{ g plant}}{0.2 \text{ g P}} \times \frac{0.05 \text{ g P}}{10^{-6} \text{ g H}_2\text{O}} \times 100 = 1\%$$

In fertilized soil with a solution concentration of 1 ppm P, mass flow contributes 20% of the total requirement. The very high P concentrations that exist temporarily in and near fertilizer bands are expected to encourage further P uptake by mass flow, as well as P diffusion. For example, P concentrations between 2 and 14 ppm have been found to occur in soil-fertilizer reaction zones. Since mass flow contributes 20% or less to P transport to the root surface, P diffusion is the primary mechanism of P transport (see Chapter 2 for review of diffusion).

Organic Soil P

Organic P represents about 50% of the total P in soils and typically varies between 15 and 80% in most soils (Table 5.3). Like OM, soil organic P decreases with depth, and the distribution with depth also varies among soils (Fig. 5.7). The P content of soil OM ranges from about 1 to 3%. Therefore, if a soil contains 4% OM in the surface 6 in., the organic P content (assume 1% of OM) is

$$2 \times 10^6 \text{ lb soil/a} - 6 \text{ in.} \times 0.01 \times 0.04 = 800 \text{ lb organic P/a} - 6 \text{ in.}$$

The quantity of organic P in soils generally increases with increasing organic C and/or N; however, the C/P and N/P ratios are more variable among soils than the C/N ratio. Soils have been characterized by their C/N/P/S ratio, which also varies among soils (Table 5.4). On the average, the C/N/P/S ratio in soil is 140:10:1.3:1.3.

[1]Transpiration ratio = weight of water transpired per unit weight of plant.

TABLE 5.3   Range in Organic P Levels in Various Soils

| Location | Organic P | |
	μg/g	Percentage of Total P
Australia	40–900	—
Canada	80–710	9–54
Denmark	354	61
England	200–920	22–74
New Zealand	120–1360	30–77
Nigeria	160–1160	—
Scotland	200–920	22–74
Tanzania	5–1200	27–90
United States	4–85	3–52

SOURCE: Stevenson, *Cycles of Soil*, p. 260, John Wiley & Sons (1986).

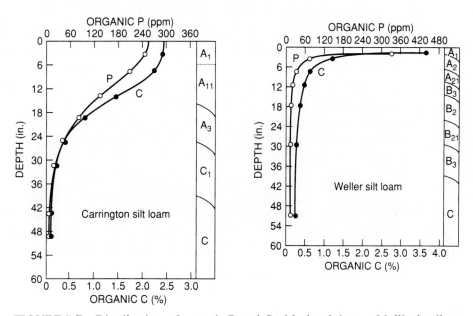

FIGURE 5.7   Distribution of organic P and C with depth in two Mollisol soils in Iowa.   *Stevenson,* Cycles of Soil, *p. 261, John Wiley & Sons, 1986.*

TABLE 5.4   Organic C, N, P, and S Ratios in Selected Soils

Location	Number of Soils	C/N/P/S
Iowa	6	110:10:1.4:1.3
Brazil	6	194:10:1.2:1.6
Scotland*		
Calcareous	10	113:10:1.3:1.3
Noncalcareous	40	147:10:2.5:1.4
New Zealand†	22	140:10:2.1:2.1
India	9	144:10:1.9:1.8

*Values for S given as total S.

†Values for subsurface layers (35–53 cm) were 105:10:3.5:1.1.

SOURCE: Stevenson, *Cycles of Soil*, p. 262, John Wiley & Sons (1986).

Many of the organic P compounds in soils have not been characterized. Most organic P compounds are esters of orthophosphoric acid ($H_2PO_4^-$) and have been identified primarily as inositol phosphates, phospholipids, and nucleic acids. The approximate proportion of these compounds in total organic P is as follows:

Inositol phosphates	10–50%
Phospholipids	1–5%
Nucleic acids	0.2–2.5%

Thus, on the average, only about 50% of organic P compounds in soils are known.

Inositol phosphates represent a series of phosphate esters ranging from monophosphate up to hexaphosphate. Phytic acid (myoinositol hexaphosphate) has six orthophosphate ($H_2PO_4^-$) groups attached to each C atom in the benzene ring (Fig. 5.8). Successive replacement of $H_2PO_4^-$ with OH^- represents the other five phosphate esters. For example, the pentaphosphate ester has five $H_2PO_4^-$ groups and one OH^-. Inositol hexaphosphate is the most common phosphate ester and comprises as much as 50% of total organic P in soils. Most of the inositol phosphates in soils are products of microbial activity and the degradation of plant residues.

Nucleic acids occur in all living cells and are produced during the decomposition of residues by soil microorganisms. Two distinct forms of nucleic acids, ribonucleic acid (RNA) and deoxyribonucleic acid (DNA), are released into soil in greater quantities than inositol phosphates, and they are broken down more quickly. Therefore, nucleic acids represent only a small portion of total organic P in soils, approximately 2.5% or less. Organic P compounds in soils called *phospholipids* are insoluble in water but are readily utilized and synthesized by soil microorganisms. Some of the most common phospholipids are derivatives of glycerol. The rate of release of phospholipids from organic sources in soils is rapid. Thus, the phospholipid content in soils is also low, about 5% or less of total organic P.

The remaining organic P compounds in soils are believed to originate from microorganisms, especially from bacterial cell walls, which are known to contain a number of very stable esters.

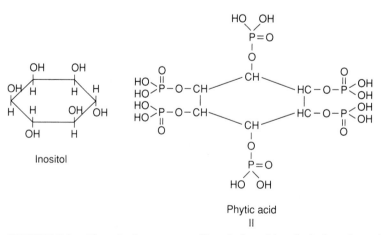

FIGURE 5.8   Chemical structure of inositol and inositol phosphate (phytic acid).

ORGANIC P TURNOVER IN SOILS   In general, P mineralization and immobilization are similar to those of N in that both reactions or processes occur simultaneously in soils and can be depicted as follows:

$$\text{Organic P} \xrightleftharpoons[\text{Immobilization}]{\text{Mineralization}} \text{inorganic P } (H_2PO_4^- / HPO_4^{2-})$$

The initial source of soil organic P is plant and animal residues, which are degraded by microorganisms to produce other organic compounds and release inorganic P (Fig. 5.1). Some organic P is resistant to microbial degradation and is most likely associated with humic acids. The inositol phosphates, nucleic acids, and phospholipids can also be mineralized in soils by a reaction catalyzed by the enzyme phosphatase:

$$
\begin{array}{ccc}
\quad\quad O & & \quad\quad O \\
\quad\quad \| & \text{Phosphatase} & \quad\quad \| \\
R{-}O{-}P{-}O^- + H_2O & \xrightarrow{\hspace{1.5cm}} & H{-}O{-}P{-}O^- + ROH \\
\quad\quad | & \text{Enzyme} & \quad\quad | \\
\quad\quad O^- & & \quad\quad O^-
\end{array}
$$

Phosphatase enzymes play a major role in the mineralization of organic P in soil. The wide range of microorganisms existing in soil are capable of mineralizing organic P. Phosphatase activity in soils increases with increasing organic C content but also is affected by pH, moisture, temperature, and other factors. Evidence of organic P mineralization can be provided by measuring changes in soil organic P during the growing season (Fig. 5.9). Organic P content decreases with crop growth and increases again after harvest.

Many factors influence the total quantity of P mineralized in soil. In most soils, total organic P is highly correlated with soil organic C; thus, P mineralization increases with increasing total organic C (Fig. 5.10). In contrast, the quantity of inorganic P immobilized is inversely related to soil organic P (Fig. 5.11). These data

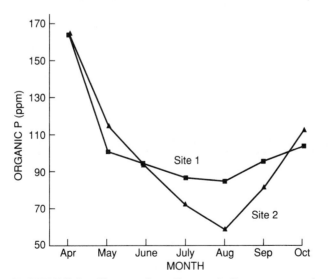

FIGURE 5.9   Changes in soil organic P over a cropping season for two locations.   *Dormaar*, Can. J. Soil. Sci., *52:107, 1972.*

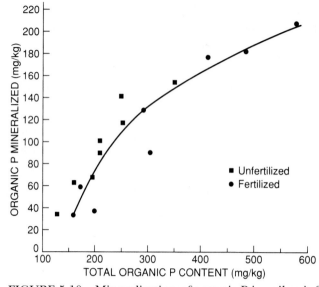

FIGURE 5.10   Mineralization of organic P in soil as influenced by total organic P.   *Sharpley, SSSAJ, 49:907, 1985.*

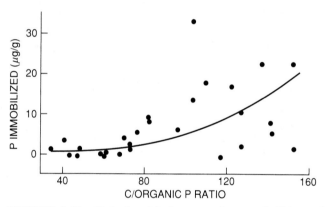

FIGURE 5.11   Relationship between inorganic P immobilization and C/P ratio in the soil.   *Enwezor, Soil Sci., 103:62, 1967.*

show that as the ratio of soil organic C to P increases (i.e., decreasing organic P), P immobilization increases.

The C/P ratio of the decomposing residues regulates the predominance of P mineralization over immobilization, just as the C/N ratio regulates N mineralization over immobilization. The following guidelines have been suggested:

C/P Ratio	Mineralization/Immobilization
<200	Net mineralization of organic P
200–300	No gain or loss of inorganic P
>300	Net immobilization of inorganic P

Expressed as % P in the degrading residue, net P immobilization occurs when % P < 0.2% and net mineralization occurs with >0.3% P. When residues are added to soil,

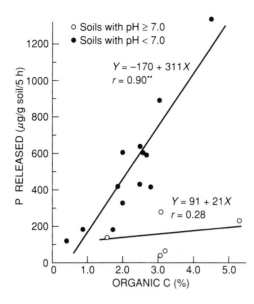

FIGURE 5.12   Influence of soil organic C and pH on P mineralization.   *Tabatabai and Dick,* Soil Biol. Biochem., *11:655, 1979.*

net P immobilization occurs during the early stages of decomposition, followed by net P mineralization as the C/P ratio of the residue decreases. Thus, P mineralization/immobilization processes are similar to those described for N (Fig. 4.1).

Other factors affecting the quantity of P mineralization/immobilization are temperature, moisture, aeration, pH, cultivation intensity, and P fertilization. The environmental effects are similar to those described for N mineralization/immobilization, since both are microbial processes (see Chapter 4).

The effect of soil pH and organic C on P mineralization is illustrated in Figure 5.12. The pH influence is related to (1) OH^- competing with $H_2PO_4^-$ or HPO_4^{2-} for bonding sites, (2) greater microbial activity at neutral pH levels, and (3) increased precipitation of Ca-P minerals at pH levels above 7.

Inorganic fertilizer P can be immobilized to organic P by microorganisms. The quantity of P immobilized varies widely, with values of 25 to 100% of applied P reported. Continued fertilizer P applications can increase the organic P content and subsequently increase P mineralization. Increases of 3 to 10 lb/a/yr in organic P with continued P fertilization are possible. In general, organic P will accumulate with P fertilization when C and N are available in quantities relative to the C/N/P ratio of the soil OM. Inorganic P will likely accumulate if C and N are limiting.

Further evidence of organic P mineralization is clearly demonstrated by the decrease in organic P with continued cultivation. When virgin soils are brought under cultivation, the OM content decreases, as described in Chapter 4. With this decrease in OM, extractable inorganic P initially increases but then decreases in a few years. In the northern Great Plains, organic C and P decreased an average of 38 and 21% after 60 and 70 years of cultivation, respectively, with no decrease in inorganic P (Table 5.5). Studies in the Midwest showed that after 25 years of cultivation, organic P mineralization reduced organic P 24% in the surface soil, which was less than the loss in organic C and N. In the southern Plains states, losses of organic P are greater because of increased soil temperature. In temperate regions, the decline in organic P with cultivation is generally less than that of organic C and N because of fewer loss mechanisms for P,

TABLE 5.5   Loss of Organic P with Continued Cultivation in Three Soils of the
Canadian Prairie

Soil Association	Native Prairie	60–70 Years of Cultivation	C or P Loss (%)
		Content	
Blaine Lake			
Organic C, mg/g	48	33	32
Total P, μg/g	823	724	12
Organic P, μg/g	645	528	18
Inorganic P, μg/g	178	196	
Sutherland			
Organic C, mg/g	38	24	37
Total P, μg/g	756	661	12
Organic P, μg/g	492	407	17
Inorganic P, μg/g	256	254	
Bradwell			
Organic C, mg/g	32	17	46
Total P, μg/g	746	527	29
Organic P, μg/g	446	315	29
Inorganic P, μg/g	300	212	29

SOURCE: Tiessen et al., *Agron. J.,* 74:831 (1982).

resulting in comparatively greater conservation of organic P. Under higher temperature and moisture regimes, equal losses of organic C, N, and P have been observed.

Measuring organic P cycling in soils is more difficult than for N because inorganic P produced through mineralization can be removed from solution by (1) P adsorption to clay and other mineral surfaces and (2) P precipitation as secondary Al^{3+}, Fe^{3+}, or Ca-P minerals. However, the quantity of P mineralized during a growing season varies widely among soils (Table 5.6). Large quantities of organic P are mineralized in tropical, high-temperature environments. In the Midwest, organic P mineralization probably contributes about 4 to 10 lb/a/yr of plant available P.

Much is yet to be learned about the immobilization and mineralization of P in soils and its relation to C, N, and S cycling. These reactions have an important bearing on nutrient management, but the extent of their influence is difficult to quantify. We can assume the following:

1. If adequate amounts of N, P, and S are added to soils to which crop residues are returned, some of the added nutrients may be immobilized in fairly stable organic C compounds.
2. Continued cropping of soils without the addition of supplemental N, P, and S will result in the mineralization of these elements and their subsequent depletion in soils.
3. If N, P, or S is present in insufficient amounts, the synthesis of soil OM may be reduced. All of these reactions presuppose the presence of adequate C and conditions conducive to the synthesis and breakdown of soil OM.

Inorganic Soil P

As organic P is mineralized to inorganic P or as P is added to soil, the inorganic P in solution not absorbed by plant roots or immobilized by microorganisms can be

TABLE 5.6   Quantities of Organic P Mineralized in a Growing Season for Several Soils

Location	Land Use	Soil	Study Period (yr)	Organic P Mineralized (kg/ha/yr)	Organic P Mineralized/Year (%/yr)
	Slightly weathered, temperate soils				
Australia	Grass	—	4	6	4
Australia	Wheat	—	55	0.3	0.3
Canada	Wheat	Silt loam	90	7	0.4
		Sandy loam	65	5	0.3
England	Grassland	Silt and sandy loam	1	7–40	1.3–4.4
	Arable	Silt and sandy loam	1	2–11	0.5–1.7
	Woodland	Silt loam	1	22	2.8
England	Cereal crop	—	—	0.5–8.5	—
England	Deciduous forest	Brown earth	1	9	1.2
	Grass	Brown earth	1	14	1.0
Iowa	Row crops	Clay loam	80	9	0.7
Maine	Potatoes	Silt loam	50	6	0.9
Minnesota	Alfalfa	Silty clay loam	60	12	1.2
Mississippi	Cotton	Silt loam	60	5	1.0
	Soybean	Silty clay loam	40	8	1.0
New Mexico	Row crops	Loam	30	2	0.4
Texas	Sorghum	Clay	60	7	1.0
	Weathered, tropical soils				
Honduras	Corn	Clay	2	6–27	5.9–11.9
		Clay	2	10–22	6.9–8.8
Nigeria	Bush	Sandy loam	1	123	24
	Cocoa	Sandy loam	1	91	28
Ghana					
Cleared shaded		Ochrosol fine	3	141	6
Tropical half shaded		Sandy loam	3	336	17
Rainforest exposed			3	396	17

SOURCE: Stewart and Sharpley, *SSSA Spec. Publ. No. 19*, p. 111 (1987).

adsorbed to mineral surfaces (labile P) or precipitated as secondary P compounds (Fig. 5.1). Surface adsorption and precipitation reactions collectively are called *P fixation* or *retention*. The extent of inorganic P fixation depends on many factors, most importantly soil pH (Fig. 5.13). In acidic soils, inorganic P precipitates as Fe/Al-P secondary minerals and/or is adsorbed to surfaces of Fe/Al oxide and clay minerals. In neutral and calcareous soils, inorganic P precipitates as Ca-P secondary minerals and/or is adsorbed to surfaces of clay minerals and $CaCO_3$.

P retention is a continuous sequence of precipitation and adsorption. With low-solution P concentrations, adsorption probably dominates, while precipitation reactions proceed when the concentration of P and associated cations in the soil solution exceeds that of the solubility product (K_{sp}) of the mineral. Where water-soluble fertilizers are applied to soils, the soil solution concentrations of P and accompanying cations are very high. Initially, P precipitation reactions occur because the solution P and accompanying cation concentrations exceed a specific mineral solubility. As the solution P level declines, P adsorption to reactive surface sites occurs. It is probable that both reactions proceed, to some extent, immediately following fertilizer P addition. Regardless of the relative contributions of adsorption and precipitation reactions, understanding these P fixation processes is important for optimum P nutrition and efficient management of fertilizer P.

PRIMARY AND SECONDARY MINERAL P SOLUBILITY   The P cycle (Fig. 5.1) illustrates that solution P levels are buffered by the release of adsorbed P from mineral surfaces (labile P), mineralization of organic P, and dissolution of solid P minerals. Ultimately, the P concentration in solution is controlled by the solubility of inorganic P minerals in soil. The most common P minerals found in acidic soils are Al- and Fe-P minerals, while Ca-P minerals predominate in neutral and calcareous soils (Table 5.7).

Mineral solubility represents the concentration of ions contained in the mineral that is maintained in solution. Each P mineral will support specific ion concentrations that depend on the solubility product of the mineral. For example, $FePO_4 \cdot 2H_2O$ will dissolve according to the following equation:

$$FePO_4 \cdot 2H_2O + H_2O \leftrightarrows H_2PO_4^- + H^+ + Fe(OH)_3 \qquad (1)$$

As solution $H_2PO_4^-$ decreases with P uptake by the crop, strengite in the soil can dissolve to resupply or maintain the $H_2PO_4^-$ concentration in solution. This reaction also shows that as H^+ increases (decreasing pH), the $H_2PO_4^-$

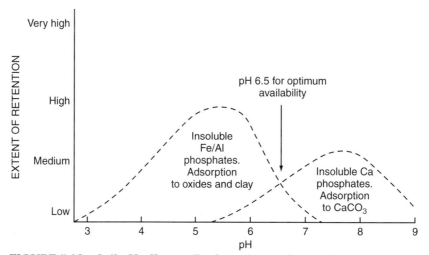

FIGURE 5.13   Soil pH effect on P adsorption and precipitation.   *Adapted from Stevenson,* Cycles of Soil, *p. 250, John Wiley & Sons, 1986.*

TABLE 5.7   Common P Minerals Found in Acidic, Neutral, and Calcareous Soils

Acidic soils*	
Variscite	$AlPO_4 \cdot 2H_2O$
Strengite	$FePO_4 \cdot 2H_2O$
Neutral and calcareous soils	
Dicalcium phosphate dihydrate (DCPD)	$CaHPO_4 \cdot 2H_2O$
Dicalcium phosphate (DCP)	$CaHPO_4$
Octacalcium phosphate (OCP)	$Ca_4H(PO_4)_3 \cdot 2.5H_2O$
β-tricalcium phosphate (βTCP)	$Ca_3(PO_4)_2$
Hydroxyapatite (HA)	$Ca_5(PO_4)_3OH$
Fluorapatite (FA)	$Ca_5(PO_4)_3F$

*Minerals are listed in order of decreasing solubility.

concentration decreases. Therefore, the specific P minerals present in soil and the concentration of solution P supported by these minerals are highly dependent on solution pH.

The relationship between the solubility of the various P minerals and soil solution pH is shown in Figure 5.14. In this diagram, the y-axis represents the concentration of $H_2PO_4^-$ or HPO_4^{2-} in soil solution. From Figure 5.5, $H_2PO_4^-$ is the predominant ion below pH 7.2 and HPO_4^{2-} is the predominant ion above pH 7.2. The x-axis expresses the soil solution pH. At pH 4.5, $AlPO_4 \cdot 2H_2O$ and $FePO_4 \cdot 2H_2O$ control the concentration of $H_2PO_4^-$ in solution. Increasing pH increases the $H_2PO_4^-$ concentration in solution because the minerals dissolve according to eq. 1, which is also depicted in the diagram as a positive slope. Increasing P availability is often observed when acidic soils are limed. Also, hydroxyapatite or fluorapatite can be used as a fertilizer in very low pH soils, as shown by their high solubility at low pH (Fig. 5.14). In contrast, they cannot be

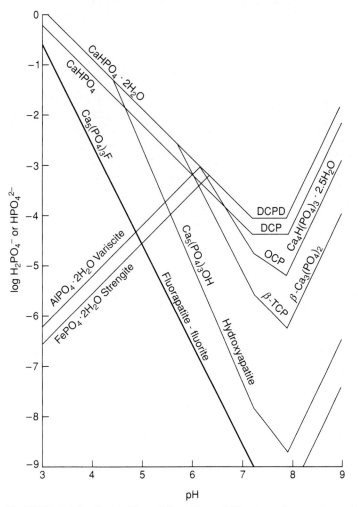

FIGURE 5.14   Solubility of Ca, Al, and Fe phosphate minerals in soils. *Lindsay,* Chemical Equilibria in Soils, *Wiley Interscience, p. 181, 1979.*

used to supply plant available P in neutral or calcareous soils because of their low solubility.

As pH is increased, the variscite and strengite solubility lines intersect several lines, representing the solubility of Ca-P minerals. For example, at pH 4.8, both strengite and fluorapatite can exist in soil, supporting $10^{-4.5}\,M\,H_2PO_4^-$ in solution. Between pH 6.0 and 6.5, the Al-P and Fe-P minerals can coexist with β-tricalcium phosphate (β-TCP), octacalcium phosphate (OCP), dicalcium phosphate (DCP), and dicalcium phosphate dihydrate (DCPD) at about $10^{-3.2}\,M$ $H_2PO_4^-$ in solution, which is about the highest solution P concentration that can exist in most unfertilized soils.

The solubility of Ca-P minerals is affected much differently than that of the Al-P and Fe-P minerals, as shown by the negative slopes of the Ca-P lines in Figure 5.14. As pH increases, the $H_2PO_4^-$ concentration decreases as the Ca-P precipitates, as described by the following equation for DCPD:

$$CaHPO_4 \cdot 2H_2O + H^+ \leftrightarrows Ca^{2+} + H_2PO_4^- + 2H_2O \qquad (2)$$

For example, assume that a soil contains β-TCP and the pH is 7.0. If pH decreases, the concentration of $H_2PO_4^-$ increases until about pH 6.0 as β-TCP dissolves. If pH decreases below 6.0, strengite and/or variscite will precipitate and the $H_2PO_4^-$ concentration in solution will decrease.

The slopes of the Ca-P lines in Figure 5.14 change above pH 7.2 because HPO_4^{2-} becomes the dominant species in solution compared with $H_2PO_4^-$. Thus, the mineral solubility lines represent only HPO_4^{2-} above pH 7.2. Above pH 7.8, the slope of the Ca-P solubility lines becomes positive, which means that as pH increases above 7.8, HPO_4^{2-} concentration increases. The change in solubility is due to the competing reaction of $CaCO_3$ solubility given by

$$\begin{array}{c} CaHPO_4 \cdot 2H_2O + H^+ \rightarrow Ca^{2+} + H_2PO_4^- + 2H_2O \\ Ca^{2+} + CO_2 + H_2O \leftrightarrows CaCO_3 + 2H^+ \end{array} \qquad (3)$$

$$\overline{CaHPO_4 \cdot 2H_2O + CO_2 \leftrightarrows H_2PO_4^- + H^+ + H_2O + CaCO_3} \qquad (4)$$

The precipitation of $CaCO_3$ in soil occurs at pH 7.8 and above. As solution Ca^{2+} decreases with $CaCO_3$ precipitation in soils (eq. 3), the DCPD will dissolve (eq. 2) to resupply solution Ca^{2+}. Therefore, when DCPD dissolves, $H_2PO_4^-$ also increases, as shown in eq. 4, which is the sum of eqs. 2 and 3. All of the Ca-P minerals listed in Table 5.7 behave similarly in soils that contain $CaCO_3$. Even though these P solubility relationships show solution P concentration increasing above pH 7.8, P availability to plants actually decreases. The HPO_4^{2-} released to solution by dissolution of Ca-P minerals will adsorb to the precipitating $CaCO_3$.

Although not readily apparent from the P solubility diagram (Fig. 5.14), P minerals that support the lowest concentration of P in solution (i.e., the lowest P solubility) are the most stable in soils. The apatite minerals, TCP, and OCP are more stable than DCPD in slightly acidic and neutral soils, for example. Therefore, the P mineral solubility relationships shown in Figure 5.14 can be used to understand the fate of fertilizer P applied to soils (see pg. 187–191).

An important fertilizer P source is monocalcium phosphate [MCP, $Ca(H_2PO_4)_2$], which is very soluble in soil. When MCP applied to soil dissolves, the concentration of $H_2PO_4^-$ in soil solution is much higher than the P concentrations supported by the minerals shown in Figure 5.14. Because the soil P minerals have lower solubility, the $H_2PO_4^-$ from the fertilizer will likely precipitate as these minerals. For example, in an acidic soil, fertilizer $H_2PO_4^-$ reacts with Fe^{3+} and Al^{3+} in solution to form $FePO_4$ and $AlPO_4$ compounds, respectively. As a result, the $H_2PO_4^-$ concentration in solution decreases once the precipitation reactions begin. In neutral and calcareous soils, fertilizer $H_2PO_4^-$ initially precipitates as DCDP and DCP within the first few weeks after application. After 3 to 5 months, OCP begins to precipitate, with β-TCP forming after 8 to 10 months. After long periods of time, apatite minerals eventually form.

Thus, after MCP is applied to soil, a series of reactions occur that decrease the elevated concentration of $H_2PO_4^-$ in solution as the insoluble minerals precipitate. These reactions in soils cannot be altered and help explain why plant recovery of fertilizer P is generally lower than recovery of soluble nutrients such as NO_3^- and SO_4^{2-}.

P ADSORPTION REACTIONS Labile inorganic P represents $H_2PO_4^-$ and/or HPO_4^{2-} adsorbed to mineral surfaces (Fig. 5.1). In acidic soils, Al and Fe oxide and hydroxide minerals are primarily involved in adsorption of inorganic P. Since the soil solution is acidic, the surface of these minerals has a net positive charge, although both positive and negative sites exist. The predominance of positive charges readily attracts $H_2PO_4^-$ and other anions. P ions adsorb to the Fe/Al oxide surface by interacting with OH^- and/or OH_2^+ groups on the mineral surface (Fig. 5.15). When the orthophosphate ion is bonded through one Al-O-P bond, the $H_2PO_4^-$ is considered *labile* and can be readily desorbed from the mineral surface to soil solution. When two Al-O bonds with $H_2PO_4^-$ occur, a stable six-member ring is formed (Fig. 5.15). Consequently, desorption is more difficult and the $H_2PO_4^-$ is considered *nonlabile*.

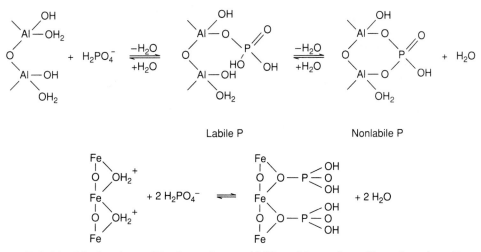

FIGURE 5.15 Mechanism of P adsorption to Al/Fe oxide surface. Phosphate bonding through one Al-O bond results in labile P; however, bonding through two Fe-O or Al-O bonds produces a stable structure that results in very little desorption of P.

In acidic soils, P adsorption also readily occurs on the broken edges of kaolinite clay minerals (see Fig. 2.7). Again, exposed OH^- groups on these edges can exchange for $H_2PO_4^-$ similarly to exchange with OH^- on the surface of Fe/Al oxides. Cations held to the surface of silicate clay minerals also influence P adsorption by developing a small positive charge near the mineral surface saturated with cations. This small positive charge attracts and holds small quantities of anions such as $H_2PO_4^-$. As discussed earlier, precipitation of Al-P minerals in acidic soils and Ca-P minerals in neutral and calcareous soils occurs at high P concentrations.

In calcareous soils, small quantities of P can be adsorbed through replacement of CO_3^{-2} on the surface of $CaCO_3$. At low P concentrations, surface adsorption predominates; however, at high P concentrations, Ca-P minerals precipitate on the $CaCO_3$ surface. Other minerals, mostly $Al(OH)_3$ and $Fe(OH)_3$, contribute to adsorption of solution P in calcareous soils.

Adsorption Equations. Several equations have been developed to describe P adsorption in soils. The Freundlich and Langmuir equations, or modifications of them, have been used most frequently. These equations are helpful for understanding the relationship between the quantity of P adsorbed per unit soil weight and the concentration of P in solution. Therefore, all of the equations have the general form

$$q = f(c)$$

where q is the quantity of P adsorbed and is a function (f) of the solution P concentration (c).

The *Freundlich equation* is represented by

$$q = ac^b$$

where a and b are coefficients that vary among soils, and q and c are the same as defined for the previous equation. This equation does not predict or include a maximum adsorption capacity and therefore works best with low solution P concentrations (Fig. 5.16). According to the Freundlich equation, energy of adsorption decreases as the amount of adsorption increases.

Since P adsorption data often exhibit a maximum P adsorption capacity at some solution P concentrations, another equation is needed to describe situations in which the adsorption sites are saturated with P. The *Langmuir equation* includes a term for the maximum P adsorption and is described by

$$q = \frac{abc}{1 + ac}$$

where q, c, and a are defined as before and b is the P adsorption maximum (Fig. 5.16). The P adsorption maximum in the Langmuir equation implies that a monolayer of P ions is adsorbed on the surface of the mineral, which occurs at relatively higher solution P concentrations than described by the Freundlich equation. The equation also shows that further increases in solution P concentration do not increase P adsorption. Although this does not

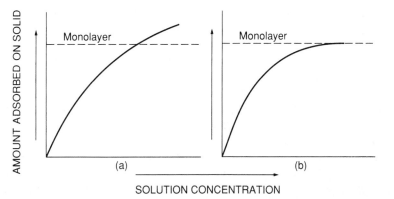

FIGURE 5.16   Graphical representation of adsorption isotherms of the Freundlich (a) and Langmuir (b) equations used to describe adsorption reactions in soils.

occur with all soils, the Langmuir equation frequently has been used to quantify P adsorption maxima.

These adsorption equations provide virtually no information about the mechanisms responsible for adsorption. They are incapable of showing whether Fe/Al oxides, silicate clays, or $CaCO_3$ dominate the adsorption reactions. In addition, the equations do not indicate whether P adsorption involves replacement of either hydroxyl, silica, or carbonate. There is interest in knowing if nonlabile P can become available to plants. Release of adsorbed P is extremely slow; it is not usually completed within hours or days. The rate of desorption decreases with time. In general, desorption appears to become very slow after about 2 days.

As might be expected, the extent of desorption depends on the nature of the adsorption complex at the surface of the Fe/Al oxides. Formation of six-membered ring structures, illustrated in Figure 5.15, prevents the desorption of P.

FACTORS INFLUENCING P RETENTION IN SOILS   Many soil physical and chemical properties influence the P solubility and adsorption reactions (P fixation) in soils. Consequently, these soil properties also affect solution P concentration, P availability to plants, and recovery of P fertilizer by crops. Understanding these influences will enable us to manage soil and fertilizer P efficiently for optimum production.

Soil Minerals. Adsorption and desorption reactions are affected by the type of mineral surfaces in contact with P in the soil solution. Fe/Al oxides have the capacity to adsorb large amounts of P in solution. Although present in most soils, they are most abundant in weathered, acidic soils. Fe/Al oxides can occur as discrete particles in soils or as coatings or films on other soil particles. They also exist as amorphous Al hydroxy compounds between the layers of expandable Al silicates. In soils with significant Fe/Al oxide contents, the less crystalline or the more amorphous the oxides, the larger their P fixation capacity because of their greater surface areas.

P is adsorbed to a greater extent by 1:1 clays (e.g., kaolinite) than by 2:1 clays (e.g., montmorillonite) because of the higher amounts of Fe/Al oxides associated with kaolinitic clays that predominate in highly weathered soils. Kaolinite has a larger number of exposed OH groups in the Al layer that can exchange with P. In addition, kaolinite develops pH-dependent charges on its edges that can adsorb P (see Fig. 2.7).

Figure 5.17 shows the influence of clay mineralogy on P adsorption. First, compare the three soils with >70% clay content. Compared with the Oxisol and Andept soils, very little P adsorption occurred in the Mollisol, composed mainly of montmorillonite, with only small amounts of kaolinite and Fe/Al oxides. The Oxisol soils contained Fe/Al oxides and exhibited considerably more P adsorption capacity compared with the Mollisol soils. Greatest P adsorption occurred with the Andept soils, composed principally of Fe/Al oxides and other minerals.

Soils containing large quantities of clay will fix more P than soils with low clay content (Fig. 5.17). In other words, the more surface area exposed with a given type of clay, the greater the tendency to adsorb P. For example, compare the three Ultisol soils containing 6, 10, and 38% clay. A similar relationship is evident in the Oxisol soils (36, 45, and 70% clay) and the Andept soils (11 and 70% clay).

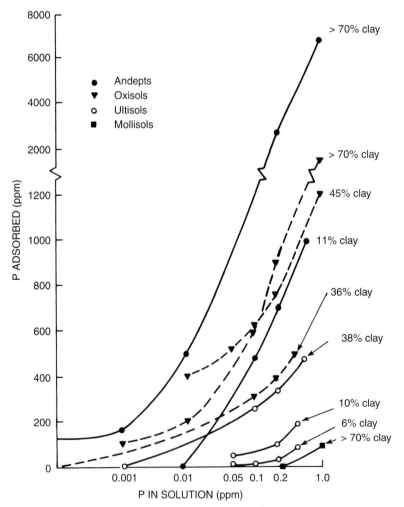

FIGURE 5.17   P adsorption influenced by clay content.   *Sanchez and Uehara, The Role of Phosphorus in Agriculture, p. 480, ASA, Madison, Wisc., 1980.*

In calcareous soils, P adsorption to $CaCO_3$ surfaces occurs; however, much of the adsorption is attributed to Fe oxide impurities. The amount and reactivity of $CaCO_3$ will influence P fixation. Impure $CaCO_3$ with large surface area exhibits greater P adsorption and more rapid precipitation of Ca-P minerals. Calcareous soils with highly reactive $CaCO_3$ and a high Ca–saturated clay content will exhibit low solution P levels, since P can readily precipitate or adsorb. In relative terms, acidic soils fix two times more P per unit surface area of soil than neutral or calcareous soils. The P adsorbed is held with five times more bonding energy in acidic soils than in calcareous soils.

To maintain a given level of solution P in soils with a high retention capacity, it is necessary to add larger quantities of P fertilizers (Fig. 5.18). In any one soil, the P concentration in solution increased with increasing P additions. Larger additions of P were required to reach a given level of solution P in fine-textured compared with coarse-textured soils. Consequently, high-clay, calcareous soils often require more fertilizer P than loam soils to optimize yields.

Soil pH. Adsorption of P by Fe/Al oxides declines with increasing pH. Gibbsite [γ-Al(OH)$_3$] adsorbs the greatest amount of P at pH 4 to 5. P adsorption by goethite (α-FeOOH) decreases steadily between pH 3 and 12 (Fig. 5.19).

P availability in most soils is at a maximum near pH 6.5 (Figure 5.13). At low pH values, the retention results largely from the reaction with Fe/Al oxides and precipitation as $AlPO_4$ and $FePO_4$. As pH increases, the activity of Fe and Al decreases, which results in lower P adsorption/precipitation and higher P concentration in solution. Above pH 7.0, Ca^{2+} can precipitate with P as Ca-P minerals

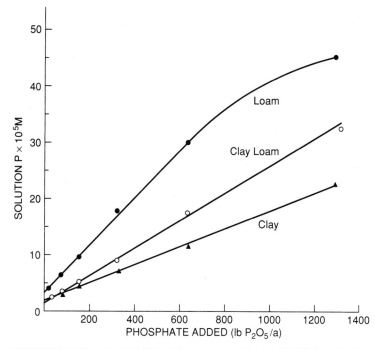

FIGURE 5.18   P solubility (the mean activity of DCP in solution) as a function of the amounts of CSP added to three calcareous soils of different texture. *Cole et al., SSSA Proc., 23:119, 1959.*

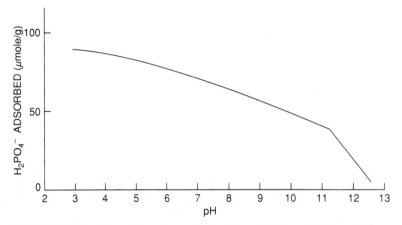

FIGURE 5.19   The adsorption of P by Fe oxide (goethite) as influenced by soil pH.   *Adapted from Hingston et al.,* Trans. 9th Int. Cong. Soil Sci., *1:1459–61, 1968.*

(Fig. 5.14) and P availability again decreases. The pH range of minimum P adsorption (pH 6.0 to 6.5) shown in Figure 5.13 also corresponds with the pH range of maximum P solubility (Fig. 5.14).

Liming acidic soils generally increases the solubility of P. Overliming can depress P solubility due to the formation of more insoluble Ca-P minerals similar to those occurring in pH > 7.0 soils that are naturally high in Ca^{2+}.

Cation Effects. Divalent cations on the CEC enhance P adsorption relative to monovalent cations. For example, clays saturated with Ca^{2+} retain greater amounts of P than those saturated with Na^+ or other monovalent ions. Current explanations for this effect of Ca^{2+} involve making positively charged edge sites of clay minerals more accessible to P anions. This action of Ca^{2+} is possible at pH values slightly less than 6.5, but in soils more basic than this, Ca-P minerals would directly precipitate from solution.

Concentration of exchangeable Al^{3+} is also an important factor in P adsorption in soils since 1 meq of exchangeable Al^{3+}/100 g of soil may precipitate up to 100 ppm P in solution. The following illustrates one of the possible ways that hydrolyzed Al^{3+} can adsorb soluble P.

Cation Exchange:

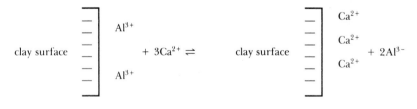

Hydrolysis:

$$Al^{3+} + 2H_2O \leftrightarrows Al(OH)_2^+ + 2H^+$$

Precipitation and/or Adsorption:

$$Al(OH)_2^+ + H_2PO_4^- \leftrightarrows Al(OH)_2H_2PO_4 \ (K_{sp} = 10^{-29})$$

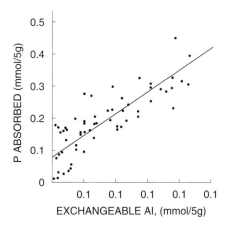

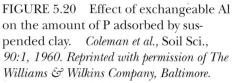

FIGURE 5.20 Effect of exchangeable Al on the amount of P adsorbed by suspended clay. *Coleman et al.,* Soil Sci., *90:1, 1960. Reprinted with permission of The Williams & Wilkins Company, Baltimore.*

Strong correlations between P adsorption and exchangeable Al^{3+} have been reported when there is appreciable hydrolysis of Al^{3+} (Fig. 5.20).

Anion Effects. Both inorganic and organic anions can compete with P for adsorption sites, resulting in decreased adsorption of added P. Weakly held inorganic anions such as NO_3^- and Cl^- are of little consequence, whereas adsorbed OH^-, $H_3SiO_4^-$, SO_4^{2-}, and MoO_4^{2-} can be competitive. The strength of bonding of the anion with the mineral surface determines the competitive ability of that anion. For example, SO_4^{2-} is unable to desorb much $H_2PO_4^-$, since $H_2PO_4^-$ is capable of forming a stronger bond than is SO_4^{2-}.

Organic anions from sources such as organic waste materials and wastewater treatment can affect the P adsorption-desorption reactions in soils. The impact of organic anions on reduction of adsorbed P is related to their molecular structure and pH. Organic anions form stable complexes with Fe and Al, which reduces adsorbed P. Oxalate and citrate can be adsorbed on soil surfaces similarly to $H_2PO_4^-$. Some of the effects of organic anions on P adsorption are partially responsible for the beneficial action of organic matter on P availability.

Extent of P Saturation. In general, P adsorption is greater in soils with little P adsorbed to mineral surfaces. As fertilizer P is added to soil and the quantity of P adsorption subsequently increases, the potential for additional P adsorption decreases. When all adsorption sites are saturated with $H_2PO_4^-$, further adsorption will not occur and recovery of applied fertilizer P should increase.

Organic Matter. Organic compounds in soils increase P availability by (1) the formation of organophosphate complexes that are more easily assimilated by plants, (2) anion replacement of $H_2PO_4^-$ on adsorption sites, (3) the coating of Fe/Al oxides by humus to form a protective cover and reduce P adsorption, and (4) increasing the quantity of organic P mineralized to inorganic P.

Organic anions produced from the decomposition of OM form stable complexes with Fe and Al, preventing their reaction with $H_2PO_4^-$. These complex ions can exchange for P adsorbed on Fe/Al oxides. The anions that are most effective in replacing $H_2PO_4^-$ are citrate, oxalate, tartrate, and malate.

Organic P compounds can move to a greater depth than can inorganic P in soil solution. Continued application of manure can result in elevated P levels at 2- to 4-ft depths. In contrast, application of the same quantity of P as inorganic fertilizer P results in much less downward movement of P.

Time and Temperature. P adsorption in soils follows two rather distinct patterns: an initial rapid reaction followed by a much slower reaction. The adsorption reactions involving exchange of P for anions on the surface of Fe/Al oxides are extremely rapid. The much slower reactions involve (1) formation of covalent Fe-P or Al-P bonds on Fe/Al oxide surfaces (Fig. 5.15) and (2) precipitation of a P compound for which the solubility product has been exceeded. These slow reactions involve a transition from more loosely bound to more tightly bound adsorbed P, which is less accessible to plants.

The initial compounds precipitated during the reaction of fertilizer P in soils are relatively unstable and will usually change with time into more stable and less soluble compounds (see p. 188). Table 5.8 shows the percentage of DCPD converted to OCP as a function of time and temperature. Conversion of 70% or more of the DCPD occurred after 10 months at 10°C and after only 4 months at 20 and 30°C.

Generally P adsorption increases with higher temperatures in soils to which fertilizer has been added. However, on soils with no added P, the amounts of P extracted are not affected by soil temperatures. P adsorption in soils of warm regions in the world is generally greater than in soils of temperate regions. These warmer climates also give rise to soils with higher contents of the Fe/Al oxides. Mineralization of P from soil OM or crop residues depends on soil biological activity, which increases with increasing temperature. Usually, mineralization rates double with each 10° increase in temperature.

Flooding. In most soils there is an increase in available P after flooding, largely due to a conversion of Fe^{3+} phosphate to soluble Fe^{2+} phosphate and hydrolysis of Al/Fe phosphate. Other mechanisms include dissolution of occluded P, increased mineralization of organic P in acidic soils, increased solubility of Ca phosphate in calcareous soils, and greater diffusion of P. These changes in P availability explain why the response to applied P by irrigated rice is usually less than the response of an upland crop grown on the same soil.

FERTILIZER P MANAGEMENT CONSIDERATIONS An important practical consequence of P adsorption and precipitation reactions is the time after application during which the plant is best able to utilize the added P. On some soils with a high fixing capacity, this period may be short, whereas with other soils it may last for months or even years. This time period will determine whether the fertilizer P should be applied at one time in the rotation or in smaller, more frequent applications. Adsorption of fertilizer P is greater in fine-textured soils because the amount of reactive mineral surface is greater than in coarse-textured soils.

Also important is the placement of P in the soil. If fertilizer P is broadcast and incorporated, the P is exposed to a greater amount of surface; hence, more fixa-

TABLE 5.8   Percentage of DCPD Hydrolyzed to OCP as a Function of Time and Temperature

| Temperature (°C) | Percentage OCP Present at | | | |
	1 Month	*2 Months*	*4 Months*	*10 Months*
10	<5	20	20	70
20	<5	40	75	100
30	<5	30	80	100

SOURCE: Sheppard and Racz, *Western Canada Phosphate Symp.*, p. 170 (1980).

tion takes place than if the same amount of fertilizer had been band applied. Band placement reduces the contact between the soil and fertilizer, with a subsequent reduction in P adsorption (see Chapter 10). Although this is not the only factor to consider in P fertilizer placement, it is very important for crops grown on low-P soils with a high P adsorption capacity. Thus, band placement generally increases the plant utilization of fertilizer P.

P Sources

Organic P

Animal and municipal wastes are excellent sources of plant available P, with manure accounting for 98% of organic P applied to cropland. The form and content of P in fresh animal waste varies greatly depending on the P content of the feed and the type of animal. Typically, inorganic P ranges from 0.3 to 2.4% of the dry weight, while organic P ranges from 0.1 to 1% (Table 5.9). Of the total P content in fresh manure, organic P represents 30 to 70%. For example, organic P may account for 40% of total P in swine manure, whereas poultry manure may contain 60% organic P. As discussed in relation to N, manure storage and handling change the nutrient content of manure. Storage usually increases inorganic P through mineralization of organic P; thus, organic P decreases with storage time. The data in Table 5.10 show 85% inorganic P after 3 or 4 months of liquid storage of swine waste, where initial inorganic P was 60 to 70% of the fresh manure dry weight. Additional information on P content in manures and manure management is provided in Chapter 10.

TABLE 5.9   P Content of Selected Animal Wastes

Animal	Total P		Inorganic P
		% of Dry Matter	
Swine	1.5–2.5		0.8–2.0
Beef cattle	0.7–1.2		0.5–0.8
Dairy cattle	0.5–1.2		0.3–1.0
Poultry	0.9–2.2		0.3–1.2
Horses	0.4–1.4		0.2–0.8

TABLE 5.10   Distribution of P among Organic or Inorganic P Compounds in Liquid Stored Swine Manure

Animal	Total P (% dry matter)	Percentage of Total P
Total inorganic P	1.5–2.0	85
Total organic P	0.2–0.3	15
Inorganic P in solution	0.01–0.20	5
Organic P in solution	0.01–0.03	<4
Microbial P	0.02–0.04	<2

SOURCE: *Van Faassen, 1987.* In V.d. Meer (ed), Animal Manure on Grassland Crops. p. 27–45. Martinus Nijhoff Publ., Netherlands.

P content in sewage sludge ranges from 2 to 4%, with most present as inorganic P (Table 10.21). Thus, 40 to 80 lb P/a would be applied per ton of material. If 80% was inorganic P and plant available during the first year, then 32 to 64 lb P/a would be needed. Because of relatively high transportation and processing expenses, application rates generally exceed 1 t/a, and therefore the total amount of P applied can greatly exceed typical crop requirements.

MICROBIAL P FERTILIZATION   The use of microorganisms to increase plant available P has been documented. Since the late 1950s, bacteria collectively called *phosphobacterins* have been soil applied to increase the P uptake and yield of crops. An average of 10% yield increases have been reported; however, other studies showed no response to phosphobacterin applied to numerous crops over a wide geographic area.

In the 1980s several fungi, in particular *Penicillium bilaii*, were shown to increase P uptake, especially in high-pH, calcareous soils. Increased solubilization of native soil mineral P and added rock phosphate (RP) have been observed. Phosphate-solubilizing organisms apparently release organic acids that may dissolve P minerals. The future use of P-solubilizing organisms is uncertain, but it is currently being evaluated in Canada and Australia.

Inorganic P

P FERTILIZER TERMINOLOGY   The terms used to describe the P content in fertilizers are *water soluble, citrate soluble, citrate insoluble, available,* and *total P* (as P_2O_5). A small sample is first extracted with water, and the amount of P contained in the filtrate represents the fraction that is *water soluble.*

The remaining water-insoluble material is extracted with 1 *N* ammonium citrate. The P content of the filtrate is the *citrate-soluble* P. The sum of the water-soluble and citrate-soluble P represents plant *available* P. The amount of P in the residue remaining from the water and citrate extractions is *citrate-insoluble* P. The sum of available and citrate-insoluble P represents the total amount present.

P CONTENT OF FERTILIZERS   The P content of fertilizers is expressed as P_2O_5 instead of as elemental P. Although attempts have been made to change from % P_2O_5 to % P, most still express P concentration in fertilizers as % P_2O_5. Similarly, the concentration of K in fertilizers is usually expressed as % K_2O instead of % K. As a matter of interest, N was formerly guaranteed as % NH_3 rather than as % N, as is now done.

The conversion between % P and % P_2O_5 is given by

$$\% \text{ P} = \% \text{ } P_2O_5 \times 0.43$$
$$\% \text{ } P_2O_5 = \% \text{ P} \times 2.29$$

The conversion factors are derived from the ratio of molecular weights of P and P_2O_5:

$$\frac{2 \times \text{M wt P}}{\text{M wt } P_2O_5} = \frac{2 \times 31}{142} = 0.43$$

For example, the % P concentration in $CaHPO_4$ (DCP) is

$$\frac{\text{M wt of P}}{\text{M wt CaHPO}_4} = \frac{31}{136} \times 100 = 23\% \text{ P}$$

Thus, the % P_2O_5 is calculated by

$$\% \text{ P}_2\text{O}_5 = 23\% \times 2.29 = 53\%$$

P FERTILIZER SOURCES   Rock phosphate (RP) is the only important raw material for P fertilizers. The general formula for RP is $Ca_{10}(PO_4)_6(X)_2$, where X is either F^-, OH^-, or Cl^-. These minerals are called *apatites;* the most common RP mined is fluorapatite (the F^- form of RP). RP contains numerous impurities; the most common ones are CO_3, Na, and Mg. Carbonate-fluorapatite (francolite) is the primary apatite mineral in the majority of phosphate rocks. The high reactivity of some phosphate rocks is due to the occurrence of francolite. The major deposits are found in the United States, Morocco, Russia, South Africa, and China. The United States produces about 40% of the world's RP, although nearly 50% of the world reserves are in Morocco. The common P fertilizers are produced from either acid- or heat-treated RP to break the apatite bond and to increase the water-soluble P content. The common commercially available P fertilizers are listed in Table 5.11.

Rock Phosphate.   After several processing and purification steps, RP contains between 11.5 and 17.5% total P (27 to 41% P_2O_5). None of the P is water soluble, although the citrate solubility varies from 5 to 17% of the total P. RP can be used directly as a P fertilizer under certain conditions; however, in most situations, P fertilizers processed from RP are more effective.

Finely ground apatitic RPs are effective only on acidic soils (pH < 6). On low-P, acidic soils, RP application can supply sufficient plant available P but *only* when RP is applied in quantities two to three times the rates of superphosphate. RP has been reported to give better residual effects than superphosphate, but whenever this was so, it was found that the rates of applied RP were much greater than for superphosphate.

Environmental conditions such as warm climates, moist soils, and long growing seasons increase the effectiveness of RP. RP is used extensively for plantation crops such as rubber, oil palm, and cacao grown on very acidic soils (pH < 5) in Southeast Asia. Ground RP is sometimes used for restoration of low-P soils on abandoned farms and on newly broken lands. For these purposes, a heavy initial application is recommended, such as 1 to 3 t/a, which may be repeated at 5- to 10-year intervals. Addition of 1,000 lb/a of reactive RP is used to rehabilitate tropical savanna land in Southeast Asia.

In situations where the reactivity of RP is inadequate for immediate crop response and the P-fixation capacity of the soil quickly renders soluble P fertilizer unavailable to plants, partially acidulated RP can increase the water-soluble P content and improve the short-term crop response to RP. Partially acidulated RP is produced by treating RP with 10 to 20% of the quantity of H_3PO_4 used for the manufacture of triple superphosphate or by reacting it with 40 to 50% of the amount of H_2SO_4 normally used in the production of single superphosphate.

TABLE 5.11   Common Commercially Available P Fertilizers

Material	Frequently Used Abbreviations	Analysis (%)				Form of P	Percentage Total P Available	Formula of Main P Compound
		N	P_2O_5	K_2O	S			
Rock phosphate	RP	—	25–40	—	—	Orthophosphate	14–65	$[Ca_3(PO_4)_2]_3 \cdot CaF_x \cdot (CaCO_3)_x \cdot (Ca(OH)_2)_x$
Single superphosphate	SSP	—	16–22	—	11–12	Orthophosphate	97–100	$Ca(H_2PO_4)_2$
Wet process phosphoric acid	—	—	48–53	—	—	Orthophosphate	100	H_3PO_4
Triple superphosphate	TSP or CSP	—	44–53	—	1–1.5	Orthophosphate	97–100	$Ca(H_2PO_4)_2$
Ammonium phosphates								
Monoammonium phosphate	MAP	11–13	48–62	—	0–2	Orthophosphate	100	$NH_4H_2PO_4$
Diammonium phosphate	DAP	18–21	46–53	—	0–2	Orthophosphate	100	$(NH_4)_2HPO_4$
Ammonium polyphosphate	APP	10–15	35–62	—	—	Mixture of ortho- and polyphosphates	100	$(NH_4)_3HP_2O_7 + NH_4H_2PO_4$ + others
Urea-ammonium phosphate	UAP or UAPP	21–34	16–42	—	—	Mixture of ortho- and polyphosphates	100	$NH_4H_2PO_4 \cdot (NH_4)_3HP_2O_7$
Ammoniated normal superphosphate	—	2–5	14–21	—	9–11	Orthophosphate	97–100	$NH_4H_2PO_4 \cdot CaHPO_4$
Ammoniated triple superphosphate	—	4–6	44–53	—	0–1	Orthophosphate	96–100	$CaHPO_4 \cdot NH_4H_2PO_4$
Potassium phosphates								
Monopotassium phosphate	—	—	51	35	—	Orthophosphate	100	KH_2PO_4
Dipotassium phosphate	—	—	41	54	—	Orthophosphate	100	K_2HPO_4
Potassium polyphosphate	—	—	51	40	—	Polyphosphate and orthophosphate	100	$K_3HP_2O_7 \cdot KH_2PO_4$ · others

SOURCE: Follett et al., *Fertilizers and Soil Amendments*, p. 131, Prentice-Hall, (1981).

Phosphoric Acid. Phosphoric acid (H_3PO_4) is manufactured by treating RP with H_2SO_4 and is called *green* or *wet process acid.* A higher concentration of H_2SO_4 is used in the wet process acid reaction than is used to produce single superphosphate. Both reactions produce gypsum ($CaSO_4 \cdot 2H_2O$), which can be used for other industrial purposes. Gypsum is also used as a soil amendment for sodic soils and as a S source. Phosphoric acid can be made by heating RP in an electric arc furnace to produce elemental P, which is then reacted with O_2 to produce P_2O_5 and subsequently with H_2O to form H_3PO_4.

Phosphoric acid made by burning is termed *white* or *furnace acid* and is used almost entirely by the nonfertilizer segment of the chemical industry. White acid has a much higher degree of purity than green acid; however, the high energy cost involved in manufacturing makes it expensive and limits its use in the fertilizer industry.

Agricultural-grade green acid, containing 17 to 24% P (39 to 55% P_2O_5), is used to acidulate RP to make TSP and is neutralized with NH_3 in the manufacture of NH_4^+ phosphates and liquid fertilizers. It can also be applied by injection in the soil or to irrigation water, particularly in alkaline and calcareous areas, but this method requires special handling and equipment. Almost all wet acid is used to manufacture other P fertilizers.

Calcium Orthophosphates. The Ca orthophosphate fertilizers—single superphosphate, triple superphosphate, and enriched superphosphates, were once the most important P sources.

The superphosphates are *neutral fertilizers* in that they have no appreciable effect on soil pH, as do H_3PO_4 and the NH_4^+-containing fertilizers. Single superphosphate (SSP) is manufactured by reacting H_2SO_4 with RP:

$$[Ca_3(PO_4)_2]_3 \cdot CaF_2 \; + \; 7H_2SO_4 \; \rightarrow \; 3Ca(H_2PO_4)_2 \; + \; 7CaSO_4 \; + \; 2HF$$

| Rock phosphate | Sulfuric acid | Monocalcium phosphate | Gypsum | Hydrofluoric acid |

SSP, 7 to 9.5% P (16 to 22% P_2O_5), is 90% water soluble and essentially all is plant available. SSP is an excellent source of P and S (12% S), but its low P analysis is a main disadvantage in its use. As a result, SSP is not commercially available in the U.S., but is in other countries. Because of the low energy requirements for production, use of H_2SO_4 by-products of other industries, and increasing demand for S and micronutrients, SSP use may increase in the future.

Triple or concentrated superphosphate (TSP or CSP), containing 17 to 23% P (44 to 52% P_2O_5), is manufactured by treating RP with H_3PO_4:

$$[Ca_3(PO_4)_2]_3 \cdot CaF_2 \; + \; 12H_3PO_4 \; + \; 9H_2O \; \rightarrow \; 9Ca(H_2PO_4)_2 \; + \; CaF_2$$

| Rock phosphate | Phosphoric acid | Water | Monocalcium phosphate | Calcium fluoride |

TSP was manufactured to increase the P content of SSP, although TSP contains very little S (0 to 1%).

TSP is an excellent source of P and was the most common source of fertilizer P used in the United States until the early 1960s, when NH_4^+ phosphates became popular (Fig. 5.21). Its high P content is an advantage because transportation, storage, and handling costs make up a large fraction of the total fertilizer cost. TSP is manufactured in granular form and is used in mixing and blending with

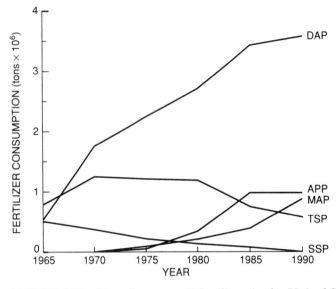

FIGURE 5.21   Use of common P fertilizers in the United States. DAP, diammonium phosphate; APP, ammonium polyphosphate; MAP, monoammonium phosphate; TSP, triple superphosphate; SSP, single superphosphate.   *USDA, 1990.*

other materials and in direct soil application. SSP and TSP can be ammoniated to produce monoammonium phosphate, or MAP ($NH_4H_2PO_4$). The reactions shown in Figure 5.22 illustrate the ammoniation of superphosphates but also reveal that excessive ammoniation produces very insoluble P (see Fig. 5.14). The total P content of the end product is decreased in proportion to the weight of the NH_3 added. Ammoniation of superphosphates offers the advantage of inexpensive N but decreases the amount of water-soluble P in the product.

Ammonium Phosphates. Ammonium phosphates are produced by reacting wet process H_3PO_4 with NH_3 (Fig. 5.23). MAP contains 11 to 13% N and 21 to 24% P (48 to 55% P_2O_5); however, the most common grade is 11-22-0 (11-52-0). DAP contains 18 to 21% N and 20 to 23% P (46 to 53% P_2O_5); the most common grade is 18-20-0 (18-46-0). Although MAP use has increased significantly in the last decade, DAP is more widely used than any other P fertilizer in the U.S. (Fig. 5.21). The increased interest in and use of NH_4 phosphates result from considerable evidence supporting increased $H_2PO_4^-$ uptake when NH_4^+ is placed with P fertilizer.

Both MAP and DAP are granular fertilizers and completely water soluble. Ammonium phosphates have the advantage of a high plant-food content, which minimizes shipping, handling, and storage costs. In addition, they can be used for formulating solid fertilizers by bulk blending or in manufacturing suspension fertilizers. MAP and DAP are also used for direct application as starter fertilizers.

Care must be taken with row or seed placement of DAP since free NH_3 can be produced, causing seedling injury and inhibiting root growth, according to the following reaction:

$$(NH_4)_2HPO_4 \rightarrow 2NH_4^+ + HPO_4^{2-} \text{ (pH 8.5)}$$

Ammoniation of normal superphosphate

$Ca(H_2PO_4)_2$ + NH_3 $\longrightarrow$ $CaHPO_4$ + $NH_4H_2PO_4$
Monocalcium   Ammonia   Dicalcium   Monoammonium
phosphate   phosphate   phosphate

$NH_4H_2PO_4$ + $CaSO_4$ + NH_3 $\longrightarrow$ $CaHPO_4$ + $(NH_4)_2SO_4$
Monoammonium Calcium Ammonia   Dicalcium   Ammonium
phosphate   sulfate   phosphate   sulfate

$2\,CaHPO_4$ + $CaSO_4$ + $2\,NH_3$ $\longrightarrow$ $Ca_3(PO_4)_2$ + $(NH_4)_2SO_4$
Dicalcium   Calcium   Ammonia   Tricalcium   Ammonium
phosphate   sulfate   phosphate   sulfate

Ammoniation of triple superphosphate

$Ca(H_2PO_4)_2$ + NH_3 $\longrightarrow$ $CaHPO_4$ + $NH_4H_2PO_4$
Monocalcium   Ammonia   Dicalcium   Monoammonium
phosphate   phosphate   phosphate

$3\,CaHPO_4$ + NH_3 $\longrightarrow$ $NH_4H_2PO_4$ + $Ca_3(PO_4)_2$
Dicalcium   Ammonia   Monoammonium   Tricalcium
phosphate   phosphate   phosphate

Ammoniation usually expressed in terms of kg NH_3 per 20 kg of P_2O_5. For normal superphosphate, the normal range is 4–6 kg NH_3 per 20 kg P_2O_5; for triple superphosphate, 3–4 kg NH_3 per 20 kg P_2O_5.

FIGURE 5.22 Ammoniation of single superphosphate (SSP) and triple superphosphate (TSP) to produce monoammonium phosphate (MAP). The reactions are carefully controlled to prevent excessive ammoniation and formation of water-insoluble tricalcium phosphate and dicalcium phosphate (DCP). *Follett et al.,* Fertilizers and Soil Amendments, *p. 122, Prentice-Hall, 1980.*

NH_3 + H_3PO_4 $\longrightarrow$ $NH_4H_2PO_4$
Ammonia   Orthophosphoric   Monoammonium
                acid   phosphate

$2\,NH_3$ + H_3PO_4 $\longrightarrow$ $(NH_4)_2HPO_4$
Ammonia   Orthophosphoric   Diammonium
                acid   phosphate

$3\,NH_3$ + $H_4P_2O_7$ $\longrightarrow$ $(NH_4)_3HP_2O_7$
Ammonia   Pyrophosphoric   Triammonium
                acid   pyrophosphate

FIGURE 5.23 Reactions of ammonia with ortho- and pyrophosphate to produce monoammonium phosphate (MAP), diammonium phosphate (DAP), and ammonium polyphosphate (APP). *Follett et al.,* Fertilizers and Soil Amendments, *p. 127, Prentice-Hall, 1981.*

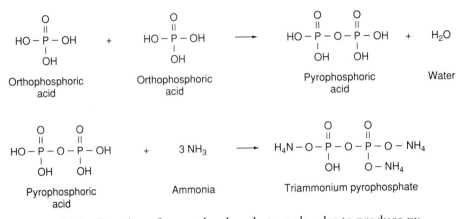

FIGURE 5.24   Reaction of two orthophosphate molecules to produce pyrophosphate. The reaction can continue to form longer chain products called
polyphosphates. Ammonium of pyro- and polyphosphates produces ammonium
polyphosphate (APP).   *Follett et al.*, Fertilizers and Soil Amendments, *p. 130,
Prentice-Hall, 1981.*

$$NH_4^+ + OH^- \rightarrow NH_3 + H_2O$$

These problems are especially common in calcareous or high-pH soils. Adequate separation of seed from DAP is usually all that is required to eliminate
seedling damage. In most cases, the N rate should not exceed 15 to 20 lb N/a as
DAP applied with the seed. Seedling injury with MAP is seldom observed except
in sensitive crops such as canola/rapeseed and flax.

The initial soil reaction pH of DAP is about 8.5, which favors NH_3 production
(see Fig. 4.35), whereas the reaction pH with MAP is 3.5. Except for the differences in reaction pH and seedling injury when applied with the seed, few agronomic differences exist between MAP and DAP. Reports of improved crop response to MAP compared with DAP on high-pH or calcareous soils are generally
not substantiated. Low-reaction pH with MAP has been claimed to increase micronutrient availability in calcareous soils, but this has not been demonstrated.

Ammonium Polyphosphate. APP is manufactured by reacting pyrophosphoric
acid, $H_4P_2O_7$, with NH_3 (Fig. 5.24). Pyrophosphoric acid is produced from dehy-
dration of wet process acid. *Polyphosphate* is a term used to describe two or more
orthophosphate ions ($H_2PO_4^-$) combined together, with the loss of one H_2O
molecule per two $H_2PO_4^-$ ions (Fig. 5.24). APP is a liquid containing 10 to 15%
N and 15 to 16% P (34 to 37% P_2O_5), with about 75 and 25% of the P present as
polyphosphate and orthophosphate, respectively. The most common APP grade
is 10-15-0 (10-34-0).

Granulation during the manufacture of APP results in a solid product of
11-24-0 (11-55-0). Granular APP can be applied directly or blended with
other granular fertilizers. Liquid APP is more popular and can be directly applied or mixed with other liquid fertilizers. Commonly, UAN and APP are
combined and subsurface band applied.

One unique property of APP is the chelation or sequestering reaction with
metal cations, which maintains higher concentrations of micronutrients in APP

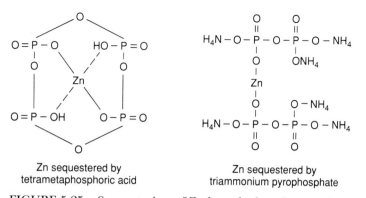

Zn sequestered by
tetrametaphosphoric acid

Zn sequestered by
triammonium pyrophosphate

FIGURE 5.25   Sequestering of Zn by polyphosphate molecules can maintain a greater Zn concentration in solution than orthophosphate.   *Follett et al., Fertilizers and Soil Amendments, p. 132, Prentice-Hall, 1981.*

than are possible with orthophosphate solutions (Fig. 5.25). For example, APP can maintain 2% Zn in solution compared with only 0.05% Zn with orthophosphate.

A granular fertilizer, urea-ammonium phosphate (UAP), is produced by reacting urea with APP. The fertilizer grade is 28-12-0 (28-28-0), containing 20 to 40% polyphosphate. UAP can be easily blended with other granular fertilizers. Like DAP, seedling damage occurs when UAP is applied with the seed.

Potassium Phosphate. Potassium phosphate is represented by two salts, KH_2PO_4 and K_2HPO_4, which have the grades 0-52-35 (22% P, 29% K) and 0-41-54 (18% P, 45% K), respectively. They are completely water soluble and find their greatest market in soluble fertilizers sold in small packets for home and garden use. Their high content of P and K makes them attractive possibilities for commercial application on a farm scale. Developments in the economics of producing these salts will determine whether they can be manufactured on a large scale for use as commercial fertilizers.

In addition to the high plant nutrient content of the K phosphates, they have other desirable characteristics. Because they contain no Cl^-, K phosphates are ideally suited for solanaceous crops such as potatoes, tomatoes, and many leafy vegetables that are sensitive to high levels of Cl^-. Their low salt index reduces the risk of injury to germinating seeds and to young seedlings when they are placed in or close to the seed row.

Behavior of P Fertilizers in Soils

The chemical characteristics of the soil and the P fertilizer source determine the soil-fertilizer reactions, which influence fertilizer P availability to plants. Many of the same factors that affect native P availability, discussed earlier, also influence fertilizer P reaction product chemistry and availability. As seen in Figure 5.1, P fertilizer added to soil initially increases solution P but subsequently solution P decreases through influences of mineral, adsorbed (labile), and organic P fractions.

The commonly used granular P fertilizers are 90 to 100% water soluble and dissolve rapidly when placed in moist soil. Water sufficient to initiate dissolution moves to the granule by either capillarity or vapor transport. A nearly saturated

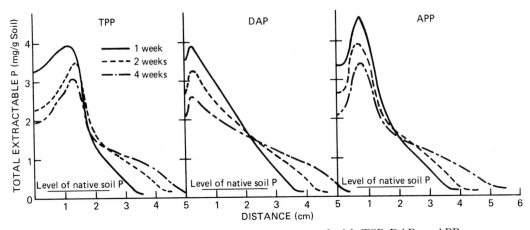

FIGURE 5.26   P distribution profiles in columns treated with TSP, DAP, or APP. *Khasawneh et al., Soil Sci. Soc. Am. J., 38:446, 1974.*

solution of the P fertilizer material forms in and around fertilizer granules or droplets (Figs. 5.26, 5.27, 5.28).

While water is drawn into the fertilizer, the fertilizer solution moves into the surrounding soil. Movement of water inward and fertilizer solution outward continues to maintain a nearly saturated solution as long as the original salt remains. Initial movement of P away from a fertilizer application site seldom exceeds 3 to 5 cm (Fig. 5.26).

Diffusion of fertilizer P reaction products away from the dissolving granule increases with increasing soil moisture content. Extensive reaction zones combined with thorough distribution of the reaction products are factors that should enhance absorption of P by plant roots encountering reaction zones.

As the saturated P solutions move into the first increments of soil, the chemical environment is dominated by the solution properties rather than by the soil properties. Solutions formed from water-soluble P fertilizers have pH values between 1 and 8 and contain from 2.9 to 6.8 mol/l of P (Table 5.12). The concentration of the accompanying cations ranges from 1.3 to 10.2 mol/l.

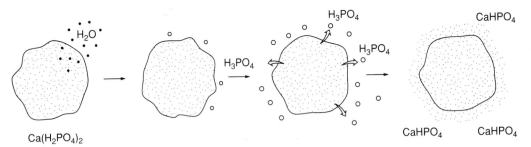

FIGURE 5.27   Reaction of a monocalcium phosphate (MCP) granule in soil. Water vapor moves toward the granule, which begins to dissolve. Phosphoric acid forms around the granule, resulting in a solution pH of 1.5. The acidic solution causes other soil minerals to dissolve, increasing the cation (and anion) concentration near the granule. With time the granule dissolves completely and the solution pH increases, with subsequent precipitation of a dicalcium phosphate (DCP) reaction product.

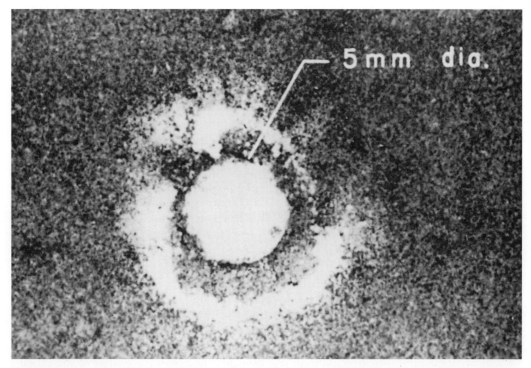

FIGURE 5.28   Distribution of monocalcium phosphate (MCP) reaction products after 14 days' reaction at 5°C in the Bradwell very fine sandy loam.   *Hinman et al., Can. J. Soil Sci., 42:229, 1962.*

TABLE 5.12   Phosphate Compounds Commonly Found in Fertilizers and Compositions of Their Saturated Solutions

		Composition of Saturated Solution			
Compound	*Formula*	*Solution Symbol*	*pH*	*P (mol/liter)*	*Accompanying Cation (mol/liter)*
Highly water soluble					
Monocalcium phosphate	$Ca(H_2PO_4) \cdot H_2O$	TPS	1.5	4.5	Ca	1.3
Monoammonium phosphate	$NH_4H_2PO_4$	MAP	3.5	2.9	NH_4	2.9
Triammonium pyrophosphate	$(NH_4)_3HP_2O_7 \cdot H_2O$	TPP	6.0	6.8	NH_4	10.2
Diammonium phosphate	$(NH_4)_2HPO_4$	DAP	8.5	3.8	NH_4	7.6
Sparingly soluble					
Dicalcium phosphate	$CaHPO_4$ $CaHPO_4 \cdot 2H_2O$	DCP/ DCPD	6.5	≈ 0.002	Ca	0.001
Hydroxyapatite	$Ca_{10}(PO_4)_6(OH)_2$	HAP	6.5	$\approx 10^{-5}$	Ca	0.001

SOURCE: Sample et al., in F. E. Khasawneh et al. (Eds.), *The Role of Phosphorus in Agriculture*, p. 275. American Society of Agronomy Madison, Wisc. (1980).

When the concentrated P solution leaves the granule, droplet, or band site and moves into the surrounding soil, the soil components are altered by the solution; at the same time, the solution's composition is changed by its contact with soil. Some soil minerals may actually be dissolved by the concentrated P solution, resulting in the release of cations such as Fe^{3+}, Al^{3+}, Mn^{2+}, K^+, Ca^{2+}, and Mg^{2+}. Cations from exchange sites may also be displaced by these concentrated solutions. P in the concentrated solutions reacts with these cations to form specific compounds, referred to as *soil-fertilizer reaction products*.

An example of fertilizer P dissolution and reaction product formation is shown in Figure 5.27. Monocalcium phosphate [MCP, $Ca(H_2PO_4)_2$] is added to soil, and water diffuses toward the granule. As the MCP dissolves, H_3PO_4 is formed, resulting in a solution pH of 1.5 near the granule (Table 5.12). Other soil minerals in contact with the H_3PO_4 may be dissolved, increasing the cation concentration near the granule. Subsequently, the solution pH will increase as the H_3PO_4 is neutralized. Within a few days or weeks, DCP and/or DCPD will precipitate as the initial fertilizer reaction product. Depending on the native P minerals initially present in the soil, OCP, TCP, HA, or $Fe/AlPO_4$ may eventually precipitate (see Fig. 5.14).

Precipitation reactions are favored by the very high P concentrations existing in close proximity to P fertilizer. Adsorption reactions are expected to be most important at the periphery of the soil-fertilizer reaction zone, where P concentrations are much lower. Although both precipitation and adsorption occur at the application site, precipitation reactions usually account for most of the P being retained in that vicinity. The precipitation of DCP at the application site of MCP is readily apparent in Figure 5.28. From 20 to 34% of the applied P will remain as this reaction product at the granule site.

Although the initial reaction products are unstable and are usually transformed with time into more stable but less water-soluble compounds, they will have a favorable influence on the P nutrition of crops. Some of the initial reaction products will provide P concentrations in solution 1,000 times those in untreated soil. The rate of change of the initial reaction products is influenced by soil properties and environmental factors. For example, after initial DCP formation (a few weeks), formation of OCP may take 3 to 5 months.

Further transformation to TCP or HA may take 1 year or longer. The residual value of fertilizer P depends on the nature and reactivity of long-term reaction products.

In acidic soils, reaction products formed from MCP include DCP and eventually $AlPO_4$ and $FePO_4$ precipitates (Fig. 5.14). If the soil is very acidic with low Ca^{2+}, then $AlPO_4$ may precipitate first. Similarly, in calcareous soils, DCP is the dominant initial reaction product. Also, in the presence of extremely large amounts of calcium carbonate, OCP may form.

Because MAP has a reaction pH of 3.5 compared with pH 8.5 for DAP, P should be more soluble near the dissolving granule. The acidic pH with MAP may temporarily reduce the rate of P reaction product precipitation in calcareous soils. Although differences in pH among the various P fertilizers (Table 5.12) cause differences in reaction product chemistry, the overall effect is temporary because the volume of soil influenced by the P granule or droplet is small. Differences in availability of P sources to crops are small compared with differences in other P management factors such as P rate and placement.

APP applied to soil reacts similarly to the granular P fertilizers. The reaction pH is 6.2, and both precipitation and adsorption of the polyphosphate and $H_2PO_4^-$ present initially, plus that formed by hydrolysis of the polyphosphate, occur with soils similar to those described earlier for $H_2PO_4^-$.

Hydrolysis or the reaction of H_2O with polyphosphate results in a stepwise breakdown, producing $H_2PO_4^-$ and various shortened polyphosphate fragments. The shortened polyphosphate fragments then undergo further hydrolysis. Hydrolysis of polyphosphates to $H_2PO_4^-$ occurs by two principal pathways, either chemically or biologically. Hydrolysis of polyphosphates proceeds very slowly in sterile solutions at room temperature; however, in soils in which both mechanisms can function, hydrolysis is usually rapid.

Several factors control hydrolysis rates in soils, with enzymatic activity provided by plant roots and microorganisms being the most important. Phosphatases associated with plant roots and rhizosphere organisms are responsible for the biological hydrolysis of polyphosphates.

Temperature, moisture, soil C, pH, and various conditions that encourage microbial and root development favor phosphatase activity and hydrolysis of polyphosphates. Temperature is the most important environmental factor influencing the rate of polyphosphate hydrolysis. The extent of hydrolysis of pyro- and polyphosphate increased substantially by elevating the temperature from 5 to 35°C (Fig. 5.29).

Polyphosphates are as effective as $H_2PO_4^-$ as sources of P for crops. Plants can absorb and utilize the polyphosphates directly. Because polyphosphates have the ability to form metal ion complexes, they may be effective in mobilizing Zn in soils in which deficiencies have been induced by high pH or high P levels (Fig. 5.25). After the addition of a high rate of pyrophosphate, only slight increases in Zn in the soil solution have been observed. This short-lived effect probably results from either sequestering of Zn by the pyrophosphate or solubilization of soil OM. Any complexing of Zn by polyphosphate can only be transitory because hydrolysis is usually very rapid.

INTERACTION OF N WITH P   Since N accounts for at least one-half of the total number of ions absorbed, it is reasonable that P uptake is influenced by the presence of fertilizer N. N promotes P uptake by plants by (1) increasing top and root growth, (2) altering plant metabolism, and (3) increasing the solubility and availability of P. Increased root mass is largely responsible for increased crop uptake of P. NH_4^+ fertilizers have a greater stimulating effect on absorption than NO_3^-.

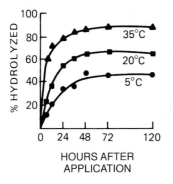

FIGURE 5.29   Effects of temperature on hydrolysis of water-soluble pyrophosphate (200 ppm). *Chang and Racz*, Can. J. Soil Sci., *57:271, 1977.*

Greater effectiveness of fertilizer P can occur when fertilizer application systems place P in close association with NH_4^+-N sources. For example, agronomic advantages, often resulting in 5 to 6 bu/a yield increases of winter wheat, can be gained by simultaneously injecting anhydrous NH_3 and APP solution into the soil. This and other application methods are discussed at greater length in Chapter 10.

EFFECT OF GRANULE OR DROPLET SIZE   The relative effectiveness of P fertilizers is influenced by the size of the granule or droplet and the water solubility of the P fertilizer. Since water-soluble P is rapidly converted to less soluble P reaction products, decreasing the contact between soil and fertilizer generally improves the plant response to P fertilizer. Increasing the granule or droplet size and/or band application of the fertilizer decreases soil-fertilizer contact and maintains a higher solution P concentration for a longer time compared with broadcast P and/or fine particle size.

SOIL MOISTURE   Moisture content of the soil influences the effectiveness and availability of applied P in various forms. When the soil water content is at field capacity, 50 to 80% of the water-soluble P can be expected to move out of the fertilizer granule within a 24-hour period. Even in soils with only 2 to 4% moisture, 20 to 50% of the water-soluble P moves out of the granule within the same time.

RATE OF APPLICATION   Water solubility may be much more important at low rates of P than at high rates. When optimum application rates cannot be used, it is important that materials of high water solubility be used for full benefit from the limited amount of fertilizer applied. The effect of the degree of water-soluble P on corn yields in Iowa is shown in Figure 5.30.

Even though fertilizer P eventually forms less soluble P compounds, the P concentration in solution increases with P application rate. With time the P concentration decreases as less soluble P compounds precipitate. The duration of elevated solution P levels depends on the rate of P fertilizer applied, the method of P placement, the quantity of P removed by the crop, and the soil properties that influence P availability.

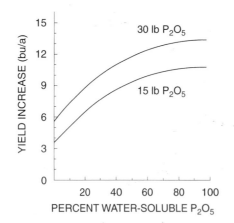

FIGURE 5.30   Effect of rate and water solubility of applied phosphate fertilizer on the yield increase of corn.   *Webb et al., SSSA Proc., 22:533, 1958.*

RESIDUAL P   In both acidic and basic soils, substantial benefits from residual P can persist for 5 to 10 years or longer. The duration of the response is, of course, influenced by the amount of residual P.

For example, in Figure 5.31, with all three rates there was a substantial increase in available P. Following a rapid decline in available P in the first year, there was a gradual decrease of 2 to 7 ppm soil P per year, depending on the P rate. At the end of the experiment, soil test P was about two, four, and eight times greater than that of the unfertilized soil. P fertilizer is usually recommended when NaHCO₃ soil test levels are below 15 ppm, and large, economical yield increases are expected from applied P when soil tests are below 10 ppm.

These data demonstrate that relatively high P rates are needed to substantially increase and maintain residual available P over a long time period. The data shown in Figure 5.32 illustrate the change in plant available P with P rates based on crop need. First, plant removal of P in the unfertilized soil caused initial soil test P to decrease substantially over the 6-year study period. Annual application of 50 kg/ha P maintained the soil test P at 2 to 3 ppm above the initial soil test level, whereas the intermediate P rate (25 kg/ha P) resulted in soil

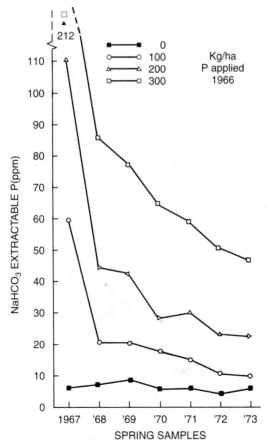

FIGURE 5.31   Effect of single applications of P on the NaHCO₃ extractable P levels in the soils while being cropped alternately with wheat and flax from 1967 to 1973.   *Spratt,* Better Crops Plant Food, *62:24, 1978.*

test levels midway between the 0 and 50 kg/ha P annual rates. Triennial application of 75 kg/ha P increased available P in the first year; however, soil test P subsequently decreased below the initial soil test level until the next triennial application. Similarly, 75 kg/ha P applied only in the first year maintained soil test P at or above the initial level during the first 3 years, followed by decreasing soil test P in subsequent years. These data illustrate the importance of soil testing for accurately determining when additional fertilizer P is needed for optimum production.

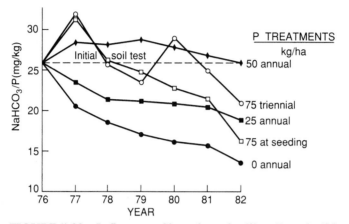

FIGURE 5.32   Influence of broadcast fertilizer P on buildup or decline in soil test P over 6 years.   *Havlin et al.,* SSSAJ, *48:332, 1984.*

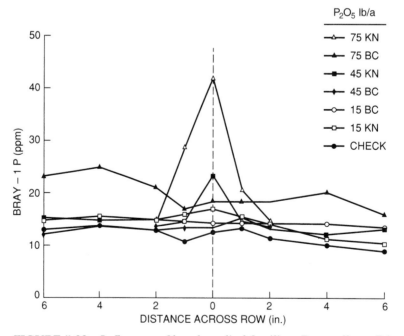

FIGURE 5.33   Influence of band-applied fertilizer P on soil test P in the band 23 months after application.   *Havlin et al.,* Proc. FFF Symposium, *p. 213, 1990.*

P placement also influences the quantity of residual fertilizer P (Fig. 5.33). On this low-P soil, soil test P for broadcast (BC) P applied at 15 and 45 lb P_2O_5/a was no different from that of the unfertilized soil, indicating that the applied fertilizer P not taken up by the crop had been converted to P compounds with a solubility similar to that of the native P minerals. However, increasing band-applied (KN) P from 15 to 75 lb/a P_2O_5 dramatically increased soil test P in the band, indicating that the solubility of the P reaction products is greater than that of the native P minerals and that they persist for several years after application.

There is some question about the need for additional P even when residual P levels are high. Low rates of P in starter fertilizers placed with or near the seed row are potentially beneficial on high-P soils when the crop is stressed by cold, wet conditions and diseases such as root rots. Although residual P contributes significantly to crop yields, additional banding of P may be required to maximize crop production (see Chapter 10).

Selected References

BARBER, S. A. 1984. *Soil Nutrient Bioavailability: A Mechanistic Approach.* John Wiley & Sons, New York.

FOLLETT, R. H., L. S. MURPHY, and R. L. DONAHUE. 1981. *Fertilizers and Soil Amendments.* Prentice-Hall, Englewood Cliffs, N.J.

KHASAWNEH, F. E. (Ed.). 1980. *The Role of Phosphorus in Agriculture.* American Society of Agronomy, Crop Science Society of America, Soil Science Society of America, Madison, Wisc.

LINDSAY, W. L. 1979. *Chemical Equilibria in Soils.* John Wiley & Sons, New York.

STEVENSON, F. J. 1986. *Cycles of Soil: Carbon, Nitrogen, Phosphorus, Sulfur, Micronutrients.* John Wiley & Sons, New York.

YOUNG, R. D., D. G. WESTFALL, and G. W. COLLIVER. 1985. Production, marketing, and use of phosphorus fertilizers, pp. 324–76. In O. P. ENGLESTAD (Ed.), *Fertilizer Technology and Use.* Soil Science Society of America, Madison, Wisc.

CHAPTER 6

Potassium

Potassium (K) is absorbed by plants in larger amounts than any other nutrient except N. Although the total K content of soil is usually many times greater than the amount taken up by a crop during a growing season, in most cases only a small fraction of it is available to plants.

Potassium is present in relatively large quantities in most soils, ranging between 0.5 and 2.5%. Total K content is lower in coarse-textured soils formed from sandstone or quartzite and higher in fine-textured soils formed from rocks high in K-bearing minerals.

Soils of the Southeast and Northwest United States have been highly leached and generally have a low-K content. In contrast, the soils of the Midwest and Western states generally have a high K content because these soils are formed from geologically young parent materials and under conditions of lower rainfall.

In tropical soils, total K content can be quite low because of the origin of the soils, high rainfall, and continued high temperatures. Unlike N and P, which are deficient in most tropical soils due to leaching and/or fixation, the need for K frequently arises only after a few years of cropping a virgin soil. From 70 to 90% of the total K is contained in the forest vegetation, and it takes only a few crops to remove the K in forest residues.

The K Cycle

Listed in increasing order of plant availability, soil K exists in four forms:

mineral	5,000–25,000 ppm
nonexchangeable	50–750 ppm
exchangeable	40–600 ppm
solution	1–10 ppm

The unavailable form accounts for 90 to 98% of the total soil K; the slowly available form, 1 to 10%; and the readily available form, 0.1 to 2%. The relationships and transformations among the various forms of K in soils are depicted in Figure 6.1. K cycling or transformations among the K forms in soils are dynamic. Because of the continuous removal of K by crop uptake and leaching, there is a continuous but slow transfer of K in the primary minerals to the exchangeable and slowly

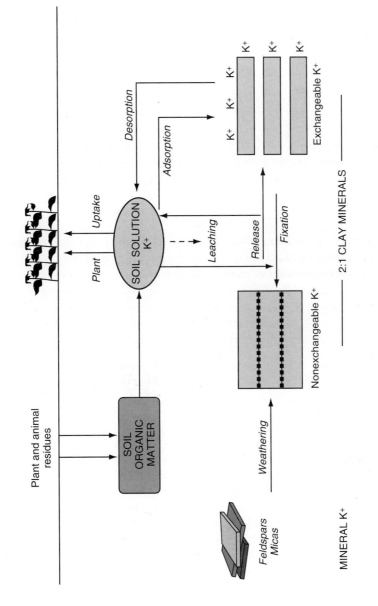

FIGURE 6.1   K equilibria and cycling in soils.

available forms. Under some soil conditions, including applications of large amounts of fertilizer K, some reversion to the slowly available form will occur.

Exchangeable and solution K equilibrate rapidly, whereas fixed K equilibrates very slowly with the exchangeable and solution forms. Transfer of K from the mineral fraction to any of the other three forms is extremely slow in most soils, and this K is considered essentially unavailable to crops during a single growing season.

Before detailing the specific components of the K cycle, we discuss K nutrition and K uptake from soils.

Functions and Forms of K in Plants

Forms

K is absorbed by plant roots as a K^+ ion, with the concentration in tissue ranging between 0.5 and 6% of dry matter.

Functions

K, unlike N, P, and most other nutrients, forms no coordinated compounds in the plant. Instead it exists solely as K^+, either in solution or bound to negative charges on tissue surfaces through radicals such as

$$R\text{-}C\text{-}O^-$$
$$\|$$
$$O$$

As a result of its strictly ionic nature, K^+ has functions particularly related to the ionic strength of solutions within plant cells.

ENZYME ACTIVATION   Enzymes are involved in many important plant physiological processes, and more than 80 plant enzymes require K for their activation. Enzyme activation is regarded as the single most important function of K. These enzymes are abundant in meristematic tissue at the growing points, both above and below ground level, where cell division takes place rapidly and where primary tissues are formed.

Starch synthetase is an enzyme involved in the conversion of soluble sugars into starch, which is a vital step in the grain-filling process. Nitrogenase is the enzyme responsible for reducing atmospheric N_2 to NH_3 in the cells of *Rhizobium* bacteria (Chapter 4). The intensity of the N_2 reduction process depends on the supply of carbohydrates. K enhances carbohydrate transport to nodules for synthesis of amino acids.

WATER RELATIONS   K provides much of the osmotic "pull" that draws water into plant roots. Plants that are K deficient are less able to withstand water stress, mostly because of their inability to make full use of available water.

Maintenance of plant turgor is essential to the proper functioning of photosynthetic and metabolic processes. The opening of stomata occurs when there is

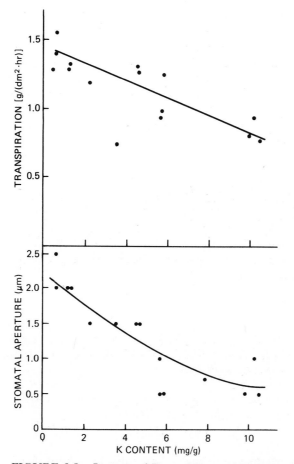

FIGURE 6.2   Improved K nutrition reduced the transpiration rate of peas due to smaller stomatal apertures.   *Brag,* Physiol. Plant., *26:254, 1972.*

an increase of turgor pressure in the guard cells surrounding each stoma, which is brought about by an influx of K. Malfunctioning of stomata due to a deficiency of K is related to lower rates of photosynthesis and less efficient use of water.

Transpiration, the loss of water through stomata, accounts for the major portion of a plant's water use. K can affect the rate of transpiration and water uptake through regulation of stomatal openings. An example of how improved K nutrition reduced the transpiration rate by stomatal closure is shown in Figure 6.2.

ENERGY RELATIONS   Plants require K for the production of adenosine triphosphate (ATP), which is produced in both photosynthesis and respiration. The amount of CO_2 assimilated into sugars during photosynthesis increases sharply with increasing K (Fig. 6.3).

TRANSLOCATION OF ASSIMILATES   Once CO_2 is assimilated into sugars during photosynthesis, the sugars are transported to plant organs, where they are stored or used for growth. Translocation of sugars requires energy in the form of ATP—which requires K for its synthesis.

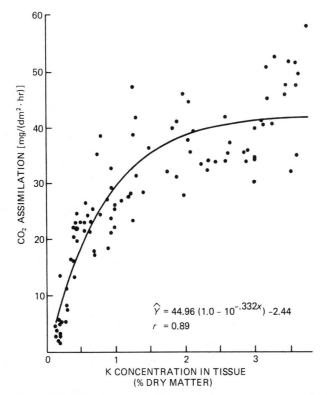

FIGURE 6.3   Adequate K in corn leaves increases photosynthesis as measured by CO_2 fixation.   *Smid and Peaslee*, Agron. J., *68:907, 1976.*

The translocation of sugar from leaves is greatly reduced in K-deficient plants. For example, normal translocation in sugarcane leaves is approximately 2.5 cm/min; however, the rate is reduced by half in K-deficient plants.

N UPTAKE AND PROTEIN SYNTHESIS   Total N uptake and protein synthesis are reduced in K-deficient plants, as indicated by a buildup of amino acids. Again, the involvement of K is through the need for ATP for both processes.

Visual Deficiency Symptoms

When K is limiting, characteristic deficiency symptoms appear in the plant. Typical K deficiency symptoms in alfalfa consist of white spots on the leaf edges, whereas chlorosis and necrosis of the leaf edges are observed with corn and other grasses (see color plates inside book cover).

Since K is mobile in the plant, visual deficiency symptoms usually appear first in the lower leaves, progressing toward the top as the severity of the deficiency increases. K deficiency can also occur in young leaves at the top of high-yielding, fast-maturing crops such as cotton and wheat.

Another symptom of insufficient K is weakening of straw in grain crops, which causes lodging in small grains and stalk breakage in corn (Fig. 6.4) and sorghum. Table 6.1 shows how seriously stalk breakage can affect production through impaired yields and harvesting losses.

FIGURE 6.4 Response of corn to K on a low-K soil. Note the poor growth and lodged condition of the crop on the right. *Courtesy of the Potash & Phosphate Institute, Atlanta, Ga.*

K stress can increase the degree of crop damage by bacterial and fungal diseases, insect and mite infestation, and nematode and virus infection. Soybeans are highly susceptible to pod and stem blight caused by the fungus *Diaporthe sojae* L. The relationship between percentage of soybean seed infected by *D. sojae* L. and K treatment is illustrated in Table 6.2. Higher rates of K as either KCl or K_2SO_4 markedly decreased the incidence of disease in each variety. Lack of K in wetland rice greatly increases the severity of foliar diseases such as stem rot, sheath blight, and brown leaf spot.

TABLE 6.1 Effect of N and K on Yields and Stalk Breakage of Corn

K_2O Applied (lb/a)	Applied (lb/a)		
	0	*80*	*160*
	Yield (bu/a)		
0	48	33	38
80	73	116	119
160	59	122	129
	Stalk Breakage (%)		
0	9	57	59
80	4	3	8
160	4	4	4

SOURCE: Schulte, *Proc. Wisconsin Fert. and Aglime Conf.,* p. 58 (1975).

TABLE 6.2   Effect of K on Soybean Seed Yield and Disease

KCl or K_2SO_4 (g/cylinder)	Seeds per Plant		Diseased Seed (%)*	
	Var. A	Var. B	Var. A	Var. B
Control	254	200	87	62
2	262	207	65	58
10	275	209	21	33
30 + 10 sidedress	264	200	13	14
	NS[†]		LSD = 6.0	

*Percentage of gray, moldy seed (*D. sojae* infected).

[†]Not significant.

SOURCE: Crittenden and Svec., *Agron. J.*, 66:697 (1974).

Forms of Soil K

Soil Solution K

Plants take up the K^+ ion from the soil solution. The concentration of solution K needed varies considerably, depending on the type of crop and the amount of growth.

Levels of solution K commonly range between 1 and 10 ppm, with 4 ppm being representative. The K concentration in soil saturation extracts usually varies from 3 to 156 ppm, and the higher figures are found in arid or saline soils. Under field conditions, soil solution K varies considerably due to the concentration and dilution processes brought about by evaporation and precipitation, respectively.

The effectiveness of soil solution K^+ for crop uptake is influenced by the presence of other cations, particularly Ca^{2+} and Mg^{2+}. It may also be desirable to consider Al^{3+} in very acidic soils and Na^+ in salt-affected soils. The activity ratio (AR_e^k)

$$\frac{\text{activity of } K^+}{\sqrt{\text{activity of } Ca^{2+} \text{ and } Mg^{2+}}} \text{ or } \frac{(ak)}{\sqrt{a_{Ca+Mg}}}$$

in a solution in equilibrium with a soil provides a satisfactory estimate of the availability of K. This ratio is a measure of the "intensity" of labile K in the soil and represents the K that is immediately available to crop roots.

Soils with similar AR_e^k values may have quite different capacities for maintaining AR_e^k while K^+ is being depleted by plant uptake or leaching. Thus, to describe the K status of soils, it is necessary to specify not only solution K (intensity) but also the way in which the intensity depends on the quantity of labile K present. The quantity-intensity relationships (Q/I) are discussed in the section "Exchangeable K," since this fraction has the principal role in replenishing solution K.

K ABSORPTION BY PLANTS   Diffusion and mass flow of K to plant roots account for the majority of K absorbed. The amount of K that can diffuse is directly related to the intensity of K in the soil solution.

Mass flow depends on the amount of water used by plants and the concentration in solution. The relative contribution of mass flow to K absorption by the plant can

TABLE 6.3   Mechanisms and Rate of K Transport in Soils

Situation	*Mechanism*	*Rate (cm/day)*
In profile	Mainly mass flow	Up to 10
Around fertilizer source	Mass flow and diffusion	≈ 0.1
Around root	Mainly diffusion	0.01–0.1
Out of clay interlayers	Diffusion	10^{-7}

SOURCE: Tinker, in G. S. Sekhon (Ed.), *Potassium in Soils and Crops,* Potash Research Institute of India, New Delhi (1978).

be estimated. For example, assume that the average K concentration in the crop is 2.5% and that the transpiration ratio is 400 g H_2O/g plant. Water moving to the root would need to contain in excess of 60 ppm K for mass flow to provide sufficient K. Since most soils contain only about one-tenth this amount, mass flow contributes only 10% of the K requirement. It is apparent, however, that mass flow could supply considerably more K to crops grown in soils naturally high in water-soluble K or where fertilizer K has elevated K in the soil solution. Involvement of mass flow in the transport of K is identified in Table 6.3.

Diffusion of K occurs in response to a concentration gradient, resulting in K transport from a zone of high concentration to one of lower concentration. It is a slow process compared with mass flow (Table 6.3). K diffusion takes place in the moisture films surrounding soil particles and is influenced by soil and environmental conditions, including moisture content, tortuosity of the diffusion path, and temperature, which influence the diffusion rates of ions such as K^+ (see Chapter 2).

K diffusion to roots is limited to very short distances in soil, usually only 1 to 4 mm from the root surface during a growing season. Diffusion in many soils accounts for 88 to 96% of K absorption by roots. The nature of K diffusion to roots can be seen from autoradiographs (Fig. 6.5). These were made using ^{86}Rb, which closely resembles K. Since K absorption occurs within only a few millimeters of the root, K that is farther away, although possibly plant available, is not positionally available.

Exchangeable K

Like other exchangeable cations, K^+ is held around negatively charged soil colloids by electrostatic attraction (Fig. 6.1). The distribution of K between negatively charged sites on soil colloids and the soil solution is a function of the kinds and amounts of complementary cations, the anion concentration, and the properties of the soil cation exchange materials.

Some of the principles of K^+ exchange from soil colloids are summarized by the two equations that follow. Consider first the reaction

$$\text{clay} \begin{bmatrix} \text{Al} \\ \text{K} \\ \text{K} \\ \text{K} \\ \text{K} \end{bmatrix} + \text{CaSO}_4 \rightarrow \begin{bmatrix} \text{Al} \\ \text{Ca} \\ \text{K} \\ \text{K} \end{bmatrix} \text{clay} + \text{K}_2\text{SO}_4$$

If a soil colloid is saturated with K^+ and a neutral salt such as $CaSO_4$ is added, some of the adsorbed K^+ will be replaced by Ca^{2+}. The amount of replacement will depend on the nature and amount of the added salt, as well as the quantity

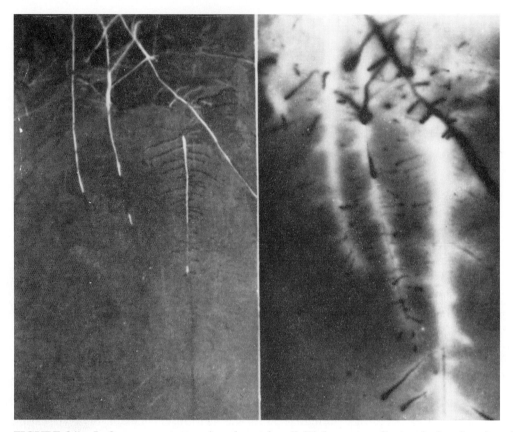

FIGURE 6.5 Left, corn roots growing through soil. Right, autoradiograph showing the effect of corn roots on ^{86}Rb distribution in the soil. Lighter areas are where ^{86}Rb concentration is reduced by root uptake of [86]Rb. *Barber, Potassium availability at the soil–root interface and factors influencing K uptake, in R. D. Munson (Ed.),* Potassium in Agriculture. *ASA, CSSA, SSSA, Madison, Wisc., 1985.*

of K^+ adsorbed on the clays. On some soils used for production of perennial crops, $CaSO_4$ is applied to encourage K^+ displacement and movement into the subsoil, where it becomes available to roots deeper in the profile. Suppose that a soil condition is represented by the following reaction:

$$
\text{clay} \begin{bmatrix} \text{Ca} \\ \text{Al} \\ \text{Al} \end{bmatrix} + \text{KCl} \rightarrow \begin{bmatrix} \tfrac{1}{2}\text{Ca} \\ \text{K} \\ \text{Al} \\ \text{Al} \end{bmatrix} \text{clay} + \tfrac{1}{2}\text{CaCl}_2
$$

This soil clay contains adsorbed Ca^{2+} and Al^{3+} to which KCl has been added. Because the Ca^{2+} is more easily replaced than the Al^{3+}, the added K^+ will replace some of the Ca^{2+} and will itself be adsorbed onto the surface of the clay. This

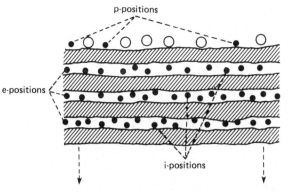

FIGURE 6.6   Binding sites for K on 2:1 clay minerals such as illite, vermiculite, and chlorite.   *Mengel and Haeder,* Potash Rev., *11:1, 1973.*

reaction illustrates an important point: the greater the degree of Ca^{2+} saturation, the greater the adsorption of K^+ from the soil solution. This is consistent with the previous example, in which the Ca^{2+} from $CaSO_4$ replaced K^+ from the colloid. Ca, when added as a neutral salt, replaces Al^{3+} only with great difficulty, and if a soil clay contains K^+, Na^+, and NH_4^+ in addition to Al^{3+}, these ions, rather than the Al^{3+}, will be replaced. In such cases, there will be a net transfer of K^+ to the soil solution. The difference between cations in ease of displacement is defined by the *lyotropic* series for exchangeable cations (see Chapter 2).

Sandy soils with a high base saturation lose less of their exchangeable K^+ by leaching than soils with a low base saturation. Liming increases base saturation thus, decreasing the loss of exchangeable K^+. Part of this effect may be due to an increase in pH-dependent CEC.

Exchangeable K^+ on soil colloids is held at three types of exchange sites, or binding positions (Fig. 6.6). The planar position (*p*) on the outside surfaces of some clay minerals such as mica is rather unspecific for K. By contrast, the edge position (*e*) and the inner position (*i*) in particular have a rather high specificity for K. Under field conditions, soil solution K concentrations are buffered more readily by K^+ held to *p* positions; however, K held on all three positions contributes to solution K.

Because of the major role of exchangeable K in replenishing soil solution K, there is much interest in defining the relationship between exchangeable K (*Q* for quantity) and the activity of soil solution K (*I* for intensity). The *Q/I* ratio is used to quantify the K status of soils.

Labile soil K (held in *p* positions) may be more reliably estimated by *Q/I* than by the measurement of exchangeable K with l-*N* NH_4OAc. Higher values of labile K indicate a greater K release into soil solution, resulting in a larger pool of labile K. Fertilizer K will also increase labile K.

Q/I measures the ability of soil to maintain the intensity of soil solution K^+ and is proportional to the CEC. A high value signifies good K-supplying power or (BC), whereas a low figure suggests a need for K fertilization. Liming can increase *Q/I*, presumably as a result of the increase in pH-dependent CEC.

When *Q/I* values are low, small changes in exchangeable K produce large differences in soil solution K^+. Potential BC is extremely small in sandy soils in which

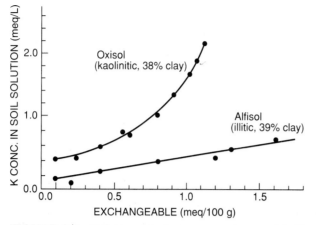

FIGURE 6.7   Relationship between exchangeable K and K concentration in the soil solution of two soils with the same clay content but different clay mineralogy. The steeper slope of the kaolinitic soil indicates less buffer capacity. *Nemeth,* unpublished.

the CEC is due mainly to OM. In such soils, intense leaching or rapid plant growth can seriously deplete available K in just a few days.

The difference in soil solution K^+ between kaolinitic and illitic clays is indicated in Figure 6.7. The steeper slope of the kaolinitic soil indicates lower BC. The effect of a lower clay content, and hence lower BC, on increasing the K^+ concentration in the soil solution around the root is shown in Figure 6.8.

In general, the relation between exchangeable and solution K^+ is a good measure of the availability of the more labile K in soils to plants. The ability of a soil to maintain the activity ratio against depletion by plant roots and leaching is governed partly by the labile K pool, partly by the rate of release of fixed K, and partly by the diffusion and transport of K ions in the soil solution.

Nonexchangeable and Mineral K

The remaining soil K is generally referred to as *nonexchangeable* and *mineral* K. Although nonexchangeable K reserves are not always immediately available, they can contribute significantly to maintenance of the labile K pool in soil. In some soils, nonexchangeable K becomes available as the exchangeable and solution K^+ are removed by cropping or lost by leaching. In other soils, release from nonexchangeable K is too slow to meet crop requirements.

The rate of K supply or release to solution and exchangeable K is largely governed by the weathering of K-bearing micas and feldspars. The K feldspars are orthoclase and microcline ($KAlSi_3O_8$) and the micas are muscovite [$KAl_3Si_3O_{10}(OH)_2$], biotite [$K(Mg,Fe)_3AlSi_3O_{10}(OH)_2$], and phlogopite [$KMg_2Al_2Si_3O_{10}(OH)_2$]. The ease with which these K minerals weather depends on their properties and the environment. As far as plant response is concerned, the availability of K in these minerals, although slight, is of the order biotite > muscovite > potassium feldspars. Feldspars have a three-dimensional crystal structure, with K located throughout the mineral lattice. K can be released from feldspars only by destruction of the mineral.

K feldspars are the largest natural reserve of K in many soils. In moderately weathered soils, there are usually considerable quantities of K feldspars. They

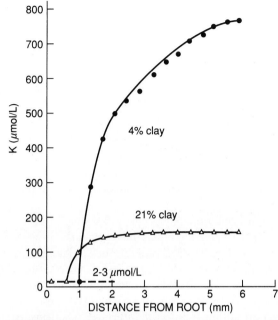

FIGURE 6.8   K concentration in the soil solution around a maize root in two soils after 3 days. The initial exchangeable K content was 0.17 meq/100 g in the loamy soil (21% clay) and 0.37 meq/100 g in the sandy soil (4% clay). Because of the lower BC of the sandy soil, the difference of concentrations in the soil solution were higher than that of exchangeable K.   *Claassen and Jungk,* Soil Sci. Soc. Am. J. *41:1322, 1982.*

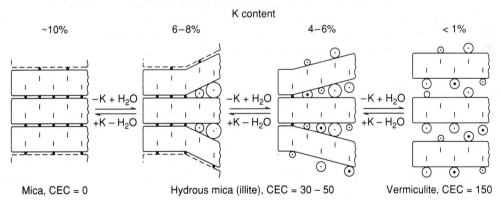

FIGURE 6.9   Schematic weathering of micas and their transformation into clay minerals: a matter of K release and fixation.   *McLean,* Potassium in Soil and Crops, *Potash Research Institute of India, New Delhi, pp. 1–13, 1979.*

often occur in much smaller amounts or may even be absent in strongly weathered soils such as those in humid tropical areas.

The micas are 2:1 layer-structured silicates composed of a sheet of Al octahedra between two sheets of Si tetrahedra (see Chapter 2). Potassium ions reside mainly between the silicate layers (Fig. 6.9). Bonding of interlayer K is stronger in dioctahedral than in trioctahedral micas; therefore, K release generally occurs more readily with biotite than with muscovite.

The gradual release of K from positions in the mica lattice results in the formation of illite (hydrous mica) and eventually vermiculite, with an accompanying gain of water or OH_3^+ and swelling of the lattice (Fig. 6.9). There is also an increase in the specific surface charge and CEC of the clay minerals formed during the weathering and transformation of mica.

K release from mica is both a cation exchange and a diffusion process, requiring time for the exchanging cation to reach the site and for the exchanged ion (K^+) to diffuse from it. A low K concentration or activity in the soil solution favors the liberation of interlayer K. Thus depletion of K by the plant or leaching may induce release of K from nonexchangeable interlayer positions. It is possible for K^+ to be progressively released and diffused from all interlayer locations, or it may come only from alternate interlayers, leading to the formation of interstratified mica-vermiculite.

K FIXATION   K fixation predominately occurs in soils high in 2:1 clays and with large amounts of illite (Fig. 6.9). Fixation of K is the result of reentrapment of K^+ ions between the layers of the 2:1 clays. The 1:1 minerals such as kaolinite do not fix K.

K ions are sufficiently small to enter the silica sheets, where they are held very firmly by electrostatic forces. NH_4^+ has nearly the same ionic radius as K^+ and is subject to similar fixation (Chapter 4). Cations such as Ca^{2+} and Na^+ have larger ionic radii than K^+ and do not move into the interlayer positions. Because NH_4^+ can be fixed by clays in a manner similar to that of K^+, its presence will alter both the fixation of added K and the release of fixed K. Just as the presence of K^+ can block the release of fixed NH_4^+, the presence of NH_4^+ can block the release of fixed K (Fig. 6.10). The NH_4^+ ions evidently are held in the interlayer positions, further trapping the K^+ ions already present.

K fixation is generally more important in fine-textured soils, which have a high fixation capacity for both K^+ and NH_4^+. Although it is not generally considered

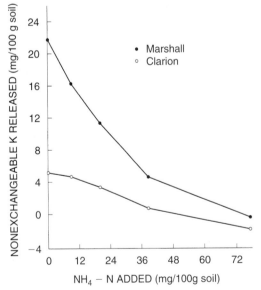

FIGURE 6.10   Nonexchangeable K released by Marshall and Clarion surface soils during a 10-day cropping period as influenced by the amount of added NH_4^+.   *Welch and Scott, SSSA Proc., 25:102, 1961.*

to be a serious factor in limiting the crop response to either applied NH_4^+ or K^+, increasing the concentration of K^+ in soils with a high fixation capacity will obviously encourage greater fixation.

Air drying some soils high in exchangeable K can result in fixation and a decline in exchangeable K. In contrast, drying of field-moist soils low in exchangeable K, particularly subsoils, will frequently increase exchangeable K. The release of K upon drying is thought to be caused by cracking of the clay edges and exposure of interlayer K, which can then be released to exchange sites.

The effects of wetting and drying on the availability of K under field conditions is difficult to quantify. They are important, however, in soil testing. Soil test procedures call for the air drying of samples before analysis. Drying can substantially modify soil test K values and subsequent recommendations for K fertilization.

The freezing and thawing of moist soils may also be important in K release and fixation. With alternate freezing and thawing, illitic soils release K, whereas in other soils, particularly those high in exchangeable K, no K release is observed. It seems probable that freezing and thawing play a significant role in the K supply of certain soils, depending on their clay mineralogy and degree of weathering.

Retention of K in less available or fixed forms is of considerable practical significance. As with P, the conversion of K to slowly available or fixed forms reduces its immediate value as a plant nutrient. However, it must not be assumed that K fixation is completely unfavorable. K fixation results in conservation of K, which can become available over a long period of time and thus is not entirely lost to plants, although plants vary in their ability to utilize slowly available K.

Loss of K by Leaching

In most soils, except those that are quite sandy or subject to flooding, K leaching losses are small. In the humid tropics, leaching is recognized as a major factor in limiting productivity. Under natural vegetation, leaching is low, in the range of 0 to 5 lb/a/yr. On cleared land after fertilizer application, 35% of the K may be leached with cropping, and much higher losses occur on bare land. In these soils emphasis should be placed on annual or split applications rather than on buildup of soil K. Monitoring with soil tests is crucial. Ultimately, leaching losses of K are important only in coarse-textured, organic, or humid tropic soils in areas of high rainfall.

The K source can influence the amount of K moved through the profile by leaching (Table 6.4). Compared with the KCl, the SO_4^{2-} and PO_4^{3-} sources exhibit greater anion adsorption to positive exchange sites. Thus, with fewer anions in solution available for leaching, fewer K^+ would be leached. Remember that solutions must be electrically neutral (negative charges equal positive charges); therefore, for every one negative charge leached, one positive charge must also be leached.

TABLE 6.4   Influence of Source on Leaching Loss of K in Turf

K Source	*"Rainfall" Applied (in.)*				
	10	*20*	*50*	*75*	*100*
			% K Lost		
Potassium chloride, KCl	17	75	91	91	94
Potassium sulphate, K_2SO_4	0	15	53	79	79
Potassium phosphate, K_3PO_4	0	0	0	18	33

SOURCE: Sartain, *Soil Sci. Fert. Sheet.*, SL52, Univ. of Florida, Gainsville, FL, (1988).

Factors Affecting K Availability

Clay Minerals

The greater the proportion of clay minerals high in K, the greater the potential K availability in a soil. For example, soils containing vermiculite, montmorillonite, or illite have more K than soils containing predominantly kaolinitic clays, which are more highly weathered and very low in K. Intensively cropped montmorillonitic soils may be low in K and require K fertilization for optimum crop production.

Cation Exchange Capacity

Finer-textured soils usually have a higher CEC and can hold more exchangeable K; however, a higher level of exchangeable K does not always mean that a higher level of K will be maintained in the soil solution. In fact, soil solution K^+ in the finer-textured soils (loams and silt loams) may be considerably lower than that in a coarse-textured (sandy) soil at any given level of exchangeable K (Fig. 6.8).

Amount of Exchangeable K

Determination of exchangeable K is a measure of K availability. Many studies show the relationship between soil test K and response to applied K. Fertilizer applications of K can be adjusted downward with increasing levels of available soil K (Fig. 6.11).

Capacity to Fix K

In general, the amount of K needed to increase exchangeable K 1 ppm may vary from 1 to 45 lb K/a or more, depending on the soil. The wide difference is related in part to the variation in K fixation potential among soils. Fortunately,

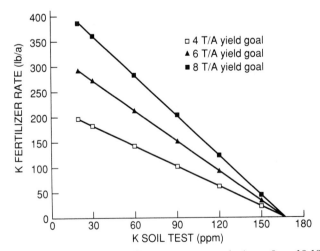

FIGURE 6.11   K fertilizer recommendations for alfalfa in the northern Plains. Fertilizer K rate decreases with increasing soil test K and decreasing alfalfa yield goal.

some of the K that is fixed may be subsequently released to crops, but the release may be too slow for high levels of crop production.

Soil Moisture

With low soil moisture, water films around soil particles are thinner and discontinuous, resulting in a more tortuous path for K^+ diffusion to roots. Increasing K levels or moisture content in the soil will accelerate K diffusion.

Soil moisture can have substantial effects on K transport in soil (Fig. 6.12). Increasing soil moisture from 10 to 28% increases total transport by up to 175%.

Soil Temperature

The effect of temperature on K uptake is due to changes in both availability of soil K and root activity. Reduced temperature slows down plant processes, plant growth, and rate of K uptake. For example, K influx into corn roots at 15°C (59°F) was only about one-half of that at 29°C (84°F) (Fig. 6.13). In the same study, the root length increase over a 6-day period was eight times greater at 29°C than at 15°C. K concentration in the shoot was 8.1% at 29°C and 3.7% at 15°C.

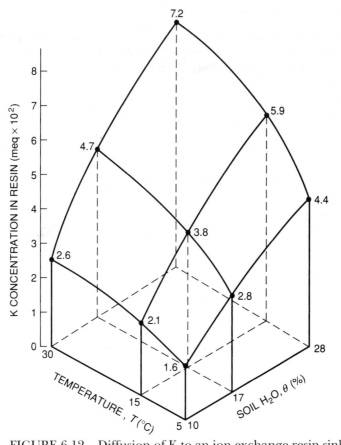

FIGURE 6.12   Diffusion of K to an ion exchange resin sink in Bozeman silt loam during 96 hours, as influenced by temperature and soil moisture. *Skogley*, Proc. 32nd Annu. Northwest Fert. Conf., *Billings, Mont., 1981.*

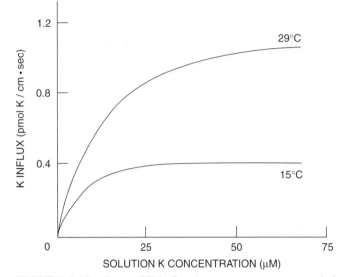

FIGURE 6.13   Rate of K influx into young corn roots is increased by higher temperature and K concentration in solution.   *Ching and Barber,* Agron, J., *71:1040, 1979.*

The supplemental K needed to increase K uptake at low temperatures overcomes some of the adverse effect that low temperature has on rate of diffusion. Providing high levels of K is a practical way of overcoming some of the problems of low temperature. Temperature effects are probably a major reason for crop responses to row-applied fertilizer for early-planted crops such as corn. The beneficial effect of K fertilization for early seedings of barley grown on soils high in available K is attributed mainly to improvement in the K supply under cool soil conditions (Table 6.5).

Soil Aeration

Respiration and the normal functioning of roots are strongly dependent on an adequate O_2 supply. Under high moisture levels or in compact soils, root growth is restricted, O_2 supply is lowered, and absorption of K and other nutrients is slowed. The inhibitory action of poor aeration on nutrient uptake is most pronounced with K (Table 6.6).

Soil pH

In very acid soils, toxic amounts of exchangeable Al^{3+} and Mn^{2+} create an unfavorable root environment for uptake of K^+ and other nutrients. When acid soils are limed, exchangeable Al^{3+} and $Al(OH)_2^+$ are converted to insoluble $Al(OH)_3$. This change removes the Al^{3+} from cation exchange competition with K^+, and it frees exchange sites so that K^+ can compete with Ca^{2+} for them. As a consequence, greater amounts of K^+ can be held by clay colloids and removed from the soil solution. Leaching losses of K will also likely be reduced.

TABLE 6.5   Effect of K Fertilization on Early
Seedings of Barley Grown on Montana Soils
High in Available K

Seeding Date	K_2O^* (lb/a)	Yield (bu/a)
April 6	0	48
	20	55
May 6	0	36
	20	42
June 3	0	30
	20	33

*N at 60 lb/a/yr and P_2O_5 at 25 lb/a/yr.
SOURCE: Dubbs, *Better Crops Plant Food*, 65:27 (1981).

TABLE 6.6   Uptake of Nutrients by Corn
Grown in Nonaerated and Aerated Cultures of a
Silt Loam Soil Containing 50% Water

Component Measured	Relative Uptake: Nonaerated/Aerated
K	0.3
N	0.7
Mg	0.8
Ca	0.9
P	1.3
Dry matter	0.6

SOURCE: Lawton, *Soil Sci. Soc. Am. J.*, 10:263 (1946).

Both Ca^{2+} and Mg^{2+} compete with K^+ for entry into plants; thus, soils high in one or both of these cations may require high levels of K for satisfactory nutrition of crops. According to the activity ratio defined earlier, K uptake would be reduced as Ca^{2+} and Mg^{2+} are increased; conversely, uptake of these two cations would be reduced as the available supply of K is increased. Thus, the availability of K is somewhat more dependent on its concentration relative to that of Ca^{2+} and Mg^{2+} than on the total quantity of K present.

Sources of K

Organic K

K in organic wastes (manures and sewage sludge) occurs predominately as soluble inorganic K^+. In animal waste, K content ranges between 0.2 and 2% or 4 to 40 lb K/t of dry matter (Table 10.17). The average K content in sewage sludge is 10 lb K/t. Therefore, waste materials can supply sufficient quantities of plant available K, depending on the rate applied. Most waste application rates are governed by the quantity of N or P applied to minimize impacts of land application of waste on surface and groundwater quality. If low waste rates are utilized on K-deficient soils, additional K may be needed.

Inorganic K Fertilizers

Deposits of soluble K salts are found well beneath the surface of the earth but also in the brines of dying lakes and seas (Fig. 6.14). Many of these deposits have high purity and lend themselves to mining of agricultural and industrial K salts, termed *potash*.

The world's largest high-grade potash deposit is in Canada. It extends 450 miles long and 150 miles wide, with a depth of 3,000 to 7,000 feet.

Like P, the K content of fertilizers is presently guaranteed in terms of its K oxide (K_2O) equivalent (Table 6.7). Converting between % K and % K_2O is accomplished by the following expressions:

$$\% \text{ K} = \% \text{ } K_2O \div 1.2$$

$$\% \text{ } K_2O = \% \text{ K} \times 1.2$$

POTASSIUM CHLORIDE (KCL)   Fertilizer-grade KCl contains 50 to 52% K (60 to 63% K_2O) and varies in color from pink or red to brown or white, depending on

FIGURE 6.14   Major potash production areas in North America.

the mining and recovery process used. There is no agronomic difference among the products. The white soluble grade is popular in the fluid fertilizer market.

KCl is by far the most widely used K fertilizer. It is used for direct application to the soil and for the manufacture of N-P-K fertilizers. When added to the soil, it readily dissolves in the soil water.

POTASSIUM SULFATE (K_2SO_4) K_2SO_4 is a white solid material containing 42 to 44% K (50 to 53% K_2O) and 17% S. It is produced by different processes, some of which involve reactions of other salts with KCl and some of which involve reactions with S or H_2SO_4.

K_2SO_4 finds its greatest use on potatoes and tobacco, which are sensitive to large applications of Cl. Its behavior in the soil is essentially the same as that of KCl, but it has the advantage of supplying S.

POTASSIUM MAGNESIUM SULFATE (K_2SO_4, $MgSO_4$)   Potassium magnesium sulfate is a double salt containing 18% K (22% K_2O), 11% Mg, and 22% S. It has the advantage of supplying both Mg and S and is frequently included in mixed fertilizers for that purpose on soils deficient in these two elements. It reacts as would any other neutral salt when applied to the soil.

POTASSIUM NITRATE (KNO_3) KNO_3 contains 13% N and 37% K (44% K_2O). Agronomically, it is an excellent source of fertilizer N and K. KNO_3 is marketed largely for use on fruit trees and on crops such as cotton and vegetables. If production costs can be lowered, it might compete with other sources of N and K for use on crops of a lower value.

POTASSIUM PHOSPHATES ($K_4P_2O_7$, KH_2PO_4, K_2HPO_4)   Several K phosphates have been produced and marketed on a limited basis. Their advantages are (1) high analysis, (2) low salt index, (3) adapted to preparation of clear fluid fertilizers high in K_2O, (4) formulation of polyphosphates, and (5) well suited for use on potatoes and other crops sensitive to excessive amounts of Cl^-.

POTASSIUM CARBONATE (K_2CO_3), POTASSIUM BICARBONATE ($KHCO_3$), AND POTASSIUM HYDROXIDE (KOH)   These salts are used primarily for the production of high-purity fertilizers for foliar application or other specialty uses. The high cost of manufacture has precluded their widespread use as commercial fertilizers.

POTASSIUM THIOSULFATE ($K_2S_2O_3$) AND POTASSIUM POLYSULFIDE (KS_x)   Analysis of these liquid fertilizers, $K_2S_2O_3$ and KS_x, is 0-0-25-17 and 0-0-22-23, respectively. $K_2S_2O_3$ is compatible with most liquid fertilizers and is well suited for foliar application and drip irrigation.

TABLE 6.7   Plant Nutrient Content of Common K Fertilizers and Other Sources

Material	N (%)	P₂O₅ (%)	K₂O (%)	S (%)	Mg (%)
Potassium chloride	—	—	60–62	—	—
Potassium sulfate	—	—	50–52	17	—
Potassium magnesium sulfate	—	—	22	22	11
Potassium nitrate	13	—	44	—	—
Potassium hydroxide	—	—	83	—	—
Potassium carbonate	—	—	<68	—	—
Potassium orthophosphates	—	30–60	30–50	—	—
Potassium polyphosphates	—	40–60	22–48	—	—
Potassium thiosulfate	—	—	25	17	—
Potassium polysulfide	—	—	22	23	—

Selected References

BARBER, S. A. 1984. *Soil Nutrient Bioavailability: A Mechanistic Approach.* John Wiley & Sons, New York.

BARBER, S. A., R. D. MUNSON, and W. B. DANCY. 1985. Production, marketing, and use of potassium fertilizers, pp. 377–410. In O. P. ENGLESTAD (Ed.), *Fertilizer Technology and Use.* Soil Science Society of America, Madison, Wisc.

FOLLETT, R. H., L. S. MURPHY, and R. L. DONAHUE. 1981. *Fertilizers and Soil Amendments.* Prentice-Hall, Englewood Cliffs, N.J.

MUNSON, R. D. (Ed.). 1985. *Potassium in Agriculture.* Soil Science Society of America, Madison, Wisc.

CHAPTER 7

Sulfur, Calcium, and Magnesium

S, Ca, and Mg are secondary macronutrients required in relatively large amounts for good crop growth. S and Mg are needed by plants in about the same quantities as P, whereas for many plant species, the Ca requirement is greater than that for P. S reactions in soil are very similar to those of N, which are dominated by the organic or microbial fraction in the soil (see Chapter 4). In contrast, Ca^{2+} and Mg^{2+} are associated with the soil colloidal fraction and behave similarly to K^+ (see Chapter 6).

Sulfur

The S Cycle

S is the 13th most abundant element in the earth's crust, averaging between 0.06 and 0.10%. The original source of soil S is the metal sulfides in rocks. As these rocks were exposed to weathering, the minerals decomposed and S^{2-} was oxidized to SO_4^{2-}. The SO_4^{2-} was then precipitated as soluble and insoluble SO_4^{2-} salts in arid or semiarid climates, absorbed by living organisms, or reduced by other organisms to S^{2-} or S^0 under anaerobic conditions. Some of the SO_4^{2-} formed during mineral weathering drains to the sea. Oceans contain approximately 2,700 ppm SO_4^{2-}, whereas natural waters range from 0.5 to 50 ppm SO_4^{2-} but may reach 60,000 ppm (6%) in highly saline lakes and sediments.

S is present in the soil in both organic and inorganic forms, although nearly 90% of the total S in most noncalcareous surface soils exists in organic forms. The inorganic forms are solution SO_4^{2-}, adsorbed SO_4^{2-}, insoluble SO_4^{2-}, and reduced inorganic S compounds. Solution plus adsorbed SO_4^{2-} represents the readily available fraction of S utilized by plants.

S cycling in the soil-plant-atmosphere continuum is shown in Figure 7.1. There are similarities between the N and S cycles in that both have gaseous components and their occurrence in soils is associated with OM (see Chapter 5). Before detailing the components of the S cycle we discuss S uptake and metabolism in plants.

217

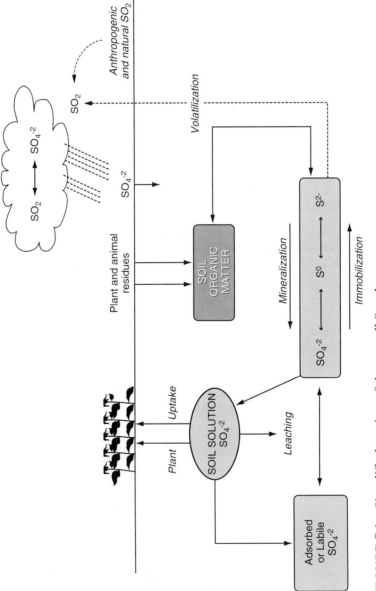

FIGURE 7.1   Simplified version of the overall S cycle.

Forms and Functions of S in Plants

FORMS   S is absorbed by plant roots almost exclusively as sulfate, SO_4^{2-}. Small quantities of SO_2 can be absorbed through plant leaves and utilized within plants, but high concentrations are toxic. Typical concentrations of S in plants range between 0.1 and 0.5%. Among the families of crop plants, S content increases in the order Gramineae < Leguminosae < Cruciferae and is reflected in the differences in S content of their seeds: 0.18–0.19%, 0.25–0.3%, and 1.1–1.7%, respectively. Much of the SO_4^{2-} is reduced in the plant to —S—S and —SH forms, although SO_4^{2-} occurs in plant tissues and cell sap.

FUNCTIONS   S is required for synthesis of the S-containing amino acids cystine, cysteine, and methionine, which are essential components of protein. Approximately 90% of the S in plants is found in these amino acids. Figure 7.2 shows the approximate cysteine and methionine content in borecole. Increasing S availability increased S content in leaves, which increased S-containing amino acids.

Plants suffering S deficiency accumulate nonprotein N in the form of NH_2 and NO_3^- (Table 7.1). It is also apparent that S fertilization improved the quality of this forage by narrowing the N/S ratio. An N/S ratio of between 9:1 and 12:1 is needed for effective use of N by rumen microorganisms. This beneficial effect of S fertilization on improving crop quality through reductions in the N/S ratio is important in animal nutrition.

Leaf tissue NO_3^- accumulates with S deficiency in vegetables and can influence food quality (Fig. 7.3). In this example, NO_3^- accumulated in lettuce only when plants exhibited visual S-deficiency symptoms (<2.5 mg/g of S).

One of the main functions of S in proteins is the formation of disulfide (—S—S—) bonds between polypeptide chains within a protein causing the protein to fold. Disulfide linkages are therefore important in determining the configuration and catalytic or structural properties of proteins.

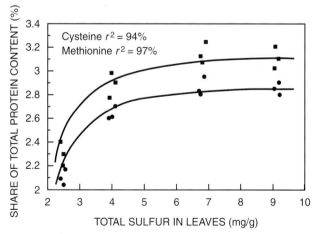

FIGURE 7.2   Relations between the S nutritional status of borecole and the concentration of cysteine and methionine in the leaf protein.   *Schung*, Sulphur in Agric., *The Sulphur Institute, Wash. D.C., 14:2–7, 1990.*

TABLE 7.1   Effects of Elemental S Application on the Yield and Quality of Orchardgrass (113 kg/ha of N Applied after Each Cutting)

Sulfur* (kg/ha)	Yield† (metric t/ha) of Cutting		Nonprotein N (%) in Cutting		Nitrate N (%) in Cutting		N/S Ratio in Cutting	
	1	3	1	3	1	3	1	3
0	3.74	1.77	1.05	1.22	0.064	0.211	21.3	21.4
23	3.72	2.55	0.64	0.85	0.037	0.184	15.3	18.7
45	3.63	2.62	0.59	0.49	0.051	0.144	14.3	14.8
90	3.40	2.89	0.51	0.44	0.037	0.137	12.2	13.4
113	3.40	2.76	0.49	0.37	0.033	0.106	10.8	10.0

*Applied in 1965 and 1967.

†Harvests taken in 1968.

SOURCE: Baker et al., *Sulphur Inst. J.,* 9(1):15 (1973).

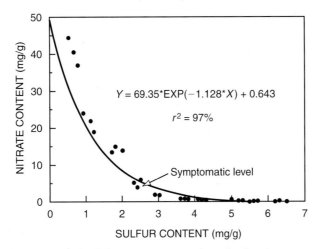

$$Y = 69.35 \cdot \mathrm{EXP}(-1.128 \cdot X) + 0.643$$

$$r^2 = 97\%$$

FIGURE 7.3   Nitrate concentrations in the dry matter of lettuce depending on the S nutritional status of the plants.   *Schung,* Sulphur in Agric., *The Sulphur Institute, Wash. D.C., 14:2–7, 1990.*

S is needed for the synthesis of coenzyme A, which is involved in the oxidation and synthesis of fatty acids, the synthesis of amino acids, and the oxidation of intermediates of the citric acid cycle.

Although not a constituent, S is required for the synthesis of chlorophyll. Table 7.2 shows the importance of adequate S nutrition in the occurrence of chlorophyll in red clover.

S is a vital part of the ferredoxins, an Fe-S protein occurring in the chloroplasts. Ferredoxin has a significant role in NO_2^- and SO_4^{2-} reduction, the assimilation of N_2 by root nodule bacteria, and free-living N-fixing soil bacteria. S occurs in volatile compounds responsible for the characteristic taste and smell of plants in the mustard and onion families.

VISUAL S DEFICIENCY SYMPTOMS   S deficiency has a pronounced retarding effect on plant growth and is characterized by uniformly chlorotic plants—

TABLE 7.2   Effect of a High Level of S Nutrition on the Chlorophyll Content of
Kenland Red Clover

Applied Sulfate (ppm S)	Chlorophyll Content (% dry weight)
0	0.49
5	0.54
10	0.50
20	1.02
40	1.18

SOURCE: Rending et al., *Agron. Abstr. Annu. Meet. Am. Soc. Agron.*, p. 109 (1968).

stunted, thin-stemmed, and spindly (see color plates inside book cover). In many
plants these symptoms resemble those of N deficiency and have undoubtedly led
to many incorrect diagnoses. Unlike N, however, S is not easily translocated from
older to younger plant parts; therefore, deficiency symptoms occur first in
younger leaves.

S-deficient cruciferous crops such as cabbage and canola/rapeseed initially develop a reddish color on the undersides of the leaves. In canola/rapeseed the
leaves are also cupped inward. As the deficiency progresses in cabbage, there is a
reddening and purpling of both upper and lower leaf surfaces; the cupped leaves
turn back on themselves, presenting flattened-to-concave surfaces on the upper
side. Paler-than-normal blossoms and severely impaired seed set also characterize
S-deficiency symptoms in rapeseed.

Forms of S in Soil

SOLUTION SO_4^{2-} SO_4^{2-} absorbed by roots reaches roots by diffusion and
mass flow. In soils containing 5 ppm or more SO_4^{2-}, all of the requirement of most
crops can be supplied by mass flow. Concentrations of 3 to 5 ppm of SO_4^{2-} in the
soil solution are adequate for the growth of many plant species, although some
crops, such as rape and alfalfa, require higher concentrations. Concentrations of
5 to 20 ppm SO_4^{2-} are common in North American soils. Sandy S-deficient soils
often contain less than 5 ppm. Except for soils in dry areas that may have accumulations of SO_4^{2-} salts, most soils contain less than 10% of total S as SO_4^{2-}.

Large seasonal and year-to-year fluctuations in SO_4^{2-} can occur, caused by the
interaction of environmental conditions on the mineralization of organic S, downward or upward movement of SO_4^{2-} in soil water, and SO_4^{2-} uptake by plants.

Sulfate content of soils is also affected by the application of S-containing fertilizers and by the SO_4^{2-} present in precipitation and irrigation waters. In localized areas near centers of industrial activity, the SO_4^{2-} content of soils can be increased by direct adsorption of SO_2 and the fallout of dry particulates.

SO_4^{2-}, like NO_3^-, can be readily leached from surface soil. The greater the
amount of percolating water, the greater the net downward movement of SO_4^{2-}
(Fig. 7.4).

Another factor influencing the loss of SO_4^{2-} is the nature of the cation in the soil
solution. Leaching losses of SO_4^{2-} are greatest when monovalent ions such as
K^+ and Na^+ predominate; next in order are the divalent Ca^{2+} and Mg^{2+} ions; leaching losses are least in acidic soils with appreciable amounts of exchangeable Al^{3+}.

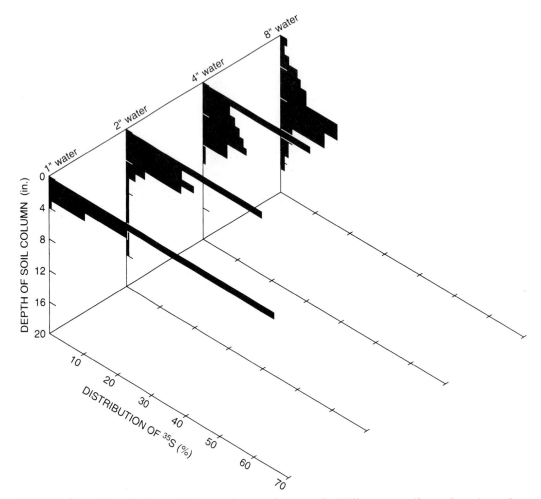

FIGURE 7.4   Distribution of S throughout columns of a Willamette soil as a function of the amount of added water.   *Chao et al.,* SSSA Proc., *26:27, 1962.*

ADSORBED SO$_4^{2-}$ Adsorbed SO$_4^{2-}$ is an important fraction in highly weathered soils in regions of high rainfall containing large amounts of Al/Fe oxides. Many Ultisol (Red-Yellow Podzol) and Oxisol (Latosol) soils contain appreciable amounts of adsorbed SO$_4^{2-}$, ranging as high as 100 ppm SO$_4^{2-}$. Adsorbed SO$_4^{2-}$ in highly weathered soils can contribute significantly to the S needs of plants because it is usually readily available, although not as rapidly available as soluble SO$_4^{2-}$.

Possible mechanisms of SO$_4^{2-}$ adsorption are

1. Anion exchange caused by positive charges on Fe/Al oxides or on the broken edges of clays, especially kaolinite, at low pH values (see Chapter 2).
2. Adsorption of SO$_4^{2-}$ by Al(OH)$_x$ complexes.
3. Amphoteric properties of soil OM develop positive charges under certain conditions.

Reserves of adsorbed SO$_4^{2-}$ in subsoils are the result of eluviation or leaching of SO$_4^{2-}$ from the upper part of soil profiles followed by their retention at lower depths. Adsorbed SO_4^{2-} can account for up to one-third of the total S in subsoils. In surface soils it usually represents less than 10% of the total S present.

Although crops can utilize adsorbed SO_4^{2-} in subsoils, they may experience S deficiency in the early growth stages until root development is sufficient to reach the subsoil. Deep-rooted crops such as alfalfa and lespedeza are unlikely to have temporary shortages of available S.

Many soil factors affect SO_4^{2-} adsorption/desorption:

1. *Clay content and type of clay mineral.* Adsorption of SO_4^{2-} increases with the clay content in soils. In general, SO_4^{2-} adsorption on H-saturated clays follows the order kaolinite > illite > montmorillonite. When saturated with Al^{3+}, adsorption is about the same for kaolinite and illite but is much lower for montmorillonite.
2. *Hydrous oxides.* Fe/Al oxides are responsible for most of the SO_4^{2-} adsorption in many soils.
3. *Soil horizon or depth.* Capacity for SO_4^{2-} adsorption is often greater in subsoils due to the presence of more clay and Fe/Al oxides.
4. *Effect of pH.* Adsorption of SO_4^{2-} in soil is favored by strongly acidic conditions, and it becomes negligible at pH > 6.5. Anion exchange capacity (AEC) increases with decreasing pH.
5. *SO_4^{2-} concentration.* Adsorbed SO_4^{2-} is in equilibrium with SO_4^{2-} in solution; therefore, increased solution SO_4^{2-} will increase adsorbed SO_4^{2-}.
6. *Effect of time.* Sulfate adsorption increases with the length of time SO_4^{2-} is in contact with the adsorbing surfaces.
7. *Presence of other anions.* Sulfate is considered to be weakly held, with the strength of adsorption decreasing in the order $OH^- > H_2PO_4^- > SO_4^{2-} > C_2H_3O_2^- > NO_3^- = Cl^-$. Phosphate will displace SO_4^{2-}, but SO_4^{2-} has little effect on $H_2PO_4^-$. Cl^- has little effect on SO_4^{2-} adsorption.
8. *Effect of cations.* The amount of SO_4^{2-} retained is affected by the associated cation or by the exchangeable cation and follows the lyotropic series: $H^+ > Ca^{2+} > Mg^{2+} > K^+ = NH_4^+ > Na^+$. Both the cation and the SO_4^{2-} from a salt may be retained, but the strength of adsorption of each will likely differ.
9. *Organic matter.* In some soils, OM may contribute to SO_4^{2-} adsorption.

Of all of these factors, the amount and type of soil colloids, pH, SO_4^{2-} concentration, and the presence of other ions in solution influence SO_4^{2-} adsorption most significantly.

SO_4^{2-} COPRECIPITATED WITH $CaCO_3$ S occurs as a coprecipitated ($CaCO_3$-$CaSO_4$) impurity with $CaCO_3$ and is an important fraction of the total S in calcareous soils. Availability of SO_4^{2-} coprecipitated with $CaCO_3$ increases with decreasing pH and particle size of the $CaCO_3$ and increasing soil moisture content.

Grinding soil samples will render the SO_4^{2-} in this fraction accessible to chemical extraction. Consequently, more S will be extracted by a particular soil-test procedure than is available under field conditions.

REDUCED INORGANIC S (S^{2-} AND S^0)   Sulfides do not exist in well-drained soils. Under anaerobic conditions in waterlogged soils, H_2S accumulates by the decay of OM. Little or no S^{2-} accumulates in oxidized soil. Sulfide accumulation is limited primarily to coastal regions influenced by seawater. In normal submerged soils well supplied with Fe, the H_2S liberated from OM is almost completely removed from solution by reaction with Fe^{2+} to form FeS, which undergoes conversion to pyrite (FeS_2). The deep color of the shore of the Black Sea is caused by the accumulation of FeS_2.

Sulfates added to waterlogged soils are reduced to H_2S. If H_2S is not subsequently precipitated by Fe and other metals, it escapes to the atmosphere. The effect of waterlogging on the production of H_2S in a rice paddy soil increases with both time and added OM (Fig. 7.5).

In some tidal marshlands, large quantities of reduced S compounds accumulate, which increase soil pH. When the areas are drained, the S compounds are oxidized to SO_4^{2-}, considerably reducing soil pH < 3.5. The general reaction for FeS_2 oxidation in soils is

$$FeS_2 + H_2O + 7/2O_2 \rightarrow Fe^{2+} + 2SO_4^{2-} + 2H^+$$

Elemental S^0 is not a direct product of SO_4^{2-} reduction in reduced soils but is an intermediate formed during chemical oxidation of S^{2-}. However, S^0 may accumulate in soils in which oxidation of reduced forms of S is interrupted by periodic flooding.

S^0 Oxidation in Soils. Elemental S^0, S^{2-}, and other inorganic S compounds can be oxidized in the soil by purely chemical means, but these are usually much slower and therefore of less importance than microbial oxidation. The rate of biological S^0 oxidation depends on the interaction of three factors: (1) the microbial population in soil, (2) characteristics of the S source, and (3) soil environmental conditions. The following discussion is especially important to plant availability of fertilizer S^0.

SOIL MICROFLORA. Chemolithotrophic S bacteria involved in S oxidation utilize energy released from S^0 oxidation and CO_2 as the C source. Their activity is described in the following equation:

$$CO_2 + S^0 + \tfrac{1}{2}O_2 + 2H_2O \rightarrow [CH_2O] + SO_4^{2-} + 2H^+$$

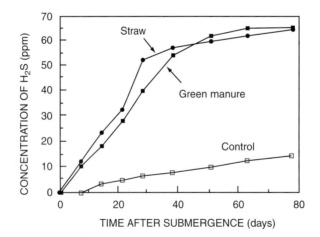

FIGURE 7.5   Effect of OM on production of H_2S in a waterlogged soil. No difference between sources of OM is observed.   *Mandal,* Soil Sci., *91:121, 1961.*

Thiobacilli are typical chemolithotrophic bacteria. Many of them are strict autotrophic aerobes, but some are facultative autotrophs.

Photolithotrophic S bacteria oxidize S^{2-} but use light for energy. The following equation summarizes their behavior:

$$CO_2 + 2H_2S \xrightarrow{\text{light}} [CH_2O] + H_2O + S^0$$

These microorganisms (*Chlorobium* and *Chromatium*) are obligate anaerobes usually found in H_2S-containing muds and stagnant waters exposed to light.

Heterotrophic fungi and bacteria are the most abundant S^0 oxidizers in some soils, where 3 to 37% of the total heterotrophic population are capable of oxidizing S^0. S^0 oxidation is greater in the rhizosphere, where there are larger and more diverse populations of S^0-oxidizing heterotrophs than beyond the rhizosphere.

The most important group of S-oxidizing organisms are the autotrophic bacteria belonging to the genus *Thiobacillus*. Variability in S^0 oxidation rates among soils is due to differences in the number of *Thiobacillus*. Addition of S^0 to soil encourages the growth of S^0-oxidizing microorganisms.

SOIL TEMPERATURE. An increase in temperature increases the S^0 oxidation rate in the soil (Fig. 7.6). Optimum temperature for the different S^0-oxidizing organisms is between 25 and 40°C. At temperatures above 55 to 60°C, the S^0-oxidizing organisms are killed.

SOIL MOISTURE AND AERATION. S^0-oxidizing bacteria are mostly aerobic, and their activity will decline if O_2 is lacking due to waterlogging. S^0 oxidation is favored by soil moisture levels near field moisture capacity (Fig. 7.7). Also evident

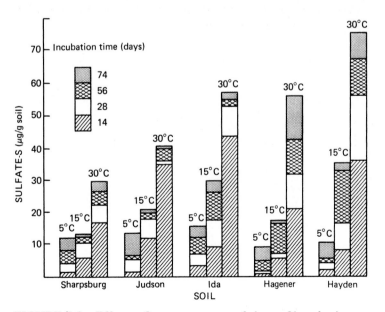

FIGURE 7.6   Effects of temperature and time of incubation on oxidation of S^0 (100 μg of S^0 per gram of soil) in soils.   *Nor and Tabatabai*, Soil Sci. Soc. Am. J., *41:739, 1979.*

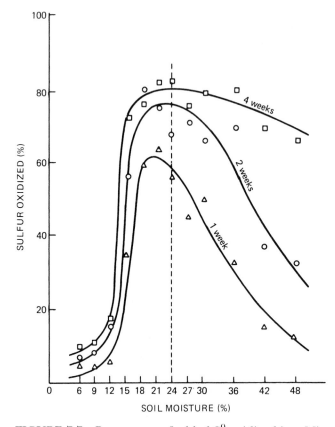

FIGURE 7.7   Percentage of added S^0 oxidized in a Miami silt loam incubated at various moisture levels after 1-, 2-, and 4-week periods. Dashed line is field moisture capacity.   *Kittams and Attoe,* Agron. J., *57:331, 1965.*

is the decline in oxidizing activity when the soils are either excessively wet or dry. Dry soils retain their ability to oxidize S^0, but a lag period can follow rewetting before they regain full capacity.

SOIL PH. Generally, microbial oxidation of S^0 occurs over a wide range in soil pH, although with some species, optimum pH can be 4.0 or lower.

ORGANIC S   The proportion of total S existing in organic forms varies considerably according to soil type and depth in the soil profile.

As described in Chapter 4, there is a close relationship between organic C, total N, and total S in soils. The C/N/S ratio in most well-drained, noncalcareous soils is approximately 120/10/1.4. Differences in the C/N/S ratios among and within soils are related to variations in parent material and other soil-forming factors, such as climate, vegetation, leaching intensity, and drainage. The N/S ratio in most soils falls within the narrow range of 6 to 8:1.

The nature and properties of the organic S fraction in soils are important since they govern the release of plant available S (Fig. 7.1). Three broad groups of S compounds in soil are recognized. These are HI-reducible S, C-bonded S, and residual or inert S. The relative importance of these three categories is shown in Table 7.3.

TABLE 7.3   Fractionation of Organic S in Surface Soils

Location*	HI-Reducible S as Percentage of Total		C-Bonded S as Percentage of Total		Residual S as Percentage of Total	
	Range	Mean	Range	Mean	Range	Mean
Quebec, Canada (3)	44–78	65	12–32	24	0–44	11
Alberta, Canada (15)	25–71	49	12–32	21	7–45	30
Australia (15)	32–63	47	22–54	30	3–31	23
Iowa, U.S. (24)	36–66	52	5–20	11	21–53	37
Brazil (6)	36–70	51	5–12	7	24–59	42

*Figures in parentheses refer to number of samples.
SOURCE: Biederbeck, in M. Schnitzer and S. U. Khan (Eds.), *Soil Organic Matter*, chap. 6. Elsevier, New York (1978).

HI-Reducible S. This fraction is composed of organic S that is reduced to H_2S by hydriodic acid. Its S is largely in the form of esters and ethers with C—O—S linkages. Examples of substances in this grouping include arylsulfates, alkylsulfates, phenolic sulfates, sulfated polysaccharides, and sulfated lipids. About 50% of the organic S occurs in this fraction, but it can range from about 27 to 59%.

Carbon-Bonded S. The S-containing amino acids, cystine and methionine, are principal components of this fraction, which accounts for about 10 to 20% of the total organic S. More oxidized S forms, including sulfoxide, sulfones, and sulfenic, sulfinic, and sulfonic acids, are also included in this fraction.

Residual S. The remaining organic S is considered to be residual. This unidentified fraction generally represents approximately 30 to 40% of the total organic S.

S MINERALIZATION AND IMMOBILIZATION

Mineralization of S is the conversion of organic S to inorganic SO_4^{2-} and is similar to mineralization of organic N (see Chapter 4). Similarly, immobilization is the conversion of SO_4^{2-} to organic S. Any factor that affects the growth of microorganisms is expected to alter the mineralization and immobilization of S.

When plant and animal residues are returned to the soil, they are digested by microorganisms, releasing some of the S as SO_4^{2-}; however, most of the S remains in organic form and eventually becomes part of the soil humus (Fig. 7.1). In contrast to the relatively rapid decomposition of fresh organic residues in soil, degradation and release of S from the large humus fraction are limited and slow.

The S supply to plants depends largely on the SO_4^{2-} released from the organic soil fraction and from plant and animal residues. Approximately 4 to 13 lb/a of S as SO_4^{2-} is mineralized each year from the organic fraction.

Factors Affecting S Mineralization and Immobilization.

S CONTENT OF OM. Mineralization of S depends on the S content of the decomposing material in much the same way that N mineralization depends on the N content. Smaller amounts of SO_4^{2-} are liberated from low-S-containing residue, which is similar to N mineralization.

S may be immobilized in soils in which either the C/S or N/S ratio is too large. At or below a C/S ratio of approximately 200/1, only mineralization of S occurs.

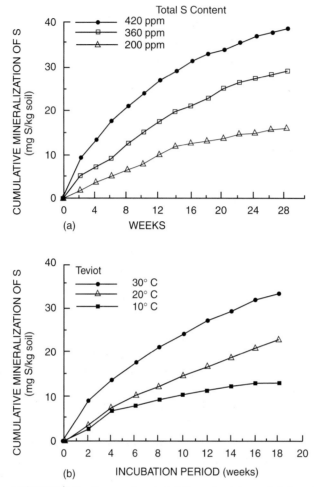

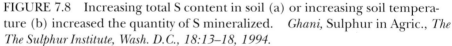

FIGURE 7.8   Increasing total S content in soil (a) or increasing soil temperature (b) increased the quantity of S mineralized.   *Ghani,* Sulphur in Agric., *The The Sulphur Institute, Wash. D.C., 18:13–18, 1994.*

Above this ratio, immobilization of SO_4^{2-} is favored, particularly if the ratio is greater than 400/1. The immobilized S is bound in soil humus, in microbial cells, and in by-products of microbial synthesis. Immobilization occurs with large C/S ratios because of conversion of C into microbial biomass, with a resultant higher need for S than when the C/S ratio is low. Fresh organic residues commonly have C/S ratios of about 50/1. Where large amounts of straw, stover, or other OM are returned to the soil, adequate N and S availability is necessary to promote rapid decomposition of the straw. Otherwise, a temporary N or S deficiency may be induced in the following crop.

SOIL TEMPERATURE. Mineralization of S is severely impeded at 10°C, increases with increasing temperatures from 20 to 40°C, and decreases at temperatures >40°C. In samples representing 12 major soil series, more S was released during incubation at 35°C than at 20°C (Figs. 7.8 and 7.9). An average Q_{10} of S mineralization of 1.9 occurred in these soils. This temperature effect on S mineralization is consistent with the relatively greater S content of soils formed in northern climates.

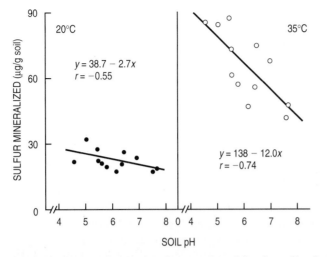

FIGURE 7.9   Relationship between total S mineralized and pH of soils incubated at 20 or 35°C.   *Tabatabai and Al-Khafaji*, Soil Sci. Soc. Am. J., *44:1000, 1980.*

SOIL MOISTURE. Mineralization of S in soils incubated at low ($<15\%$) and high ($>40\%$) moisture levels is reduced compared with the optimum moisture content of 60% of field moisture-holding capacity.

Gradual moisture changes in the range between field capacity and wilting point have little influence on S mineralization. However, drastic differences in soil moisture conditions can produce a flush of S mineralization in some soils. Increased availability of S due to soil wetting and drying may explain the observations of increased plant growth after dry periods in S-deficient soils.

SOIL PH. The effect of pH on S mineralization is not clear. Rates of S mineralization in 12 Iowa soils were found to be negatively correlated with soil pH (Fig. 7.9). In soils from other regions, the amount of S released is directly proportional to pH up to a value of 7.5. Near-neutral soil pH is normally expected to encourage microbial activity and S mineralization.

PRESENCE OR ABSENCE OF PLANTS. Soils generally mineralize more S in the presence of growing plants than in their absence. This phenomenon has been explained by stimulation of microbial activity in the "rhizosphere" brought about by the excretion of amino acids and sugars by plant roots. Appreciable immobilization of added SO_4^{2-} has been observed in uncropped soils. This finding suggests that SO_4^{2-} applied to fallowed soils could be immobilized.

TIME AND CULTIVATION. As with N, when soil is first cultivated, its S content declines rapidly. With time an equilibrium level is reached that is characteristic of the climate, cultural practices, and soil type. Before reaching this point, the rate of S mineralization gradually diminishes and becomes inadequate to meet plant needs. The C/N/S ratios of virgin soils are larger than those of the corresponding cultivated surface soils. Reduction of this ratio on cultivation suggests that S is relatively more resistant to mineralization than C and N or that the losses of organic C and N are proportionately greater than those of S.

SULFATASE ACTIVITY. As much as 50% of the total S in surface soils may be present as organic SO_4^{2-} esters. Sulfatase enzymes that hydrolyze these esters and release $SO_4{}^{2-}$ may be important in the mineralization process. The general action of sulfatases is

$$R \cdot O \cdot SO_3{}^- + H_2O \overset{\text{sulfatase}}{\longleftrightarrow} R \cdot OH + HSO_4{}^-$$

The HI-reducible ester sulfates are considered to be the natural substrates for sulfatase enzymes in soil.

SULFUR VOLATILIZATION   Volatile S compounds are produced through microbial transformations under both aerobic and anaerobic conditions. Where volatilization occurs, the volatile S compounds are dimethyl sulfide (CH_3SCH_3), carbon disulfide (CS_2), methyl mercaptan (CH_3SH), and/or dimethyl disulfide (CH_3SSCH_3). CH_3SCH_3 can account for 55 to 100% of all S volatilized.

In low-OM soils, S volatilization is negligible and generally increases with increasing OM content. The amount of S volatilized represents <0.05% of the total S present in soil and is relatively insignificant under field conditions.

Like NH_3, volatile S compounds also evolve from intact plants. Losses in plants can range from about 0.3 to 6.0% of the total S content of crops.

Volatile S compounds released by intact plants may affect the palatability and acceptability of forage plants to grazing animals. S losses from forages when they are dried in haymaking or pelleting might also influence quality and palatability.

Another source of soil S is the atmosphere. In regions in which coal and other S-containing products are burned, SO_2 is released into the air and is later brought back to earth in precipitation. Plants may also absorb SO_2 by diffusion into the leaves, which is metabolized by the plant. However, exposure to as little as 0.5 ppm of SO_2 for 3 hr can cause visible injury to the foliage of sensitive vegetation.

Liberation of SO_2 into the atmosphere by industrial nations was approximately 100 million tons in 1990. Most of this amount resulted from the combustion of fossil fuels, but industrial processes such as ore smelting, petroleum refining, and others contributed about 20% of the total amount emitted. The amount of S deposited in rainfall in the United States ranges from 1 lb/a per year in rural areas to 100 lb/a near industrial areas. S emissions are partly responsible for the acid rainfall and snowfall in industrialized regions. Strong acids including H_2SO_4 have lowered the pH of precipitation in much of these areas to between 4 and 5. Direct adsorption is an important way that SO_2 enters soils.

Because of the growing concern over air pollution, legislation requires the cleaning and scrubbing of most gases. Many industries are required to clean effluent gases, reducing the S content in precipitation. Although the emission of S compounds into the atmosphere by industrial activity has received unfavorable publicity, the fact remains that nearly 70% of the S compounds in the atmosphere are due to natural processes. Volatile S compounds are released in large quantities from volcanic activity, from tidal marshes, from decaying OM, and from other sources.

PRACTICAL ASPECTS OF S TRANSFORMATIONS   Crops grown on coarse-textured soils are generally more susceptible to S deficiency, because these soils often have low OM contents and are subject to $SO_4{}^{2-}$ leaching. Leaching losses of $SO_4{}^{2-}$ can be especially high on coarse-textured soils under conditions of high rainfall. Under such conditions, $SO_4{}^{2-}$-containing fertilizers may have to be applied more frequently than on fine-textured soils and under lighter rainfall. In more humid regions, a fertilizer containing both $SO_4{}^{2-}$ and S^0 may be required to extend the period of S availability to crops.

TABLE 7.4   Tentative Classification of Crops According to Their S Fertilizer Requirement

Crop	Fertilizer Required in Deficient Areas* (kg S/ha)	Crop	Fertilizer Required in Deficient Areas* (kg S/ha)
Group I (high)		Group III (low)	
Cruciferous forages	40–80	Sugar beet	15–25
Lucerne	30–70	Cereal forages	10–20
Rapeseed	20–60	Cereal grains	5–20
Group II (moderate)		Peanuts	5–10
Coconuts	50		
Sugarcane	20–40		
Clovers and grasses	10–40		
Coffee	20–40		
Cotton	10–30		

*Figures cited for the high end of the range apply where the potential yield is high, low available S soil, and there is considerable loss in effectiveness of applied S. Figures cited for the low end refer to the opposite situation.

SOURCE: Spencer, in K. D. McLachlan (Ed.), *Sulphur in Australasian Agriculture*. Sydney Univ. Press (1975).

Added S can be immobilized in some soils, particularly those that have a high C/S or N/S ratio. In contrast, S mineralization is favored in soils with a low C/S or N/S ratio. Again, S availability generally increases with increasing OM content. Crops grown on soils that have <1.2–1.5% OM often require S fertilization.

The S requirements of crops vary widely (Table 7.4). The actual amount of S needed depends on the balance between additions of S by precipitation, air, irrigation water, crop residues, fertilizers, and manures and losses through crop removal, leaching, and erosion.

Grasses are better able to utilize SO_4^{2-} than legumes. In grass-legume meadows the grasses can absorb available SO_4^{2-} at a faster rate than the legumes. Unless adequate S availability is maintained, the legumes will be forced out of the mixture, because S is required for N fixation by the *Rhizobia*.

S Sources

ORGANIC S   Because of the lower S requirement of most crops compared with N, most animal and municipal wastes contain sufficient quantities of plant available S. Typical S content in these organic wastes ranges between 0.2 and 1.5%, or 5 and 25 lb/t dry weight of S. With typical application rates ranging between 2 and 20 t/a, S applications would range between 10 and 250 lb/a of S.

INORGANIC S   Sulfate materials applied to the soil surface and moved into the profile with rainfall or irrigation are immediately plant available unless immobilized by microbes degrading high C/S or N/S residues. Studies comparing the effectiveness of SO_4^{2-} sources (Table 7.5) suggest that one source of SO_4^{2-} is generally equal to any other (provided that the accompanying cation is not Zn,

TABLE 7.5   S-Containing Fertilizer Materials

Material	Formula	Plant Nutrient Content (%)					S Content (lb/ton)
		N	P_2O_5	K_2O	S	Other	
Ammonium nitrate-sulfate	$NH_4NO_3 \cdot (NH_4)_2SO_4$	30	0	0	5		100
Ammonium phosphate	MAP (crude)	11	48	0	2.2		44
Ammonium phosphate-sulfate	MAP, DAP + $(NH_4)_2SO_4$	16.5	20.5	0	15.5		310
		13	39		7		140
Ammonium polysulfide solution	NH_4S_x	20	0	0	45		800
Ammonium sulfate	$(NH_4)_2SO_4$	21	0	0	24.2		484
Ammonium thiosulfate solution	$(NH_4)_2S_2O_3$	12	0	0	26		520
Ferrous sulfate	$FeSO_4 \cdot H_2O$	0	0	0	18.8	32.8 (Fe)	376
Gypsum (hydrated)	$CaSO_4 \cdot 2H_2O$	0	0	0	18.6	32.6 (CaO)	372
Magnesium sulfate (Epsom salt)	$MgSO_4 \cdot 7H_2O$	0	0	0	13	9.8 (Mg)	260
Potassium sulfate	K_2SO_4	0	0	50	17.6		352
Potassium-magnesium sulfate	$K_2SO_4 \cdot 2MgSO_4$	0	0	22	22	11 (Mg)	440
Potassium thiosulfate	$K_2S_2O_3$	0	0	25	17		
Potassium polysulfide	KS_x	0	0	22	23		
Sulfuric acid (100%)	H_2SO_4	0	0	0	32.7		654
Sulfur	S^0	0	0	0	100		2,000
Sulfur (granular with additives)		0–7	0	0	68–95		1,360–1,900
Sulfur dioxide	SO_2	0	0	0	50		1,000
Superphosphate, single	$Ca(H_2PO_4)_2 + CaSO_4 \cdot 2H_2O$	0	20	0	13.9		278
Superphosphate, triple	$Ca(H_2PO_4)_2 + CaSO_4 \cdot 2H_2O$	0	46	0	1.5		30
Urea-sulfur	$CO(NH_2)_2 + S$	36–40	0	0	10–20		200–400
Urea-sulfuric acid	$CO(NH_2)_2 \cdot H_2SO_4$	10–28	0	0	9–18		
Zinc sulfate	$ZnSO_4 \cdot H_2O$	0	0	0	17.8	36.4 (Zn)	356

SOURCE: Bixby and Beaton, *Tech. Bull. 17*. The Sulphur Institute, Washington, D.C. (1970).

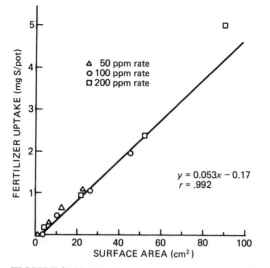

FIGURE 7.10   Influence of surface area of applied S^0 on the uptake of S by canola.   *Janzen et al.*, Proc. Alberta Soil Sci. Workshop, *p. 229, Edmonton, Alberta, February 23–24, 1982.*

Cu, or Mn, which must be applied sparingly) and that the factor determining the selection should be the cost per unit of S applied.

Elemental S^0   Elemental S^0 is a yellow, water-insoluble solid. When S^0 is finely ground and mixed with soil it is oxidized to SO_4^{2-} by soil microorganisms (see the discussion on pp. 224). The effectiveness of S^0 in supplying S to plants compared with SO_4^{2-} depends on several factors, including particle size, rate, method, and time of application; S^0-oxidizing characteristics of the soil; and environmental conditions. S^0 oxidation rates increase as particle size is reduced. As a general rule, 100% of the S^0 material must pass through a 16-mesh screen, and 50% of that should, in turn, pass through a 100-mesh screen. The finer the S^0 particle size, the greater the surface area and the faster the SO_4^{2-} formation. Because of the inverse relationship between surface area and particle diameter, the oxidation rate increases exponentially with decreasing particle diameter. Thus, increasing the S^0 surface area results in increased SO_4^{2-} availability to crops (Fig. 7.10).

When S^0 is finely ground and mixed with soil possessing a high oxidizing capacity, it is usually just as effective as other sources. Finely divided S^0 should be worked into the soil as far ahead of planting as possible.

Application of heavier rates of S^0 will increase the surface area exposed to S^0-oxidizing organisms, which should increase plant available S.

Placement of S^0 can often affect its oxidation rate, with broadcast incorporation being superior to banding. Uniform distribution of S^0 particles throughout the soil will provide greater exposure of S^0 particles to oxidizing microorganisms and minimize any potential problems caused by excessive acidity.

Dispersible, Granular S^0 Fertilizers   Water dispersible, granular S^0 fertilizers, such as S-bentonite (approximately 90% S) and micronized granular S (95% S) have several important advantages, including their high analysis resulting in savings in costs of transportation and handling; provision of S^0 of different particle sizes with varying degrees of controlled availability to plants; low susceptibility to leaching losses in areas of high and intense rainfall; and excellent durable physical forms that are well-suited for direct application or blending with most common granular fertilizers excepting those containing NO_3-N. S-bentonite pastilles are manufactured by adding bentonite to molten S^0 while micronized granular S^0 consists of 100% < 74 μm (-200 mesh) sized particles bound together with a water soluble binder.

Dispersion of S^0-bentonite pastilles into more readily oxidized finely divided S^0 occurs gradually in soil following wetting and swelling of the bentonite component. Micronized, granular S^0 disperses rapidly and completely upon wetting in soil. Dispersion of both S sources, particularly S^0-bentonite, is enhanced by exposure at the soil surface to precipitation and freezing/thawing before soil incorporation. Thus, fertilization practices involving broadcast applications are usually more dependable than banding.

Because of the uncertainty of adequate formation of SO_4^{2-} from S^0-bentonite in the first growing season after application, it should be applied well in advance of planting, preferably allowing for a period of exposure at the soil surface before incorporation. When applied just before seeding of high S requiring crops and on severely deficient soils, some SO_4^{2-} should also be provided.

S^0 Suspensions   The addition of finely ground S^0 to water containing 2 to 3% attapulgite clay results in a suspension containing 40 to 60% S. These suspensions can be applied directly to the soil or combined with suspension fertilizers.

Ammonium Thiosulfate [$(NH_4)_2S_2O_3$, or ATS]   ATS is a clear liquid containing 12% N and 26% S and is a popular S-containing product. ATS is compatible with N solutions and complete (N-P-K) liquid mixes, which are neutral to slightly acidic in pH.

ATS can be applied to the soil directly, in mixtures, or to both sprinkler and open-ditch irrigation systems. When applied to the soil, ATS forms colloidal S and $(NH_4)_2SO_4$. The SO_4^{2-} is immediately available, whereas the S^0 must be oxidized to SO_4^{2-}, thus extending the availability to the crop. Potassium thiosulfate (KTS) behaves similarly as ATS.

Ammonium Polysulfide (NH_4S_x)   Ammonium polysulfide is a red to brown to black solution having a H_2S odor. It contains approximately 20% N and 45% S. In addition to use as a fertilizer, it is used for reclaiming high-pH soils and for treatment of irrigation water to improve water penetration into the soil.

Ammonium polysulfide is recommended for mixing with anhydrous NH_3, aqua NH_3, and UAN solutions. The simultaneous application of ammonium polysulfide and anhydrous NH_3 is popular in some areas for providing both N and S. Normally, it is considered incompatible with phosphate-containing liquids. This material has a low vapor pressure, and it should be stored at a pressure of 0.5 psi to prevent loss of NH_3 and subsequent precipitation of S^0. Potassium polysulfide (0-0-22-23) has been used on a limited basis in sprinkler and flood irrigation systems for salt removal and to supply K.

Urea-Sulfuric Acid   Two typical grades used as acidifying amendments, as well as sources of N, contain 10% N and 18% S and 28% N and 9% S, respectively. They can be applied directly to the soil or added through sprinkler systems.

Because these urea–sulfuric acid formulations have pH values between 0.5 and 1.0, the equipment used must be made from stainless steel and other noncorrosive materials. Workers must wear protective clothing.

FERTILIZER USE GUIDELINES   For purposes of convenience, recommendations for the use and proper application of common S-containing fertilizers are summarized in Table 7.6.

Calcium

Ca in acidic, humid-region soils occurs largely in the exchangeable form and as primary minerals. In most of these soils, Ca^{2+}, Al^{3+}, and H^+ ions dominate the exchange complex. As with any other cation, the exchangeable and solution forms are in dynamic equilibrium (Fig. 7.11). If the activity of solution Ca^{2+} is decreased by leaching or plant removal, Ca^{2+} will desorb to resupply solution Ca^{2+}. Other cations, such as H^+ and/or Al^{3+}, occupy the exchange sites left by the desorbed Ca^{2+}. Conversely, if solution Ca^{2+} is increased, the equilibrium shifts in the opposite direction, with subsequent adsorption of some of the Ca^{2+} by the exchange complex.

The fate of solution Ca^{2+} is less complex than that of K^+. It may be (1) lost in drainage waters, (2) absorbed by organisms, (3) adsorbed onto the CEC, or (4) reprecipitated as a secondary Ca compound, particularly in arid climates.

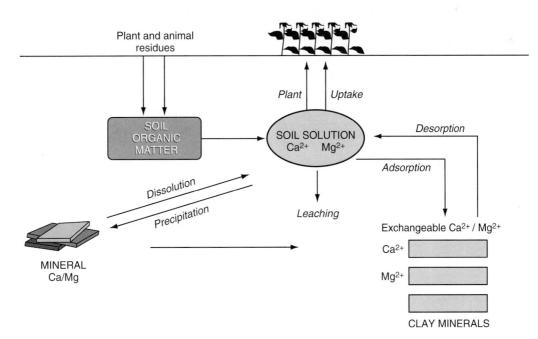

FIGURE 7.11   Simple representation of Ca and Mg equilibrium and transformations in soil.

TABLE 7.6 Recommendations for Use of Fertilizers Containing S

Solid Materials	Recommended Use	Remarks
Ammonium phosphate–S⁰ Urea-S⁰ Dispersible, granular S⁰; flake S and porous granular S	For direct application and bulk blends, apply materials several months before the beginning of the growing season; fall applications should be encouraged and allowances made for dispersion before incorporation of broadcast applications	If used in starter fertilizer or shortly before the beginning of the growing season, some readily available SO_4^{2-} should be included; dispersion of water-degradable granular S products at the soil surface before incorporation will greatly improve their agronomic effectiveness; where feasible, incorporate into soil 4 or 5 months before planting; when applied just in advance of planting or on severely S-deficient soils, some readily available SO_4^{2-} should be included.
Ammonium sulfate	For direct application and to some extent for bulk blending; should be effective at almost any time	Tends to segregate in bulk blends unless physical properties are improved by granulation; where significant leaching losses are expected, apply shortly before planting
Ammonium nitrate–sulfate; ammonium phosphate–sulfate; ordinary superphosphate; potassium sulfate; potassium magnesium–sulfate	For direct application and bulk blends; should be effective at almost any time	Where significant leaching losses of SO_4^{2-} are expected, apply shortly before planting or the beginning of the growing season
Calcium sulfate (gypsum)	For direct application; should be effective at almost any time	Difficulties may be encountered in application (dustiness, clogging)

TABLE 7.6 Recommendations for Use of Fertilizers Containing S—*Continued*

Solid Materials	Recommended Use	Remarks
Ammonium thiosulfate Potassium thiosulfate	For direct application and blending with most fluid fertilizer products; can be broadcast before planting or applied in starter fertilizers; can be topdressed on certain growing crops; can also be added through open-ditch and sprinkler irrigation systems	Can be blended with all neutral fluid phosphate products now available, all nitrogen solutions (except anhydrous ammonia), and most micronutrient solutions
Ammonium polysulfide Potassium polysulfide	For direct application and blending with other N solutions; frequently injected into soil; broadcast spray applications are possible following dilution with water; single preplant applications are effective; repeated applications at low rates are often made to growing crops through open-ditch irrigation systems	Ammonium polysulfide is generally not considered suitable for mixing with phosphate-containing fluids
Sulfuric acid	For mixing with wet process ammonium polyphosphate and anhydrous ammonia in the preparation of clear liquid blends	Sulfuric acid has been applied directly to crops such as onions and garlics for weed control purposes
Suspensions containing elemental S⁰	For direct application and for simultaneous application with other fertilizers, the suspensions should be applied several months before the beginning of the growing season	If used in starter fertilizer or shortly before the beginning of the growing season, readily available SO_4 should be included (15 to 20% of total S applied)
Suspensions containing sulfate salts	Should be effective at almost any time	Where significant leaching losses are expected, apply shortly before planting or the beginning of the growing season

SOURCE: Bixby and Beaton, *Tech. Bull. 17*. The Sulphur Institute, Washington, D.C. (1970).

Forms and Functions of Ca in Plants

Ca is absorbed by plants as Ca^{2+} from the soil solution and is supplied to the root surface by mass flow and root interception. Ca deficiency is uncommon but can occur in highly leached and unlimed acidic soils. In soils abundant in Ca^{2+}, excessive accumulation in the vicinity of roots can occur.

Ca^{2+} concentration in plants ranges from 0.2 to 1.0%. Ca is important in the structure and permeability of cell membranes. Lack of Ca^{2+} causes a breakdown of membrane structures, with resultant loss in retention of cellular diffusible compounds. Ca enhances uptake of NO_3 and therefore is interrelated with N metabolism. Ca^{2+} provides some regulation of cation uptake. For example, studies have shown that K^+ and Na^+ uptake are about equal in the absence of Ca^{2+}, but in its presence K^+ uptake greatly exceeds Na^+ uptake.

Ca is essential for cell elongation and division, and Ca^{2+} deficiency manifests itself in the failure of terminal buds of shoots and apical tips of roots to develop, which causes plant growth to cease.

In corn, Ca^{2+} deficiency prevents the emergence and unfolding of new leaves, the tips of which are almost colorless and are covered with a sticky gelatinous material that causes them to adhere to one another.

In fruits and vegetables, the most frequent indicator of Ca^{2+} deficiency consists of disorders in the storage tissues. Examples of Ca^{2+} disorders are blossom-end rot in tomato and bitter pit of apples.

Finally, Ca^{2+} is generally immobile in the plant. There is very little translocation of Ca^{2+} in the phloem, and for this reason there is often a poor supply of Ca^{2+} to fruits and storage organs. Downward translocation of Ca^{2+} is also limited in roots, which usually prevents them from entering low-Ca soils. Conditions impairing the growth of new roots will reduce root access to Ca^{2+} and induce deficiency. Problems related to inadequate Ca^{2+} uptake are more likely to occur with plants that have small root systems than with those possessing more highly developed rooting systems.

Special attention must be given to the Ca^{2+} requirements of certain crops, including peanuts, tomatoes, and celery, which are often unable to obtain sufficient Ca^{2+} from soils supplying adequate Ca^{2+} for most other crops. Proper Ca^{2+} supply is important for tree fruits and other crops such as alfalfa, cabbage, potatoes, and sugar beets, which are known to have high Ca^{2+} requirements.

Ca in Soil

The Ca concentration of the earth's crust is about 3.5%; however, the Ca^{2+} content in soils varies widely. Sandy soils of humid regions contain very low amounts of Ca^{2+}, whereas Ca^{2+} normally ranges from 0.7 to 1.5% in noncalcareous soils of humid temperate regions; however, highly weathered soils of the humid tropics may contain as little as 0.1 to 0.3% Ca. Ca levels in calcareous soils vary from less than 1% to more than 25%.

Ca in soils originated in the rocks and minerals from which the soil was formed. Anorthite ($CaAl_2Si_2O_8$) is the most important primary source of Ca, although pyroxenes and amphiboles are also fairly common in soils. Small amounts of Ca may also originate from biotite, apatite, and certain borosilicates.

Arid-region soils are generally high in Ca content because of low rainfall (Chapter 3). Calcite ($CaCO_3$) is the dominant source of Ca in semiarid- and arid-region soils. Dolomite $[CaMg(CO_3)_2]$ may also be present in association with $CaCO_3$. In some arid-region soils gypsum $(CaSO_4 \cdot 2H_2O)$ may be present.

Ca in the soil solution of temperate-region soils ranges from 30 to 300 ppm. In higher-rainfall areas, soil solution Ca^{2+} concentrations vary from 5 to 50 ppm. A level of 15 ppm soil solution Ca^{2+} is adequate for crops.

Ca concentrations in the soil higher than necessary for proper plant growth normally have little effect on Ca^{2+} uptake, because Ca^{2+} uptake is genetically controlled. Although the Ca^{2+} concentration of the soil solution is about 10 times greater than that of K^+, its uptake is usually lower than that of K^+. Plants' capacity for Ca^{2+} uptake is limited because it can be absorbed only by young root tips in which the cell walls of the endodermis are still unsuberized.

As a general rule, coarse-textured, humid-region soils formed from rocks low in Ca minerals are low in Ca. Fine-textured soils formed from rocks high in Ca are much higher in both exchangeable and total Ca. However, in humid regions, even soils formed from limestones are frequently acidic in the surface layers because of the removal of Ca and other cations by excessive leaching. As water containing dissolved CO_2 percolates through the soil, the H^+ formed displaces Ca^{2+} (and other basic cations) on the exchange complex $(CO_2 + H_2O \leftrightarrows H^+ + HCO_3^-)$. If there is considerable percolation of such water through the soil profile, soils gradually become acidic. Where leaching occurs, Na^+ is lost more readily then Ca^{2+} (see the lyotropic series in Chapter 2); however, since exchangeable and solution Ca^{2+} are much greater than Na^+ in most soils, the quantity of Ca^{2+} lost is also much greater. Ca is often the dominant cation in drainage waters, springs, streams, and lakes. Leaching of Ca^{2+} ranges from 75 to 200 lb/a per year. Since Ca^{2+} is adsorbed on the CEC, losses by erosion may be considerable in some soils.

In soils not containing $CaCO_3$, $CaMg(CO_3)_2$, or $CaSO_4$, the amount of soil solution Ca^{2+} depends on the amount of exchangeable Ca^{2+}. Soil factors of the greatest importance in determining the Ca^{2+} availability to plants are the following:

1. Total Ca supply.
2. Soil pH.
3. CEC.
4. Percentage of Ca^{2+} saturation on CEC.
5. Type of soil colloid.
6. Ratio of Ca^{2+} to other cations in solution.

Total Ca in very sandy, acidic soils with low CEC can be too low to provide sufficient available Ca^{2+} to crops. On such soils supplemental Ca may be needed to supply Ca^{2+} and correct the acidity. High H^+ activity (low soil pH) impedes Ca^{2+} uptake. For example, much higher Ca^{2+} concentrations are required for soybean root growth as the pH is lowered from 5.6 to 4.0 (Table 7.7).

In acidic soils, Ca is not readily available to plants at low saturation. For example, a low-CEC soil having only 1,000 ppm exchangeable Ca^{2+} but representing a high % Ca^{2+} saturation might well supply plants with more Ca^{2+} compared with 2,000 ppm exchangeable Ca^{2+} with a low % Ca saturation on a high-CEC soil. In other words, as the % Ca^{2+} saturation decreases in proportion to the total CEC, the amount of Ca^{2+} absorbed by plants also decreases.

High Ca^{2+} saturation indicates a favorable pH for plant growth and microbial activity and will usually mean low concentrations of exchangeable Al^{3+} in acidic

TABLE 7.7   Effect of Ca Concentration and pH in Subsurface Nutrient Solution on Soybean
Taproot Elongation in the Nutrient Solution

		Experiment 2				Experiment 3	
pH	Ca Concentration Added (ppm)	Taproot Elongation Rate* (mm/hr)	Taproot Harvest Length† (mm)	Oven Dry Wt/mm (mg)	pH	Ca Concentration Added (ppm)	Taproot Elongation Rate (mm/hr)
5.6	0.05	2.66	461	0.20	4.75	0.05	0.11
	0.50	2.87	453	0.23		0.50	0.91
	2.50	2.70	455	0.32			
4.5	0.05	0.04	24	0.54	4.0	2.50	0.44
	0.50	1.36	270	0.26		5.00	1.26
	2.50	2.38	422	0.31			

*Elongation rate during first 4 hr in solution.

†Harvested 7 1/2 days after entering the solution.

SOURCE: Lund, *Soil Sci. Soc. Am. J.,* 34:457 (1970).

soils and Na^+ in sodic soils. Many crops respond to Ca applications when % Ca^{2+} saturation is <25%.

The type of clay influences Ca^{2+} availability; 2:1 clays require higher Ca^{2+} saturation than 1:1 clays. Specifically, montmorillonitic clays require a >70% Ca^{2+} saturation for adequate Ca availability, whereas kaolinitic clays are able to supply sufficient Ca^{2+} at 40 to 50% Ca^{2+} saturation.

Increasing the Al^{3+} concentration in the soil solution reduces Ca^{2+} uptake (Fig. 7.12). Whereas Ca^{2+} uptake is depressed by NH_4^+, K^+, Mg^{2+}, Mn^{2+}, and Al^{3+}, its absorption is increased when plants are supplied with NO_3^-. A high level of NO_3^- nutrition stimulates organic anion synthesis and the resultant accumulation of cations, particularly Ca^{2+}.

Ca Sources

Ca is present as a component of the materials supplying other nutrients, particularly P. Single superphosphate (SSP) and triple superphosphate (TSP) contain 18 to 21 and 12 to 14% Ca, respectively. Synthetic chelates such as CaEDTA contain approximately 3 to 5% Ca, while some of the natural complexing substances used as micronutrient carriers contain 4 to 12% Ca. Chelated Ca can also be foliarly applied to crops. Phosphate rocks contain about 35% Ca, and when applied at high rates to acidic tropical soils, substantial amounts of Ca are supplied. Animal and municipal wastes contain approximately 2 to 5% Ca by dry weight and thus are excellent sources.

The primary sources are liming materials such as $CaCO_3$, $CaMg(CO_3)_2$, and others that are applied to neutralize soil acidity. In situations in which Ca is required without the need for correcting soil acidity, gypsum is used.

Gypsum ($CaSO_4 \cdot 2H_2O$) deposits are found at several locations in North America, and large amounts of by-product gypsum are produced in the manufacture of phosphoric acid (see Chapter 5). Gypsum has little effect on soil pH; hence, it may have some value on crops that demand an acidic soil yet need considerable Ca. It is widely used on the sodic soils in arid climates (Chapter 3).

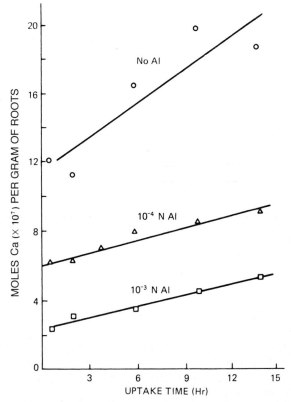

FIGURE 7.12 Influence of $AlCl_3$ on the rate of Ca uptake from 10^{-3} N $CaCl_2$ by excised wheat roots as determined by ^{45}Ca measurements. *Johnson and Jackson*, Soil Sci. Soc. Am. J., *28:381, 1964.*

Magnesium

Mg occurs predominantly as exchangeable and solution Mg^{2+} (Fig. 7.11). The absorption of Mg by plants depends on the amount of solution Mg^{2+}, soil pH, the % Mg saturation on the CEC, the quantity of other exchangeable ions, and the type of clay. Mg in soil solution may be (1) lost in percolating waters, (2) absorbed by living organisms, (3) adsorbed on the CEC, or (4) reprecipitated as a secondary mineral, predominantly in arid climates.

Forms and Functions of Mg in Plants

Mg is absorbed by plants as Mg^{2+} from the soil solution and, like Ca^{2+}, is supplied to plant roots by mass flow and diffusion. Root interception contributes much less Mg^{2+} to uptake than Ca^{2+}. The quantity of Mg^{2+} taken up by plants is usually less than that of Ca^{2+} or K^+.

Mg^{2+} concentration in crops varies between 0.1 and 0.4%. Mg^{2+} is a primary constituent of chlorophyll, and without chlorophyll the autotrophic green plant would fail to carry on photosynthesis (Fig. 4.2). Chlorophyll usually accounts for about 15 to 20% of the total Mg^{2+} content of plants.

Mg also serves as a structural component in ribosomes, stabilizing them in the configuration necessary for protein synthesis. As a consequence of Mg^{2+} deficiency, the proportion of protein N decreases and that of nonprotein N generally increases in plants.

Mg is associated with transfer reactions involving phosphate-reactive groups. Mg is required for maximal activity of almost every phosphorylating enzyme in carbohydrate metabolism. Most reactions involving phosphate transfer from adenosine triphosphate (ATP) require Mg^{2+}. Since the fundamental process of energy transfer occurs in photosynthesis, glycolysis, the citric acid or Krebs cycle, and respiration, Mg^{2+} is important throughout plant metabolism.

Because of the mobility of plant Mg^{2+} and its ready translocation from older to younger plant parts, deficiency symptoms often appear first on the lower leaves. In many species, shortage of Mg^{2+} results in interveinal chlorosis of the leaf, in which only the veins remain green. In more advanced stages the leaf tissue becomes uniformly pale yellow, then brown and necrotic. In other species, notably cotton, the lower leaves may develop a reddish-purple cast, gradually turning brown and finally necrotic (see color plates inside book cover).

GRASS TETANY   Low Mg content of forage crops, particularly grass forages, can be a problem in some areas. Cattle consuming low-Mg forages may suffer from hypomagnesemia, or grass tetany, which is an abnormally low level of blood Mg. High rates of NH_4^+ or K^+ fertilizers may depress the Mg^{2+} level in plant tissue. For example, the Mg content of young corn plants is markedly reduced when NH_4^+ rather than NO_3^- is applied. Because grass tetany often occurs in the spring, the N may still be in the NH_4^+ form, particularly if cool weather has prevailed. In addition, the high protein content of ingested forages (and other feeds) will depress the absorption of Mg by the animal.

Levels of soil Mg may be increased through the use of dolomitic limestone, if liming is advisable, or through the use of Mg-containing fertilizers. Also, the inclusion of legumes in the forage program is advisable because these plants have a higher Mg content than grasses do. Cattle can also be fed an Mg salt to help prevent grass tetany.

Although hypomagnesemia can be the result of excessive N or K fertilization, it seems more reasonable to class it as a Mg deficiency and to treat it accordingly.

Mg in Soil

Mg constitutes 1.93% of the earth's crust; however, the Mg^{2+} content of soils ranges from 0.1% in coarse, sandy soils in humid regions to 4% in fine-textured, arid, or semiarid soils formed from high-Mg parent materials. Mg in the soil originates from the weathering of rocks containing the minerals biotite, dolomite, hornblende, olivene, and serpentine. It is also found in the secondary clay minerals chlorite, illite, montmorillonite, and vermiculite. Substantial amounts of epsomite ($MgSO_4 \cdot 7H_2O$) and bloedite [$Na_2Mg(SO_4)_3 \cdot 4H_2O$] may occur in arid or semiarid soils.

The Mg concentration of soil solutions is typically 5 to 50 ppm in temperate-region soils, although Mg^{2+} concentrations between 120 and 2,400 ppm have been observed. Mg^{2+}, like Ca^{2+}, can be leached from soils, and Mg losses of 5 to

TABLE 7.8   Percentage of Exchangeable Mg Displaced by Various K Salts in the First Week of Leaching

	Soil Type and Initial Exchangeable Mg (meq %)		
	Taupo Sandy Silt (0.56 meq %)	Te Kopuru Sand (0.59 meq %)	Patea Sand (0.22 meq %)
	% displaced		
KCl (0.06 g)	12.1	6.8	31.4
K_2SO_4	4.3	6.1	30.8
$KHCO_3$ or K_2CO_3	<0.1	<0.1	1.6
KH_2PO_4	<0.1	<0.1	2.6

SOURCE: Hogg., *New Zealand J. Sci.*, 5:64 (1962).

60 lb/a have been observed. The amounts lost depend on the interaction of several factors, including the Mg content of soil, rate of weathering, intensity of leaching, and uptake by plants. Leaching of Mg^{2+} is often a severe problem in sandy soils, particularly following the addition of fertilizers such as KCl and K_2SO_4 (Table 7.8). Very little Mg displacement occurs when equivalent amounts of K are applied as either CO_3^{2-}, HCO_3^-, or $H_2PO_4^-$. Apparently, Mg^{2+} desorption and leaching in coarse-textured soils are enhanced by the presence of soluble Cl^- and SO_4^{2-}. As with Ca^{2+}, erosion losses can be considerable in some soils.

Mg in clay minerals is slowly weathered out by leaching and exhaustive cropping. Vermiculite has a high Mg content, and it can be a significant source of Mg in soils. Conditions in which Mg is likely to be deficient include acidic, sandy, highly leached soils with low CEC; calcareous soils with inherently low Mg levels; acidic soils receiving high rates of liming materials low in Mg; high rates of NH_4^+ or K^+ fertilization; and crops with a high Mg demand.

Excess Mg can occur in certain situations in which soils are formed on serpentine bedrock or are influenced by groundwaters high in Mg. Normal Ca nutrition can be disrupted when exchangeable Mg^{2+} exceeds Ca^{2+}.

Coarse-textured soils in humid regions exhibit the greatest potential for Mg deficiency. These soils normally contain small amounts of total and exchangeable Mg^{2+}. Soils are probably deficient when they contain less than 25 to 50 ppm exchangeable Mg^{2+}.

Exchangeable Mg normally accounts for 4 to 20% of the CEC of soils, but in soils derived from serpentine rock, exchangeable Mg^{2+} can exceed Ca^{2+}. The critical Mg saturation for optimum plant growth coincides closely with this range, but in most instances, % Mg saturation should not be less than 10%.

Reduced Mg^{2+} uptake in many strongly acidic soils is caused by high levels of exchangeable Al^{3+}. Al saturation of 65 to 70% is often associated with Mg deficiency. Mg deficiencies can also occur in soils with high ratios of exchangeable Ca/Mg, where this ratio should not exceed 10/1 to 15/1. On many humid-region, coarse-textured soils the continued use of high-calcic liming materials may increase the Ca/Mg ratio and induce Mg deficiency on certain crops.

High levels of exchangeable K can interfere with Mg uptake by crops. Generally, the recommended K/Mg ratios are <5/1 for field crops, 3/1 for vegetables and sugar beets, and 2/1 for fruit and greenhouse crops.

Competition between NH_4^+ and Mg^{2+} can also lower the Mg^{2+} availability to crops. Ammonium-induced Mg^{2+} stress is greatest when high rates of NH_4^+ fertilizers are applied to low exchangeable Mg^{2+} soils. This interaction may contribute to grass tetany problems. The mechanism of this interaction probably involves the H^+ released when NH_4^+ is absorbed by roots, as well as the direct effect of NH_4^+.

Mg Sources

In contrast to Ca, few primary nutrient fertilizers contain Mg, with the exception of $K_2SO_4 \cdot MgSO_4$ (see Chapter 6). Dolomite is commonly applied to low-Mg acidic soils. $K_2SO_4 \cdot MgSO_4$ and $MgSO_4$ (Epsom salts) are the most widely used materials in dry fertilizer formulations (Table 7.5). Other materials containing Mg are magnesia (MgO, 55% Mg), magnesium nitrate [$Mg(NO_3)_2$, 16% Mg], magnesium silicate (basic slag, 3 to 4% Mg; serpentine, 26% Mg), magnesium chloride solution ($MgCl_2 \cdot 10H_2O$, 8 to 9% Mg), synthetic chelates (2 to 4% Mg), and natural organic complexing substances (4 to 9% Mg).

$MgSO_4$, $MgCl_2$, $Mg(NO_3)_2$, and synthetic and natural Mg chelates are well suited for application in clear liquids and foliar sprays. Mg deficiency of citrus trees in California is frequently corrected by foliar applications of $Mg(NO_3)_2$. In some tree-fruit growing areas, $MgSO_4$ solutions are foliar applied to maintain levels, and in seriously deficient orchards several annual applications are necessary.

$K_2SO_4 \cdot MgSO_4$ is the most widely used Mg additive in suspensions. A special suspension grade (100% passing through a 20-mesh screen) of this material is available commercially. Mg content in animal and municipal wastes is similar to S content and can therefore be used to supply sufficient Mg.

Selected References

BARBER, S. A. 1984. *Soil Nutrient Bioavailability: A Mechanistic Approach.* John Wiley & Sons, New York.

BEATON, J. D., R. L. FOX, and M. B. JONES. 1985. Production, marketing, and use of sulfur products, pp. 411–54. In O. P. ENGLESTAD (Ed.), *Fertilizer Technology and Use.* Soil Science Society of America, Madison, Wisc.

FOLLETT, R. H., L. S. MURPHY, and R. L. DONAHUE. 1981. *Fertilizers and Soil Amendments.* Prentice-Hall, Englewood Cliffs, N.J.

MORTVEDT, J. J., and F. R. FOX. 1985. Production, marketing, and use of calcium, magnesium, and micronutrient fertilizers, pp. 455–82. In O. P. ENGLESTAD (Ed.), *Fertilizer Technology and Use.* Soil Science Society of America, Madison, Wisc.

TABATABAI, M. A. (Ed.). 1986. *Sulfur in Agriculture.* No. 27. ASA, CSSA, Soil Science Society of America, Madison, Wisc.

CHAPTER 8

Micronutrients

Micronutrients are just as important in plant nutrition as the major nutrients; they simply occur in plants and soils in much smaller concentrations. Plants grown on micronutrient-deficient soils can exhibit similar reductions in plant growth and yield as major nutrients. Like the major nutrients, micronutrients occur in four major forms in soil: (1) primary and secondary minerals, (2) adsorbed to mineral and organic matter surfaces, (3) organic and microbial biomass, and (4) solution (Fig. 8.1). Depending on the micronutrient, some forms are more important than others in supplying or buffering plant available micronutrients in the soil solution. Understanding the relationships and dynamics among these forms is essential for eliminating micronutrient stress in plants grown on micronutrient-deficient soils.

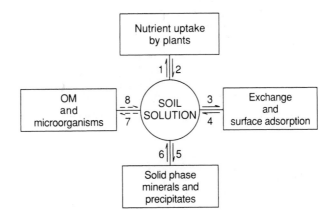

FIGURE 8.1   Relationships between the various forms of micronutrients in soils. Reactions 1 and 2 represent plant absorption and exudation, respectively; reactions 3 and 4 represent adsorption and desorption, respectively; reactions 5 and 6 represent precipitation and dissolution, respectively; and reactions 7 and 8 represent immobilization and mineralization, respectively. All of these processes interact to control concentration of micronutrients in soil solution. *Lindsay,* Micronutrients in Agriculture, *p. 42, American Society of Agronomy, 1972.*

Iron

Fe Cycle

Soil solution concentration and plant available Fe are governed predominantly by the organic fraction in soils (Fig. 8.2). Primary and secondary minerals dissolve to provide Fe incorporated into the microbial biomass and complexed by organic compounds in the soil solution. Adsorbed Fe has little influence on soil solution and plant available Fe.

FE IN PLANTS   Fe is absorbed by plant roots as Fe^{2+} and Fe^{3+}. The chemical properties of Fe make it an important part of oxidation-reduction reactions in both soils and plants. Because Fe can exist in more than one oxidation state, it accepts or donates electrons according to the oxidation potential of the reactants. The transfer of electrons between the organic molecule and Fe provides the potential for many of the enzymatic transformations. Several of these enzymes are involved in chlorophyll synthesis, and when Fe is deficient chlorophyll production is reduced, which results in the characteristic chlorosis symptoms of Fe stress.

Fe is a structural component of porphyrin molecules: cytochromes, hemes, hematin, ferrichrome, and leghemoglobin. These substances are involved in oxidation-reduction reactions in respiration and photosynthesis. As much as 75% of the total cell Fe is associated with the chloroplasts, and up to 90% of the Fe in leaves occurs with lipoprotein of the chloroplast and mitochondria membranes.

Fe in chloroplasts reflects the presence of cytochromes for performing various photosynthetic reduction processes and of ferrodoxin as an electron acceptor.

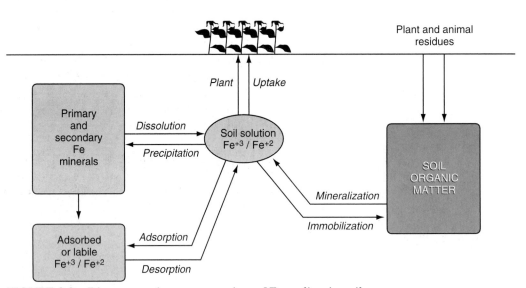

FIGURE 8.2   Diagrammatic representation of Fe cycling in soils.

Ferredoxins are Fe-S proteins and are the first stable redox compound of the photosynthetic electron transport chain. Reduction of O_2 to water during respiration is the most common function of Fe-containing compounds. Fe is also an important part of the enzyme nitrogenase, which is essential to the N_2 fixation in N-fixing microorganisms. Fe may also be capable of partial substitution for Mo as the metal cofactor necessary for the functioning of NO_3^- reductase in soybeans.

The sufficiency range of Fe in plant tissue is normally between 50 and 250 ppm. In general, when Fe contents are 50 ppm or less in the dry matter, deficiency is likely to occur. Fe deficiency symptoms show up first in the young leaves of plants, because Fe does not readily translocate from older tissues to the tip meristem; as a result, growth ceases. The young leaves develop an interveinal chlorosis, which progresses rapidly over the entire leaf. In severe cases the leaves turn entirely white. Fe-deficiency symptoms are illustrated in the color plates (see inside book cover). Fe toxicity can be observed under certain conditions. For example, in rice grown on poorly drained or submerged soils, a condition known as *bronzing* is associated with >300 ppm Fe levels in rice leaves at tillering.

FE IN SOIL

Mineral Fe. Fe comprises about 5% of the earth's crust and is the fourth most abundant element in the lithosphere. Common primary and secondary Fe minerals are olivene [$(Mg,Fe)_2SiO_4$], siderite ($FeCO_3$), hematite (Fe_2O_3), goethite ($FeOOH$), magnetite (Fe_3O_4), and limonite [$FeO(OH) \cdot nH_2O + Fe_2O_3 \cdot nH_2O$]. Iron can be either concentrated or depleted during soil development; thus, Fe concentration in soil varies widely, from 0.7 to 55%. Most of this soil Fe is found in primary minerals, clays, oxides, and hydroxides.

The solubility of the common Fe minerals in soil is very low, only 10^{-6} to 10^{-24} M Fe^{3+} in solution, depending on pH (Fig. 8.3). The mineral denoted by "soil Fe" represents an amorphous $Fe(OH)_3$ precipitate, which appears to control the solution Fe^{3+} concentration in most soils.

Soil Solution Fe. Compared with other cations in soils, the Fe^{3+} concentration in solution is very low. In well-drained, oxidized soils, the solution Fe^{2+} concentration is less than that of the dominant Fe^{3+} species in solution (Fig. 8.4). Soluble Fe^{2+} increases significantly when soils become waterlogged. The following equation describes the pH-dependent relationship for Fe^{3+}:

$$Fe(OH)_3(soil) + 3H^+ \leftrightarrow Fe^{3+} + 3H_2O$$

For every unit increase in pH, Fe^{3+} concentration decreases 1,000-fold. In contrast, Fe^{2+} decreases 100-fold for each unit increase in pH, which is similar to the behavior of other divalent metal cations (Fig. 8.4).

Over the normal pH range in soils, total solution Fe is not sufficient to meet plant requirements for Fe, even in acidic soils, where Fe deficiencies occur less frequently than in high-pH and calcareous soils (Fig. 8.5). Obviously, another mechanism that increases Fe availability to plants exists; otherwise, crops grown on almost all soils would be Fe deficient.

CHELATE DYNAMICS. Numerous natural organic compounds in soil, or synthetic compounds added to soils, are able to complex, or *chelate*, Fe^{3+} and other micronutrients. The concentration of Fe in solution and the quantity of Fe trans-

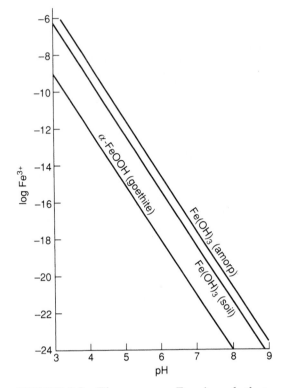

FIGURE 8.3   The common Fe minerals that control solution Fe in soils. *Lindsay,* Chemical Equilibria in Soils, *John Wiley & Sons, New York, 1979.*

ported to the root by mass flow and diffusion can be greatly increased through *complexation* of Fe with natural organic chelating compounds in the soil.

Chelate is a term derived from a Greek word meaning *claw.* Chelates are soluble organic compounds that bond with metals such as Fe, Zn, Cu, and Mn, increasing their solubility and their supply to plant roots. Natural organic chelates in soils are products of microbial activity and degradation of soil OM and plant residues. Root exudates are also capable of complexing micronutrients.

Many of the natural organic chelates have not been identified; however, compounds such as citric and oxalic acids have chelating properties (Table 8.1). Examples of the molecular structure of several synthetic chelates are shown in Figure 8.6.

The dynamics of chelation in increasing solubility and transport of micronutrients is illustrated in Figure 8.7. During active plant uptake, the concentration of chelated Fe or other micronutrients is greater in the bulk solution than at the root surface; thus, chelated Fe diffuses to the root surface in response to the concentration gradient. At the root surface the Fe^{3+} "unhooks," or dissociates from, the chelate by a mechanism not well understood. After Fe^{3+} dissociates from the chelate, the "free" chelate will diffuse away from the root back to the "bulk" solution, again because of a concentration gradient (free chelate concentration near the root > free chelate in bulk solution). The free chelate subsequently complexes another Fe^{3+} ion from solution. As the unchelated Fe^{3+} concentration decreases in solution because of chelation, additional Fe is desorbed from mineral surfaces or Fe minerals dissolve to resupply solution Fe. The chelate-micronutrient "cycling"

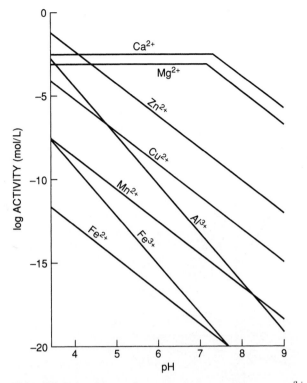

FIGURE 8.4   The influence of pH on solution Fe^{3+} concentration relative to other cations.   *Lindsay,* Chemistry in Soil Environment, *p. 189. American Society of Agronomy, Madison, Wisc., 1981.*

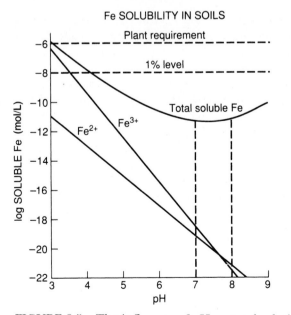

FIGURE 8.5   The influence of pH on total solution Fe concentration and its relationship to Fe required by plants.   *Lindsay,* Plant Root and Its Environment, *p. 508, Univ. Press of Virginia, 1974.*

TABLE 8.1   Chemical Formula for Several Common Synthetic and Natural Chelates

Name	Formula	Abbreviation
Ethylenediaminetetraacetic acid	$C_{10}H_{16}O_8N_2$	EDTA
Diethylenetriaminepentaacetic acid	$C_{14}H_{23}O_{10}N_3$	DTPA
Cyclohexanediaminetetraacetic acid	$C_{14}H_{22}O_8N_2$	CDTA
Ethylenediaminedi-*o*-hydroxyphenlyacetic acid	$C_{18}H_{20}O_6N_2$	EDDHA
Citric acid	$C_6H_8O_7$	CIT
Oxalic acid	$C_2H_2O_4$	OX
Pyrophosphoric acid	$H_4P_2O_7$	P_2O_7

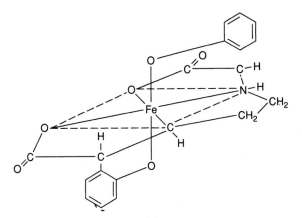

(a)

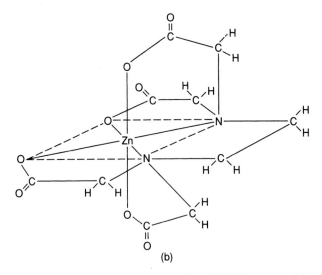

(b)

FIGURE 8.6   The structure of Fe-EDDHA (a) and Zn-EDTA (b).   *Follett et al.,*
Fertilizers and Soil Amendments, Prentice-Hall, Englewood Cliffs, N.J., 1981.

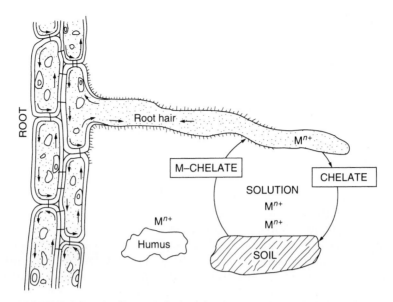

FIGURE 8.7   Cycling of chelated micronutrients (M) in soils.   *Lindsay,* Plant Root and Its Environment, *p. 517, Univ. Press of Virginia, 1974.*

is an extremely important mechanism in soils that greatly contributes to plant available Fe and other micronutrients.

For example, diffusion of Fe to sorghum roots was encouraged by the higher concentration of soluble Fe through chelation with EDDHA (Fig. 8.8).

Factors Affecting Fe Availability

SOIL pH AND BICARBONATE   Fe deficiency is most often observed on high-pH and calcareous soils in arid regions, but it may also occur on acidic soils that are very low in total Fe. Irrigation waters and soils high in bicarbonate (HCO_3^-)

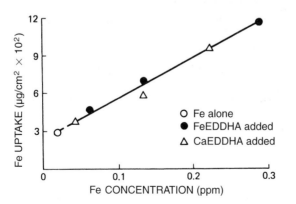

FIGURE 8.8   Uptake of Fe by sorghum roots as a function of Fe in the soil solution and chelate treatment.   *O'Connor et al.,* Soil Sci. Soc. Am. J., *35:407, 1971.*

may aggravate Fe deficiencies, probably because of the high pH levels associated with HCO_3^- accumulation. The pH of most soils containing $CaCO_3$ falls in the range 7.3 to 8.5, which coincides with the greatest incidence of Fe deficiency and the lowest solubility of soil Fe (Fig. 8.5). Bicarbonate ion can be formed in calcareous soils by the following reaction:

$$CaCO_3 + CO_2 + H_2O \rightarrow Ca^{2+} + 2HCO_3^-$$

Although the presence of lime alone does not necessarily induce Fe deficiency, its interaction with certain soil environmental conditions is related to Fe deficiency.

EXCESSIVE WATER AND POOR AERATION   The preceding reaction is promoted by the accumulation of CO_2 in excessively wet and poorly drained soils. Consequently, any compact, heavy-textured, calcareous soil is potentially Fe deficient. Iron chlorosis is often associated with cool, rainy weather when soil moisture is high and soil aeration is poor. Also, root development and nutrient absorption are reduced under these cool, wet conditions, which contributes to Fe stress. Lime-induced chlorosis often disappears when these soils are allowed to dry. Flooding and submergence of soils in which HCO_3^- formation is of no concern can improve Fe availability by increasing Fe^{2+} concentrations.

ORGANIC MATTER   Although lime-induced Fe deficiency occurs in wet soils, semiarid-region calcareous soils that are low in OM are often low in plant available Fe. This deficiency occurs especially on eroded portions of the field where the OM-rich topsoil has been removed, exposing calcareous subsoils. Land leveling for irrigation can also expose calcareous, low-OM subsoils.

Additions of OM to well-drained soils can improve Fe availability. Organic materials such as manure may supply chelating agents that aid in maintaining the solubility of micronutrients. Improved structure of fine-textured soils resulting from applications of organic manures should also increase Fe availability because of better soil aeration.

INTERACTIONS WITH OTHER NUTRIENTS   Metal cations can interact with Fe to induce Fe stress in plants. Fe deficiencies can result from an accumulation of Cu after extended periods of Cu fertilization. Pineapples in Hawaii exhibited Fe chlorosis when grown on soils high in Mn, and other plants growing on soils developed from serpentine exhibited Fe deficiency because of excess Ni. Fe deficiencies on soybeans can occur because of a low Fe/(Cu + Mn) ratio in plants. In addition to Fe deficiencies caused by excess Cu, Mn, Zn, and Mo, Fe-P interactions have been observed in some plants, probably related to precipitation of Fe-P minerals.

Plants receiving NO_3^- are more likely to develop Fe stress than those nourished with NH_4^+. When a strong acid anion (NO_3^-) is absorbed and replaced with a weak acid (HCO_3^-), the pH of the root zone increases, particularly in low buffered systems, which decreases Fe availability. Thus, Fe solubility and availability are favored by the acidity that develops when NH_4^+ is utilized by plants.

PLANT FACTORS   Although diffusion of both Fe^{3+} and Fe^{2+} to the root occurs, the Fe^{3+} is reduced to Fe^{2+} before absorption. Plant genotypes differ in their ability to take up Fe. Table 8.2 rates plants according to their sensitivity or

TABLE 8.2   Sensitivity of Crops to Low Levels of Available Fe in Soil*

Sensitive	*Moderately Tolerant*	*Tolerant*
Berries	Alfalfa	Alfalfa
Citrus	Barley	Barley
Field beans	Corn	Corn
Flax	Cotton	Cotton
Forage sorghum	Field beans	Flax
Fruit trees	Field peas	Grasses
Grain sorghum	Flax	Millet
Grapes	Forage legumes	Oats
Mint	Fruit trees	Potatoes
Ornamentals	Grain sorghum	Rice
Peanuts	Grasses	Soybeans
Soybeans	Oats	Sugar beets
Sundangrass	Orchard grass	Vegetables
Vegetables	Ornamentals	Wheat
Walnuts	Rice	
	Soybeans	
	Vegetables	
	Wheat	

*Some crops are listed under two or three categories because of variations in soil, growing conditions, and differential response varieties of a given crop.

tolerance to low levels of available Fe. Some crops appear in more than one category because of variations in soil, growing conditions, and differential response of varieties of a given crop. Fe-efficient varieties should be selected for conditions in which Fe deficiencies are likely to occur.

The ability of plants to absorb and translocate Fe appears to be a genetically controlled adaptive process that responds to Fe deficiency or stress. Roots of Fe-efficient plants alter their environment to improve the availability and uptake of Fe. Some of the biochemical reactions and changes enabling Fe-efficient plants to tolerate and adapt to Fe stress are as follows:

1. Excretion of H^+ ions from roots.
2. Excretion of various reducing or chelating compounds from roots.
3. Rate of reduction (Fe^{3+} to Fe^{2+}) increases at the root.
4. Increase in organic acids, particularly citrate, in the root sap.
5. Adequate transport of Fe from roots to tops.
6. Less accumulation of P in roots and shoots, even in the presence of relatively high P in the growth medium.

Fe Sources

ORGANIC FE   Most animal wastes contain small quantities of plant available Fe, typically ranging from 0.02 to 0.1% (0.4 to 2 lb/t). Although sufficient plant available Fe can be provided through manure application at appropriate rates, the major benefit of organic waste application is increased OM and associated natural chelation properties. Enhanced Fe chelation may supply sufficient plant

TABLE 8.3   Some Sources of Fertilizer Fe

Source	Formula	Percentage of Fe
Ferrous sulfate	$FeSO_4 \cdot 7H_2O$	19
Ferric sulfate	$Fe_2(SO_4)_3 \cdot 4H_2O$	23
Ferrous oxide	FeO	77
Ferric oxide	Fe_2O_3	69
Ferrous ammonium phosphate	$Fe(NH_4)PO_4 \cdot H_2O$	29
Ferrous ammonium sulfate	$(NH_4)_2SO_4 \cdot FeSO_4 \cdot 6H_2O$	14
Iron ammonium polyphosphate	$Fe(NH_4)HP_2O_7$	22
Iron chelates	NaFeEDTA	5–14
	NaFeEDDHA	6
	NaFeDTPA	10
Natural organic materials	—	5–10

SOURCE: Mortvedt et al. (Eds.), *Micronutrients in Agriculture,* p. 357, Soil Science Society of America, Madison, Wisc. (1972).

available Fe, even if the manure contained no Fe. In contrast, municipal waste can contain as much as 5% Fe (100 lb/t) (Table 8.3).

INORGANIC FE   Fe chlorosis is one of the most difficult micronutrient deficiencies to correct in the field. Table 8.3 lists the Fe-containing materials that are commonly used to treat Fe deficiencies. In general, soil applications of inorganic Fe sources are not effective in correcting Fe deficiency because of the rapid precipitation of very insoluble $Fe(OH)_3$ (Fig. 8.3). For example, when $FeSO_4 \cdot 7H_2O$ and Fe-EDDHA were added to soil, only 20% of the $FeSO_4 \cdot 7H_2O$ was DTPA extractable after just 1 week, compared with 70% FeEDDA after 7 weeks and 26% after 14 weeks (Fig. 8.9).

Fe deficiencies are corrected mainly with foliar application of Fe. One application of a 2% $FeSO_4$ solution at a rate of 15 to 30 gal/a is usually sufficient to alleviate mild chlorosis. However, several applications 7 to 14 days apart may be needed to remedy more severe Fe deficiencies. Injections of Fe salts directly into trunks and limbs of fruit tree species such as pears and plums have been very effective in controlling Fe chlorosis. Treatments in California orchards typically consist of pressure injection at 200 psi of between 1 to 2 pints to 1 to 2 quarts per tree of 1–2% $FeSO_4$ solutions.

With the exception of $FeSO_4$, perhaps the most widely used Fe sources are the synthetic chelates (Table 8.3). These materials are water soluble and can be applied to the soil or foliage. Chelated Fe is protected from the usual soil reactions, which result in formation of insoluble $Fe(OH)_3$.

The choice of which chelate to use as a fertilizer or for soil testing depends on (1) the specific micronutrient and (2) the stability of the chelate in the soil (Fig. 8.10). The chelate EDDHA will strongly complex Fe and is stable over the entire pH range. For example, when Fe-EDTA, Fe-DTPA, and Fe-EDDHA were added to a soil, the EDDHA chelate provided more plant available Fe than the other chelates but only partially corrected the Fe deficiency in sorghum (Fig. 8.11). Since Fe-EDDHA is the most stable Fe chelate, it is the preferred chelate fertilizer source, especially on acidic soils, although Fe-DTPA has also been used (Fig. 8.10). Un-

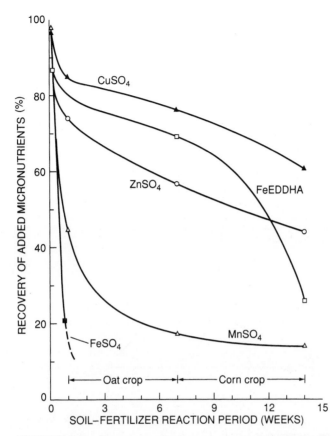

FIGURE 8.9   Recovery of micronutrients in soils fertilized with various inorganic micronutrient fertilizers.   *Follett and Lindsay,* Soil Sci. Soc. Am. J., *35:600–602, 1971.*

fortunately, Fe chelates are not cost-effective for soil application, except on high-value crops or plants.

Local acidification of small portions of the root zone can be effective in correcting Fe deficiencies in calcareous and high-pH soils. Several S products, such as S^0, ammonium thiosulfate, sulfuric acid, sulfur dioxide, and ammonium polysulfide, will lower soil pH and increase solution Fe concentration.

Complexing with polyphosphate fertilizers also increases the plant availability of Fe, but Fe-EDDHA is more effective than polyphosphate at the same Fe rates.

Zinc

Zn Cycle

Soil solution concentration and plant available Zn are governed predominantly by solution pH and Zn adsorbed on clay and organic surfaces in soils (Fig. 8.12). Primary and secondary minerals dissolve to initially provide Zn to the soil solution, which is then adsorbed onto CEC and incorporated into the microbial biomass

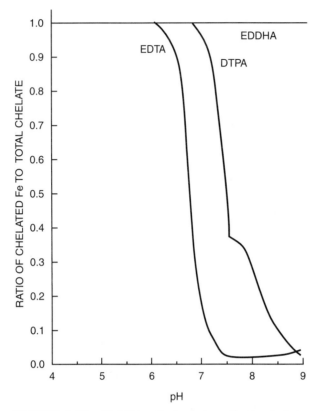

FIGURE 8.10   Stability of selected synthetic chelates with Fe in soils.   *Norvell,* Micronutrients in Agriculture, *p. 126, 1972.*

and complexed by organic compounds in the soil solution. Like Fe, chelated Zn is important to the transport of Zn to root surfaces for uptake.

ZN IN PLANTS   Plant roots absorb Zn as Zn^{2+} and as a component of synthetic and natural organic complexes. Zn is involved in many enzymatic activities, but it is not known whether it acts as a functional, structural, or regulatory cofactor. Zn is important in the synthesis of tryptophane, a component of some proteins and a compound needed for the production of growth hormones (auxins) such as indoleacetic acid. Reduced growth hormone production in Zn-deficient plants causes the shortening of internodes and smaller-than-normal leaves.

Zn deficiency can be identified by distinctive visual symptoms that appear most frequently in the leaves. Sometimes the deficiency symptoms also appear in the fruit or branches or are evident in the overall development of the plant. Symptoms common to many crops include the following:

- Occurrence of light-green, yellow, or white areas between the veins of leaves, particularly the older, lower leaves.
- Death of tissue in these discolored, chlorotic leaf areas.
- Shortening of the stem or stalk internodes, resulting in a bushy, rosetted appearance of the leaves.

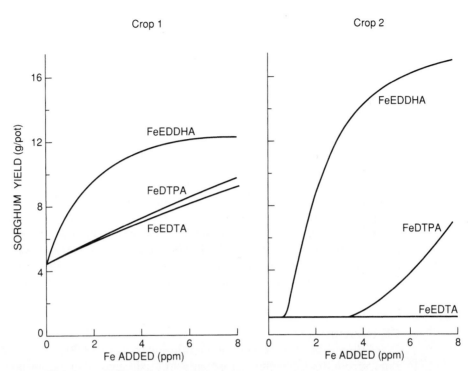

FIGURE 8.11   Effectiveness of selected synthetic Fe chelates in supplying Fe to Fe-deficient sorghum.   *Lindsay,* Plant Root and Its Environment, *p. 511, Univ. Press of Virginia, 1974.*

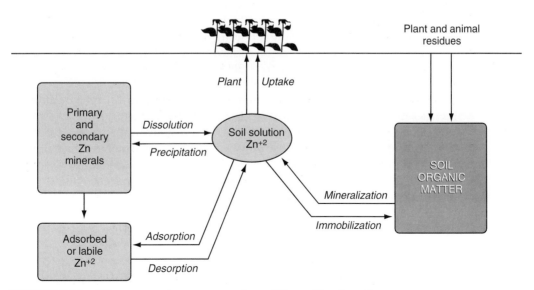

FIGURE 8.12   Diagrammatic representation of Zn cycling in soils.

- Small, narrow, thickened leaves. Often the leaves are malformed by continued growth of only part of the leaf tissue.
- Early loss of foliage.
- Malformation of the fruit, often with little or no yield.

Zn deficiency causes the characteristic little leaf and rosetting or clustering of leaves at the top of fruit tree branches, which have become mainly bare. In corn and sorghum, Zn deficiency is called *white bud,* and in cotton it is known as *little leaf.* The deficiency is referred to as *mottle leaf* or *frenching* in citrus crops and is described as *fern leaf* in Russet Burbank potato. Zn deficiency symptoms are shown in color plates (see inside book cover). Zn concentration in plants ranges between 25 and 150 ppm. Deficiencies of Zn are usually associated with concentrations of less than 20 ppm, and toxicities will occur when the Zn leaf concentration exceeds 400 ppm.

Zn deficiencies are widespread in the United States and throughout the world, especially in the rice cropland of Asia. Soil conditions most associated with Zn deficiencies are acidic sandy soils low in Zn; neutral, basic, or calcareous soils; fine-textured soils; soils high in available P; some organic soils; and subsoils exposed by land leveling or by wind and water erosion.

Zn in Soil

Mineral Zn. Zinc content of the lithosphere is about 80 ppm, and Zn in soils ranges from 10 to 300 ppm and averages approximately 50 ppm. The igneous rocks contain about 70 ppm, while sedimentary rocks (shale) contain more Zn (95 ppm) than limestone (20 ppm) or sandstone (16 ppm). Franklinite ($ZnFe_2O_4$), smithsonite ($ZnCO_3$), and willemite (Zn_2SiO_4) are common Zn-containing minerals (Fig. 8.13). Zn mineral solubility in soils often resembles the solubility represented by "soil Zn."

Soil Solution Zn. Zinc in the soil solution is very low, ranging between 2 and 70 ppb, with more than half of the Zn^{2+} in solution complexed by OM. Above pH 7.7, $ZnOH^+$ becomes the most abundant species (Fig. 8.14). Zinc solubility is highly pH dependent and decreases 100-fold for each unit increase in pH. This relationship is represented by the following formula:

$$Soil\text{-}Zn + 2H^+ \leftrightarrows Zn^{2+}$$

Thirtyfold reductions of Zn concentration in soil solutions for every unit of pH increase in the pH range 5 to 7 typically have been observed.

Diffusion is the dominant mechanism for transporting Zn^{2+} to plant roots. Complexing agents or chelates from root exudates or from decomposing organic residues facilitate the diffusion of Zn^{2+} to a root, as previously described (Fig. 8.7). Diffusion of chelated Zn^{2+} can be significantly greater than that of unchelated Zn^{2+} (Fig. 8.15).

Factors Affecting Zn Availability

Soil pH   The availability of Zn^{2+} decreases with increased soil pH, as demonstrated earlier (Fig. 8.14). Most pH-induced Zn deficiencies occur in neutral and

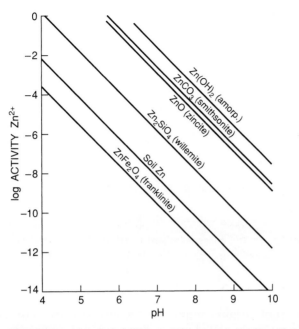

FIGURE 8.13   Solubilities of the common Zn minerals in soils.   *Lindsay, Chemical Equilibria in Soils, John Wiley & Sons, New York, 1979.*

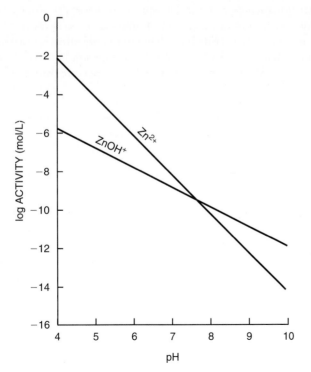

FIGURE 8.14   The common Zn species in soil solution as influenced by pH. *Lindsay, Chemical Equilibria in Soils, John Wiley & Sons, New York, 1979.*

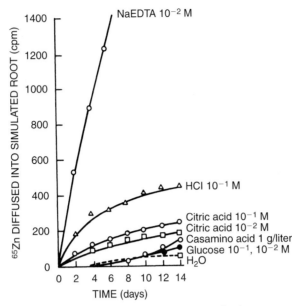

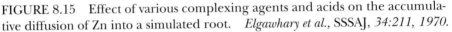

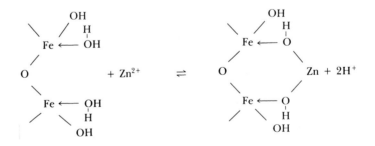

FIGURE 8.15   Effect of various complexing agents and acids on the accumulative diffusion of Zn into a simulated root.   *Elgawhary et al., SSSAJ, 34:211, 1970.*

calcareous soils, although not all of these soils exhibit Zn deficiency because of increased availability from chelation of Zn^{2+} (Fig. 8.7).

At high pH, Zn precipitates as insoluble amorphous soil Zn, $ZnFe_2O_4$, and/or $ZnSiO_4$, which reduces Zn^{2+} in soils (Fig. 8.13). Liming acidic soils, especially ones low in Zn, will reduce uptake of Zn^{2+}, which is related to the pH effect on Zn^{2+} solubility. Zn adsorption on the surface of $CaCO_3$ could also reduce solution Zn^{2+}. Adsorption of Zn^{2+} by clay minerals, Al/Fe oxides, OM, and $CaCO_3$ increases with increasing pH.

ZN ADSORPTION   The mechanism of Zn^{2+} adsorption on oxide surfaces is likely that shown in the following:

Such adsorption is considered an extension of the oxide surface resulting in retention of Zn^{2+}. Adsorption of Zn^{2+} also occurs where Zn^{2+} is less firmly held and can be replaced by other cations, such as Ca^{2+} and Mg^{2+}.

Adsorption of Zn^{2+} by bentonite, illite, and kaolinite clay minerals is directly related to the CEC of clays. Zn reversibly bound by clay minerals is exchangeable and thus can be desorbed to solution.

Zinc is strongly adsorbed by magnesite ($MgCO_3$) and to an intermediate degree by dolomite [$CaMg(CO_3)_2$] and calcite ($CaCO_3$). In magnesite and

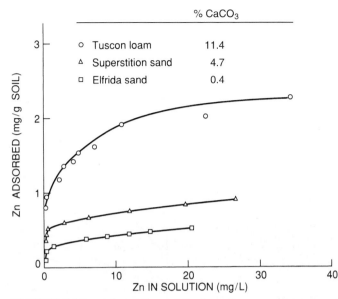

FIGURE 8.16   Adsorption of Zn by calcareous Arizona soils.   *Udo et al.,* Soil Sci. Soc. Am. J., *34:405, 1970.*

dolomite, it appears that Zn is actually adsorbed into the crystal surfaces at sites in the lattice normally occupied by Mg atoms.

Zn adsorption by carbonates is partly responsible for reduced Zn^{2+} availability in calcareous soils. Analysis of the Zn^{2+} adsorption curves in Figure 8.16 reveals that $CaCO_3$ content was the principal factor contributing to Zn adsorption.

ORGANIC MATTER   As previously discussed, Zn^{2+} forms stable complexes with soil OM components. The humic and fulvic acid fractions are prominent in Zn adsorption. Three classes of reactions of OM with Zn and other micronutrients have been distinguished:

1. Immobilization by high molecular weight organic substances such as lignin.
2. Solubilization and mobilization by short-chain organic acids and bases.
3. Complexation by initially soluble organic substances that then form insoluble salts.

The action of OM on Zn^{2+} can be expected to vary depending on the characteristics and amounts of the organic materials involved. When reactions 1 and/or 3 prevail (Fig. 8.1), availability of Zn will be reduced; this occurs in Zn-deficient peats and humic gley soils. On the other hand, formation of soluble chelated Zn compounds will enhance availability by keeping Zn^{2+} in solution (Fig. 8.7). Substances present in or derived from freshly applied organic materials have the capacity to chelate Zn^{2+}; however, the increased solution Zn^{2+} is not always reflected in enhanced Zn uptake by plants.

INTERACTION WITH OTHER NUTRIENTS   Other metal cations, including Cu^{2+}, Fe^{2+}, and Mn^{2+}, inhibit Zn^{2+} uptake, possibly because of competition for the same carrier site. The antagonistic effect of several cations, especially Cu^{2+} and Fe^{2+}, on Zn uptake by rice is shown in Figure 8.17.

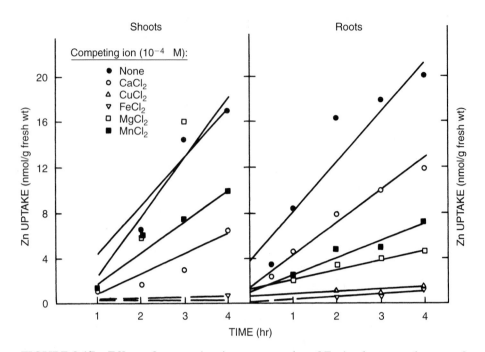

FIGURE 8.17 Effect of competing ions on uptake of Zn in shoots and roots of rice seedlings immersed in $5 \times 10^{-3} \ M \ ^{65}ZnCl_2$. *Giordano et al.,* Plant Soil, *41:637, 1974.*

High P availability can induce Zn deficiency, commonly in soils that are marginally Zn deficient. When plants are Zn deficient, their ability to regulate P accumulation is severely impaired. As a consequence, P is absorbed by roots and transported to plant tops in such excess that it becomes toxic and produces symptoms resembling Zn deficiency, in spite of adequate Zn concentrations in plant tops.

Serious concerns have been expressed about inducing Zn deficiencies, particularly in sensitive crops, as a result of building up available P in soil. The problem can be corrected or avoided by ensuring adequate Zn availability.

The popularly held belief that P-Zn reactions in soil, such as the formation of insoluble $Zn_3(PO_4)_2 \cdot 4H_2O$, are responsible for P-induced Zn deficiency should be discounted. Solubility of this compound is sufficiently high that it will readily provide Zn to plants. Mycorrhizae increase the micronutrient uptake by many plants; however, P fertilization can suppress mycorrhizal uptake of Zn and induce Zn deficiency of crops.

FLOODING   When soils are submerged, the concentration of many nutrients increases, but not with Zn. In acidic soils, Zn deficiency may be attributed to the increase in pH under reducing conditions and subsequent precipitation of franklinite ($ZnFe_2O_4$) or sphalerite (ZnS). Decreasing pH in submerged, calcareous soils would usually increase Zn solubility. However, the higher the soil pH and the poorer the aeration, the greater the Zn deficiency.

CLIMATIC CONDITIONS   Zn deficiencies are generally more pronounced during cool, wet seasons and often disappear in warmer weather. Climatic conditions during early spring that can contribute to Zn deficiency are poor light, low

TABLE 8.4   Sensitivity of Crops to Low Levels of Available Zn

High Sensitivity	Mild Sensitivity	Low Sensitivity
Beans, lima beans, and peas	Alfalfa	Asparagus
Castor beans	Barley	Carrots
Citrus	Clovers	Forage grasses
Corn	Cotton	Mustard and other crucifers
Flax	Potatoes	Oats
Fruit trees (deciduous)	Sorghum	Peas
Grapes	Sugar beets	Peppermint
Hops	Tomatoes	Rye
Onions	Wheat	Safflower
Pecans		
Pine		
Rice		
Soybeans		
Sudangrass		

temperature, and excessive moisture. Increasing soil temperature increases the availability of Zn to crops by increasing solubility and diffusion of Zn^{2+}.

PLANT FACTORS   Species and varieties of plants differ in their susceptibility to Zn deficiency (Table 8.4). Corn and beans are very susceptible to low Zn. Fruit trees in general, and citrus and peach in particular, are also sensitive.

Cultivars differ in their ability to take up Zn, which may be caused by differences in Zn translocation and utilization, differential accumulations of nutrients that interact with Zn, and differences in plant roots to exploit for soil Zn.

Zn Sources

ORGANIC ZN   Most animal wastes contain small quantities of plant available Zn, typically ranging from 0.01 to 0.05% (0.2 to 1 lb/t). With large manure application rates, sufficient plant available Zn can be provided. The primary benefit of organic waste application is increased OM and associated natural chelation properties that increase Zn concentration in soil solution and plant availability. Zn content in municipal waste varies greatly depending on the source, with an average Zn content of 0.5%, or 10 lb/t.

INORGANIC ZN   Zinc sulfate ($ZnSO_4$), containing about 35% Zn, is the most common Zn fertilizer source, although use of synthetic Zn chelates has increased (Table 8.5). Inorganic Zn sources are satisfactory fertilizers because they are very soluble in soils.

Fertilizer Zn rates depend on the crop, Zn source, method of application, and severity of Zn deficiency. Rates usually range from 3 to 10 lb/a with inorganic Zn and from 0.5 to 2.0 lb/a with a chelate or an organic Zn source. For most field and vegetable crops, 10 lb/a is recommended in clay and loam soils and 3 to 5 lb/a in sandy soils. In most cropping situations, applications of 10 lb/a of Zn can be effective for 3 to 5 years.

TABLE 8.5   Some Sources of Fertilizer Zn

Source	Formula	Percentage of Zn
Zinc sulfate monohydrate	$ZnSO_4 \cdot H_2O$	35
Zinc oxide	ZnO	78
Zinc carbonate	$ZnCO_3$	52
Zinc phosphate	$Zn_3(PO_4)_2$	51
Zinc chelates	$Na_2ZnEDTA$	14
Natural organics	—	1–5

SOURCE: Mortvedt et al. (Eds.), *Micronutrients in Agriculture,* p. 371, Soil Sci. Soc. Am. Madison, Wisc. (1972).

Because of limited Zn mobility in soils, broadcast Zn should be thoroughly incorporated into the soil; however, band application may be more effective, especially in fine-textured and very low Zn soils. The efficiency of band-applied Zn can be improved by the presence of acid-forming N fertilizers.

With perennial crops preplant soil applications of Zn are effective at rates between 20 and 100 lb/a. Soil applications are of only limited value after these crops have been established.

Foliar applications are used primarily for tree crops. Sprays containing 10 to 15 lb/a of Zn are usually applied to dormant orchards, whereas 2–3 lb/a can be foliar applied to growing crops. Damage to foliage can be prevented by adding lime to the solution or by using less soluble materials such as ZnO or $ZnCO_3$. Other methods include seed coatings, root dips, and tree injections. The former treatment may not supply enough Zn for small-seeded crops, but dipping potato seed pieces in a 2% ZnO suspension is satisfactory.

Foliar applications of chelates and natural organics are particularly suitable for quick recovery of Zn-deficient seedlings. These Zn sources can be used in high-analysis liquid fertilizers because of their high solubility and compatibility. Chelates such as ZnEDTA are mobile and can be soil applied; however, high cost usually limits their use.

Table 8.6 illustrates the effectiveness of foliar-applied $ZnSO_4$ and ZnEDTA in increasing the Zn content of several crops. In general, Zn chelates are more effective than inorganic Zn at similar rates of application. Foliar-applied Zn is more effective than soil-applied Zn.

TABLE 8.6   Comparison of Different Zn Sources and Methods of Application on Zn Content in Leaf Tissues of Selected Crops*

			Foliar Applied		
		Soil-Applied $ZnSO_4$	$ZnSO_4$		$ZnEDTA$
Crop	Control	(20 kg/ha)	(0.5 kg/ha)	(1.0 kg/ha)	(0.42 kg/ha)
			mg Zn kg^{-1}		
Alfalfa	22	37	39	50	43
Ryegrass	18	28	46	61	63
Wheat	17	21	31	41	51
Barley	21	30	43	43	54

*Zn application rates shown in parentheses.

SOURCE: Gupta, *Can. J. Soil Sci.,* 69:473 (1989).

Copper

Cu Cycle

The relationships and cycling between Cu fractions in soils are very similar to those described for Fe and Zn. Soil solution Cu concentration and plant available Cu are governed predominantly by solution pH and Cu adsorbed on clay and organic surfaces in soils (Fig. 8.18). Primary and secondary minerals dissolve to initially provide Cu to the soil solution, which is then adsorbed onto CEC and incorporated into the microbial biomass and complexed by organic compounds in the soil solution. A significant "pool" of organically complexed Cu which is in equilibrium with soil solution Cu can contribute to Cu^{2+} diffusion to plant roots.

CU IN PLANTS   Cu is absorbed by plants as the cupric ion, Cu^{2+}, and may be absorbed as a component of either natural or synthetic organic complexes. Its normal concentration in plant tissue ranges from 5 to 20 ppm. Deficiencies are probable when Cu levels in plants fall below 4 ppm in the dry matter.

Symptoms of Cu deficiency vary with the crop (see color plates inside book cover). In corn the youngest leaves become yellow and stunted, and as the deficiency becomes more severe, the young leaves pale and the older leaves die back. In advanced stages, dead tissue appears along the tips and edges of the leaves in a pattern similar to that of K deficiency. Cu-deficient small-grain plants lose color

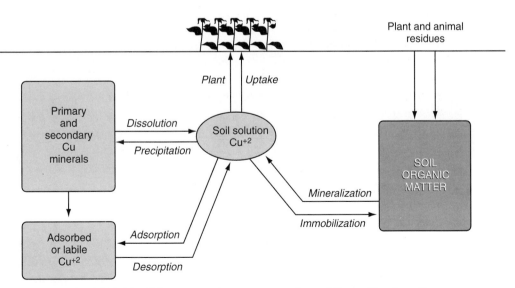

FIGURE 8.18   Diagrammatic representation of Cu cycling in soils.

in the younger leaves, which eventually break, and the tips die. Stem melanosis and take-all root rot diseases occur in certain wheat varieties when Cu is deficient. Also, ergot infection is associated with Cu deficiency in some wheat and barley varieties. In many vegetable crops the leaves lack turgor. They develop a bluish-green cast, become chlorotic, and curl, and flower production fails to take place.

Cu in its reduced form readily binds and reduces O_2. In the oxidized form the metal is readily reduced, and protein-complexed Cu has a high redox potential. These properties of Cu are exploited by enzymes that create complex polymers such as lignin and melanin. Cu is unique in its involvement in enzymes, and it cannot be replaced by any other metal ion.

Toxicity symptoms include reduced shoot vigor, poorly developed and discolored root systems, and leaf chlorosis. The chlorotic condition in shoots superficially resembles Fe deficiency. Toxicities are uncommon, occurring in limited areas of high Cu availability; after additions of high-Cu materials such as sewage sludge, municipal composts, pig and poultry manures, and mine wastes; and from repeated use of Cu-containing pesticides.

CU IN SOIL

Mineral Cu. Cu concentration in the earth's crust averages about 55 to 70 ppm. Igneous rocks contain 10 to 100 ppm Cu, while sedimentary rocks contain between 4 and 45 ppm Cu. Cu concentration in soils ranges from 1 to 40 ppm and averages about 9 ppm. Total soil Cu may be 1 or 2 ppm in deficient soils.

Malachite ($Cu_2(OH)_2CO_3$) and cupric ferrite ($CuFe_2O_4$) are the important Cu-containing primary minerals (Fig. 8.19). Secondary Cu minerals include ox-

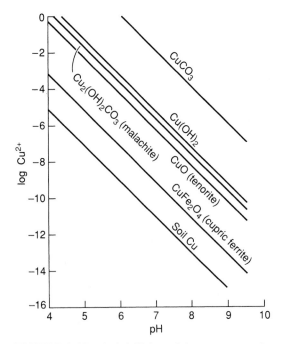

FIGURE 8.19   Solubilities of the common Cu minerals in soils.   *Lindsay, Chemical Equilibria in Soils, John Wiley & Sons, New York, 1979.*

ides, carbonates, silicates, sulfates, and chlorides, but most of them are too soluble to persist. The line indicated by "soil Cu" represents the solubility of Cu in most soils and is very close to that of $CuFe_2O_4$ (Fig. 8.19).

Soil Solution Cu. The Cu concentration in soil solution is usually very low, ranging between 10^{-8} and 10^{-6} M (Fig. 8.20). The dominant solution species are Cu^{2+} at pH < 7 and $Cu(OH)_2^0$ at pH > 7. Hydrolysis reactions of Cu ions are shown in the following equations:

$$Cu^{2+} + H_2O \leftrightarrows CuOH^+ + H^+$$

$$CuOH^+ + H_2O \leftrightarrows Cu(OH)_2^0 + H^+$$

Solubility of Cu^{2+} is pH dependent, and it increases 100-fold for each unit decrease in pH (Fig. 8.20).

Cu is supplied to plant roots by diffusion of organically bound, chelated Cu, similar to chelated Fe diffusion in soil (Fig. 8.7). Organic compounds in the soil solution are capable of chelating solution Cu^{2+}, which increases the solution Cu^{2+} concentration above that predicted by Cu mineral solubility.

Adsorbed Cu. Cu^{2+} is specifically or chemically adsorbed by layer silicate clays; OM; and Fe, Al, or Mn oxides. With the exception of Pb^{2+}, Cu^{2+} is the most strongly adsorbed of all of the divalent metals on Fe/Al oxides. The mechanism of adsorption by oxides, unlike the electrostatic attraction of Cu^{2+} on the

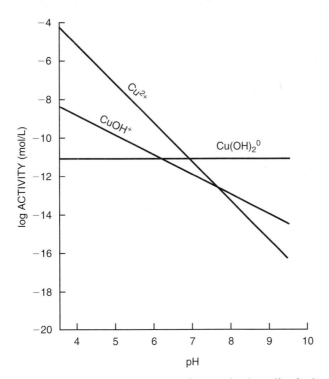

FIGURE 8.20　The common Cu species in soil solution as influenced by pH. *Lindsay,* Chemical Equilibria in Soils, *John Wiley & Sons, New York, 1979.*

CEC of clay particles, involves formation of Cu-O-Al or Cu-O-Fe surface bonds (Fig. 8.21). This chemisorption process is thus controlled by the quantity of surface OH^- groups.

Cu adsorption increases with increasing pH due to (1) increased pH-dependent sites on clay and OM, (2) reduced competition with H^+, and (3) a change in the hydrolysis state of Cu in solution. As the pH is raised, hydrolysis of Cu^{2+} adsorbed on the CEC decreases exchangeable Cu^{2+} and increases chemisorbed Cu (i.e., decreasing H^+ shifts equilibrium to the right in Fig. 8.21).

Occluded and Coprecipitated Cu. A significant fraction of soil Cu is occluded or buried in various mineral structures, such as clay minerals and Fe, Al, and Mn oxides. Cu is capable of isomorphous substitution in octahedral positions of crystalline silicate clays. It is present as an impurity within $CaCO_3$ and $MgCO_3$ minerals in arid soils and within $Al(OH)_3$ and $Fe(OH)_3$ in acidic soils.

Organic Cu. Most of the soluble Cu in surface soils is organically complexed and is more strongly bound to OM than any other micronutrient. The Cu^{2+} ion is directly bonded to two or more organic functional groups, chiefly carboxyl or phenol (Fig. 8.22). Humic and fulvic acids contain multiple binding sites, primarily carboxyl groups, for Cu. In most mineral soils, OM is intimately associated with clay, probably as a clay-metal-organic complex (Fig. 8.23).

At soil OM levels up to 8%, both organic and mineral surfaces are involved in Cu adsorption, while at higher concentrations of OM, binding of Cu takes place

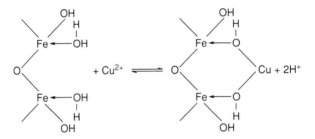

FIGURE 8.21   Chemisorption of Cu^{2+} with surface hydroxyls on $Fe(OH)_3$.

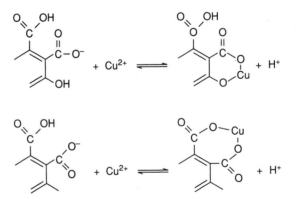

FIGURE 8.22   Mechanism of Cu complexed by organic matter.   *Stevenson and Ardakani, p. 90.* Micronutrients in Agriculture, *American Soc. Agron. Madison, Wisc., 1972.*

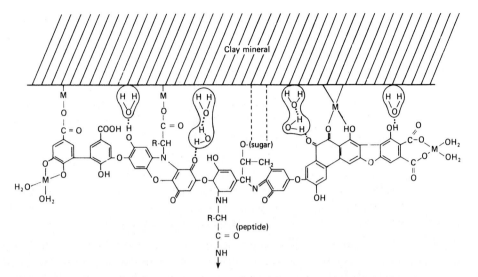

FIGURE 8.23 Schematic diagram of the clay-OM-metal (M) complex. *Stevenson and Fitch, in J. F. Loneragan (Ed.), Copper in Soils and Plants, p. 70. Academic Press, New York, 1981.*

mostly on organic surfaces. For soils having similar clay and OM contents, the contribution of OM to the complexing of Cu is highest when the predominant clay mineral is kaolinite and lowest with montmorillonite.

Factors Affecting Cu Availability

TEXTURE   Cu content in the soil solution is usually lower in excessively leached podzolic sands and calcareous sands than in other soil types.

SOIL pH   The concentration of soil solution Cu decreases with increasing pH, and its supply to plants is reduced because of decreased solubility and increased adsorption.

INTERACTIONS WITH OTHER NUTRIENTS   There are numerous interactions involving Cu. Applications of N-P-K fertilizers can induce Cu deficiencies. Furthermore, increased growth resulting from the application of N or other nutrients may be proportionally greater than Cu uptake, which dilutes Cu concentration in plants. Increasing the N supply to crops can reduce mobility of Cu in plants, since large amounts of N in plants impede translocation of Cu from older leaves to new growth. High concentrations of Zn, Fe, and P in soil solution can also depress Cu absorption by plant roots and may intensify Cu deficiency.

PLANT FACTORS   Crops vary greatly in response to Cu (Table 8.7). Among small-grain species, rye has exceptional tolerance to low levels of soil Cu and will be healthy, whereas wheat fails completely without the application of Cu. Rye can extract up to twice as much Cu as wheat under the same conditions. The usual order of sensitivity of the small grains to Cu deficiency in the field is wheat > barley >

TABLE 8.7   Sensitivity of Crops to Low Levels of Available Cu

High Sensitivity		Mild Sensitivity		Low Sensitivity	
Wheat	Carrots	Barley	Rye		Rapeseed
Flax	Lettuce	Oats	Canola		Potatoes
Canary Seed	Beets	Corn	Beans		Peas
Rice	Spinach	Timothy	Soybeans		Lupine
Alfalfa	Citrus	Clover	Forage grasses		
Sudangrass	Onion				

oats > rye. Varietal differences in tolerance to low Cu are important, and sometimes they can be as large as those among crop species.

The genotypic differences in the Cu nutrition of plants are related to (1) differences in the rate of Cu absorption by roots, (2) better exploration of soil through greater root length per plant or per unit area, (3) better contact with soil through longer root hairs, (4) modification of Cu availability in soil adjacent to roots by root exudation, (5) acidification or change in redox potential, (6) more efficient transport of Cu from roots to shoots, and/or (7) lower tissue requirement for Cu.

Severe Cu deficiency in crops planted in soils with high C/N residues is related to (1) reactions of Cu with organic compounds originating from decomposing straw, (2) competition for available Cu by stimulated microbial populations, and (3) inhibition of root development and the ability to absorb Cu. If the soil-available Cu is low, manure added to a field may accentuate the problems. OM from manure, straw, or hay can tie up Cu, making it unavailable to plants.

Cu Sources

ORGANIC CU   Although most animal wastes contain small quantities of plant available Cu (0.002 to 0.03% (0.04 to 0.6 lb/t), elevated Cu levels occur in swine manure because of high Cu content in the feed. As a result, continued application of swine manure can build up soil test Cu levels to toxicity in sensitive crops such as peanuts. Thus, with most manures, average application rates provide sufficient plant available Cu. As with Fe and Zn, the primary benefit of organic waste application is increased OM and associated natural chelation properties that increase Cu concentration in soil solution and plant availability. Cu content in municipal waste varies greatly depending on the source; its average Cu content is 0.1%, or 2 lb/t.

INORGANIC CU   The usual Cu source is $CuSO_4 \cdot 5H_2O$, although CuO, mixtures of $CuSO_4$ and $Cu(OH)_2$, and Cu chelates are also used (Table 8.8). $CuSO_4$ contains 25% Cu, is soluble in water, and is compatible with most fertilizer materials. Copper ammonium phosphate can be either soil or foliar applied. It is only slightly soluble in water but can be suspended and sprayed onto the plants. It contains 30% Cu and, like the other metal ammonium phosphates, is slowly available.

TABLE 8.8   Cu Compounds Used as Fertilizers

Source	Formula	Percentage of Cu
Copper sulfate	$CuSO_4 \cdot 5H_2O$	25
Copper sulfate monohydrate	$CuSO_4 \cdot H_2O$	35
Copper acetate	$Cu(C_2H_3O_2)_2 \cdot H_2O$	32
Copper ammonium phosphate	$Cu(NH_4)PO_4 \cdot H_2O$	32
Copper chelates	$Na_2Cu\ EDTA$	13
Organics	—	<0.5

Soil and foliar applications are both effective, but soil applications are more common, with Cu rates of 1 to 20 lb/a needed to correct deficiencies. Effectiveness is increased by thoroughly mixing Cu fertilizers into the root zone or by banding them in the seed row. Band-applied Cu is not recommended because of possible root injury. Additions of Cu can be ineffective when root activity is restricted by excessively wet or dry soil, root pathogens, toxicities, and deficiencies of other nutrients. Residual Cu fertilizer availability from as little as several pounds per acre can persist for 2 or more years, depending on the soil, crop, and Cu rate.

Application of Cu in foliar sprays is confined mainly to emergency treatment of deficiencies identified after planting. In some areas, however, Cu is included in regular foliar spraying programs. Cu chelates (CuEDTA) can be used as a foliar Cu fertilizer; however, soil application is more effective (Table 8.9).

These data also show that fall-applied $CuSO_4$ is more effective than spring applications to barley, likely because the material has more time to dissolve and move into the root zone.

Manganese

Mn Cycle

Mn exists as solution Mn^{2+}, exchangeable Mn^{2+}, organically bound Mn, and various Mn minerals. The equilibrium among these forms determines Mn availability to plants (Fig. 8.24).

TABLE 8.9   Effect of Various Cu Fertilizer Products on the Yield of Barley

Copper Product	Time and Method of Application	Rate of Application (kg/ha)	Average Yield of Barley	
			kg/ha	bu/a
$CuSO_4$	Fall ⎫	10	3,070	57
	Spring ⎬ broadcast/ incorporate	10	2,370	44
	Spring ⎭	5	1,890	35
CuEDTA	Spring ⎫	2	2,960	55
	Spring ⎬ broadcast/ incorporate	1	2,700	50
	Foliar ⎭	1	2,100	39

SOURCE: Manitoba Agric. Agdex No. 541 MG#1853, (1990).

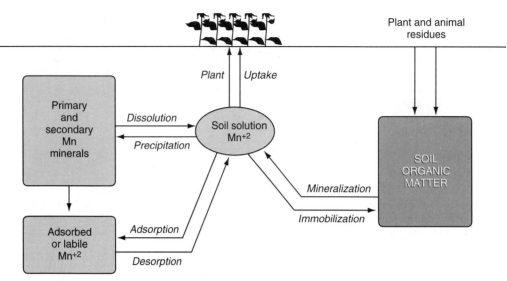

FIGURE 8.24   Diagrammatic representation of Mn cycling in soils.

The major processes in this cycle are Mn oxidation-reduction and complexing solution Mn^{2+} with natural organic chelates. The continuous cycling of OM significantly contributes to soluble Mn. Factors influencing the solubility of soil Mn include pH, redox, and complexation. Soil moisture, aeration, and microbial activity influence redox, while complexation is affected by OM and microbial activity.

MN IN PLANTS   Mn is absorbed by plants as Mn^{2+}, as well as in molecular combinations with certain natural and synthetic complexing agents. Mn concentration in plants typically ranges from 20 to 500 ppm. Concentrations of Mn in upper plant parts below 15 to 20 ppm are considered deficient.

The involvement of Mn in photosynthesis, particularly in the evolution of O_2, is well known. It also takes part in oxidation-reduction processes and in decarboxylation and hydrolysis reactions. Mn can substitute for Mg^{2+} in many of the phosphorylating and group-transfer reactions.

Although it is not specifically required, Mn is needed for maximal activity of many enzyme reactions in the citric acid cycle. In the majority of enzyme systems, Mg is as effective as Mn in promoting enzyme transformations. Mn influences auxin levels in plants, and it seems that high concentrations of this micronutrient favor the breakdown of indoleacetic acid. For satisfactory Mn nutrition of crops, soil solution and exchangeable Mn should be 2 to 3 ppm and 0.2 to 5 ppm, respectively.

Like Fe, Mn is a relatively immobile element, and deficiency symptoms usually show up first in the younger leaves. In broad-leaved plants the visual symptoms appear as an interveinal chlorosis. Leaves from Mn-deficient soybean plants are shown

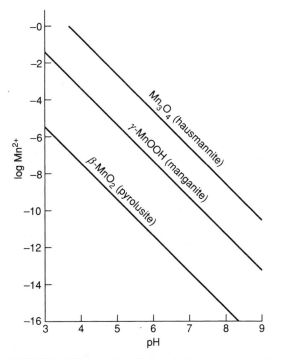

FIGURE 8.25   Solubilities of the common Mn minerals in well-aerated soils. Reducing conditions will increase Mn solubility.   *Lindsay.* Chemical Equilibria in Soils, *John Wiley & Sons, New York, 1979.*

in the color plates (see inside book cover). Mn deficiency of several crops has been described by such terms as *gray speck of oats, marsh spot of peas,* and *speckled yellows of sugar beets.* Wheat plants low in Mn are often more susceptible to root-rot diseases.

Plants are injured by excessive amounts of Mn. Crinkle leaf of cotton is Mn toxicity observed in highly acidic red and yellow soils of the southern United States. Mn toxicity can occur in crops growing on extremely acidic soils. Liming will readily correct this problem.

MN IN SOIL

Mineral Mn. Mn concentration in the earth's crust averages 1,000 ppm, and Mn is found in most Fe-Mg rocks. Mn, when released through weathering of primary rocks, will combine with O_2 to form secondary minerals, including pyrolusite (MnO_2), hausmannite (Mn_3O_4), and manganite ($MnOOH$). Pyrolusite and manganite are the most abundant (Fig. 8.25).

Total Mn in soils generally ranges between 20 and 3,000 ppm and averages about 600 ppm. Mn in soils occurs as various oxides and hydroxides coated on soil particles, deposited in cracks and veins, and mixed with Fe oxides and other soil constituents.

Soil Solution Mn. The principal species in solution is Mn^{2+}, which decreases 100-fold for each unit increase in pH, similar to the behavior of other divalent metal cations (Fig. 8.26). The concentration of Mn^{2+} in solution is predominantly controlled by MnO_2 (Fig. 8.25). Concentration of Mn^{2+} in the soil solution of acidic and neutral soils is commonly in the range of 0.01 to 1 ppm, with

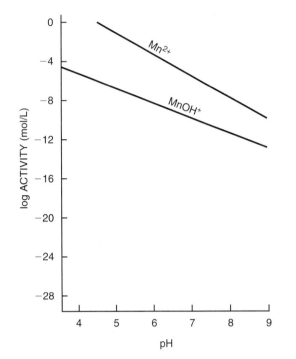

FIGURE 8.26   The common Mn species in soil solution as influenced by pH. *Lindsay.* Chemical Equilibria in Soils, *John Wiley & Sons, New York, 1979.*

organically complexed Mn^{2+} comprising about 90% of solution Mn^{2+}. Plants take up Mn^{2+}, which moves to the root surface by diffusion of principally chelated Mn^{2+}, as described for Fe (Fig. 8.7).

Mn in the soil solution is greatly increased under acidic, low-redox conditions. In extremely acidic soils, Mn^{2+} solubility can be sufficiently great to cause toxicity problems in sensitive plant species (Fig. 8.26).

Because of its mobility, Mn^{2+} can leach from soils, particularly from acidic podzols. The frequent occurrence of Mn deficiency in poorly drained mineral and organic soils is often attributed to low Mn levels resulting from leaching of soluble Mn^{2+}.

Factors Affecting Mn Availability

SOIL PH   Since Mn^{2+} in solution varies with pH, management practices that change soil pH influence Mn^{2+} availability and uptake. Liming very acidic soils decreases solution and exchangeable Mn^{2+}, due to Mn^{2+} precipitation as MnO_2 (Fig. 8.25). On the other hand, low Mn availability in high-pH and calcareous soils and in overlimed, poorly buffered, coarse-textured soils can be overcome by acidification through the use of acid-forming N or S materials. High pH also favors the formation of less available organic complexes of Mn. Activity of the soil microorganisms that oxidize soluble Mn to unavailable forms reaches a maximum near pH 7.

EXCESSIVE WATER AND POOR AERATION   Soil waterlogging will reduce O_2 and lower redox potential, which increases soluble Mn^{2+}, especially in acidic soils. Mn availability can be increased by poor aeration in compact soils and by local accumulations of CO_2 around roots and other soil microsites. The resulting low-redox conditions will render Mn more available without appreciably affecting the redox potential or pH of the bulk soil.

ORGANIC MATTER   Availability of Mn^{2+} can be strongly influenced by reactions with OM. The low availability of Mn in high-OM soils is attributed to the formation of unavailable chelated Mn^{2+} compounds. It may also be held in unavailable organic complexes in peats or muck soils. Additions of natural organic materials such as peat moss, compost, and wheat and clover straw have increased the solution and exchangeable Mn.

INTERACTION WITH OTHER NUTRIENTS   High levels of Cu, Fe, or Zn can reduce Mn uptake by plants. Addition of acid-forming NH_4^+ to soil will enhance Mn uptake. Neutral KCl, NaCl, and $CaCl_2$ applied to acidic soils also can increase the Mn availability to and concentration in plants. The relative order of the salt effect on increasing available Mn is $KCl > KNO_3 > K_2SO_4$. The effect of KCl on Mn uptake can be so strong that it produces toxicity symptoms in sensitive crops.

CLIMATIC EFFECTS   Wet weather favors the presence of Mn^{2+}, whereas warm, dry conditions encourage the formation of less available oxidized forms of Mn. Dry weather either induces or aggravates Mn deficiency, particularly in fruit trees. Wet weather is one of the conditions usually associated with a high incidence of gray speck of oats, a Mn deficiency disorder. Increasing soil temperature during the growing season improves Mn uptake, presumably because of greater plant growth and root activity.

PLANT FACTORS   Several plant species exhibit differences in sensitivity to Mn deficiency (Table 8.10). These differences in the response of Mn-efficient and Mn-inefficient plants are due to internal factors rather than to the effects of the plants on the soil. Reductive capacity at the root may be the factor restricting Mn uptake and translocation. There may also be significant differences in the amounts and properties of root exudates generated by plants, which can influence Mn^{2+} availability. It is possible that plant characteristics possessed by Fe-efficient plants may similarly influence Mn uptake in plants tolerant of Mn stress.

TABLE 8.10   Sensitivity of Crops to Low Levels of Available Mn in Soil*.

High Sensitivity		Moderate Sensitivity		Low Sensitivity	
Alfalfa	Soybeans	Barley	Potatoes	Barley	Rye
Citrus	Sugar beets	Corn	Rice	Corn	Soybeans
Fruit trees	Wheat	Cotton	Rye	Cotton	Vegetables
Oats		Field beans	Soybeans	Field beans	Wheat
Onions		Fruit trees	Vegetables	Fruit trees	
Potatoes		Oats	Wheat	Rice	

*Some crops are listed under two or three categories because of variation in soil, growing conditions, and differential response of varieties of a given crop.

TABLE 8.11   Sources of Mn Used for Fertilizer

Source	Formula	% Mn
Manganese sulfate	$MnSO_4 \cdot 4H_2O$	26–28
Manganous oxide	MnO	41–68
Manganese chloride	$MnCl_2$	17
Natural organic	—	<0.2
Manganese chelates	MnEDTA	5–12

Mn Sources

ORGANIC MN   The Mn concentration in most animal wastes is similar to Zn, ranging between 0.01 and 0.05% (0.2 and 1 lb/t). Thus, with most manures, average application rates will provide sufficient plant available Mn. As with Fe, Zn, and Cu, the primary benefit of organic waste application is increased OM and associated natural chelation properties that increase Mn concentration in soil solution and plant availability. As with the other micronutrients, Mn content in municipal waste varies greatly depending on the source. On the average, Mn content is about half the Cu content (0.05%, or 1 lb/t).

INORGANIC MN   Manganese sulfate ($MnSO_4$) is widely used for correction of Mn deficiency and may be soil or foliar applied (Table 8.11). In addition to inorganic Mn fertilizers, natural organic complexes and chelated Mn are available and are usually foliar applied.

Manganese oxide is only slightly water soluble, but it is usually a satisfactory source of Mn. MnO must be finely ground to be effective. Rates of Mn application range from 1 to 25 lb/a; higher rates are recommended for broadcast application, while lower rates are foliar applied. Band-applied Mn is generally more effective than broadcast Mn, and band treatments are usually about one-half the broadcast rates. Oxidation to less available forms of Mn is apparently delayed with band-applied Mn. Applications at the higher rates may be required on organic soils. Band application of Mn in combination with N-P-K fertilizers is commonly practiced.

Broadcast application of Mn chelates and natural organic complexes is not normally advised because soil Ca or Fe can replace Mn in these chelates, and the freed Mn is usually converted to unavailable forms. Meanwhile, the more available chelated Ca or Fe probably accentuates the Mn deficiency. Lime or high-pH-induced Mn deficiencies can be rectified by acidification resulting from the use of S or other acid-forming materials.

Boron

B Cycle

B exists in four major forms in soil: in rocks and minerals, adsorbed on clay surfaces and Fe/Al oxides, combined with OM, and as boric acid ($H_3BO_3^0$) and

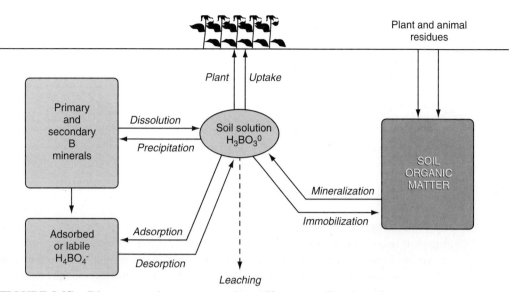

FIGURE 8.27   Diagrammatic representation of boron cycling in soils.

$B(OH)_4^-$ in the soil solution (Fig. 8.27). Understanding B cycling between the solid and solution phases is very important because of the narrow range in solution concentration that separates deficiency and toxicity in crops.

B IN PLANTS   Most of the B absorbed by plants is undissociated boric acid (H_3BO_3). Much smaller amounts of other forms, such as $B_4O_7^{2-}$, $H_2BO_3^-$, HBO_3^{2-}, and BO_3^{3-}, may be present, but they generally do not contribute significantly to plant needs. Although it is required for higher plants and some algae and diatoms, B is not needed by animals, fungi, or microorganisms.

Plants require B for a number of growth processes:

- New cell development in meristematic tissue.
- Proper pollination and fruit or seed set.
- Translocation of sugars, starches, N, and P.
- Synthesis of amino acids and proteins.
- Nodule formation in legumes.
- Regulation of carbohydrate metabolism.

B plays an essential role in the development and growth of new cells in the plant meristem. Since it is not readily translocated from older to actively growing tissues, the first visual deficiency symptom is cessation of terminal bud growth, followed by death of the young leaves. In B-deficient plants the youngest leaves become pale green, losing more color at the base than at the tip. The basal tissues break down and, if growth continues, the leaves have a one-sided or twisted appearance. Flowering and fruit development are also restricted by a shortage of B. Sterility and severely impaired seed set are late-season symptoms of B deficiency in both B-sensitive (rapeseed and clover) and B-insensitive (wheat) crops.

B-deficiency symptoms often appear in the form of thickened, wilted, or curled leaves; a thickened, cracked, or water-soaked condition of petioles and stems; and a discoloration, cracking, or rotting of fruit, tubers, or roots (see color plates inside book cover). Internal cork of apple is caused by a deficiency of this element, and a lack of B in citrus fruits results in uneven thickness of the peel, lumpy fruit, and gummy deposits in the fruit. The breakdown of internal tissues in root crops gives rise to darkened areas referred to as *brown heart* or *black heart*. Some conifers exhibit striking B deficiency symptoms including distorted branches and main stems, resin bleeding, and death of major branches. Under open growing conditions the growth form will be bushy or shrubby.

The B concentration in monocotyledons and dicotyledons generally varies between 6 and 18 ppm and 20 and 60 ppm, respectively. Levels of B in mature leaf tissue of most crops are usually adequate if over 20 ppm. B toxicity to plants is uncommon in most arable soils unless it has been added in excessive amounts in fertilizers. In arid regions, however, B toxicity may occur naturally or may develop because of a high B content in irrigation waters.

B IN SOIL

Mineral B. B is the only nonmetal among the micronutrients. It occurs in low concentrations in the earth's crust and in most igneous rocks (~10 ppm). Among sedimentary rocks, shales have the highest B concentrations (up to 100 ppm), present mainly in the clay minerals. The total concentration of B in soils varies between 2 and 200 ppm and frequently ranges from 7 to 80 ppm. Less than 5% of the total soil B is available to plants.

Tourmaline, a borosilicate, is the main B-containing mineral found in soils. It is insoluble and resistant to weathering; consequently, release of B is quite slow. An increasing frequency of B deficiencies suggests that it is incapable of supplying plant requirements under prolonged heavy cropping. B in soils of arid climates is usually sufficient because of greater mineral stability in a low weathering environment. B can substitute for Al^{3+} and/or Si^{4+} ions in silicate minerals. Following its adsorption on clay surfaces, B will slowly diffuse into interlayer positions.

Soil Solution B. Undissociated $H_3BO_3^0$ is the predominant species expected in soil solution at pH values ranging from 5 to 9. At pH > 9.2 $H_2BO_3^-$ can hydrolyze to $H_4BO_4^-$. B can be transported from the soil solution to absorbing plant roots by both mass flow and diffusion.

Adsorbed B. B adsorption and desorption can buffer solution B, which helps to reduce B leaching losses. It is a major form of B in alkaline, high-B soils. The main B adsorption sites are (1) broken Si—O and Al—O bonds at the edges of clay minerals, (2) amorphous hydroxide structures, and (3) Fe/Al oxy and hydroxy compounds. Increasing pH, clay content, and OM and the presence of Al compounds favor $H_4BO_4^-$ adsorption. B-adsorption capacities generally follow the order mica > montmorillonite > kaolinite.

Organically Complexed B. OM represents a large potential source of plant available B in soils, which increases with increasing OM. The B-OM complexes are probably

$$=C-O \quad \backslash B-OH \quad \text{or} \quad H^+ \left| \begin{array}{c} =C-O \quad \diagdown \quad O-C= \\ =C-O \diagup \quad B \diagup \quad \diagdown O-C= \end{array} \right|^-$$

Factors Affecting B Availability

SOIL PH   B normally becomes less available to plants with increasing soil pH, decreasing dramatically above pH 6.3 to 6.5 (Table 8.12). Liming strongly acidic soils frequently causes a temporary B deficiency in susceptible plants. The severity of the deficiency depends on the moisture status of the soil, the crop, and the time elapsed following liming.

The reduction in B availability following liming is caused mainly by B adsorption on freshly precipitated $Al(OH)_3$, with maximum adsorption at pH 7. Alternatively, moderate liming can be used to depress B availability and plant uptake on soils high in B. It should be noted that heavy liming does not always lead to greater B adsorption and reduced plant uptake. Higher pH resulting from liming of soils high in OM may encourage OM decomposition and release of B.

ORGANIC MATTER   The greater availability of B in surface soils compared with subsurface soils is related to the greater quantities of OM in surface soil. Applications of OM to soils can increase the B concentration in plants and even cause phytotoxicity.

SOIL TEXTURE   Coarse-textured, well-drained, sandy soils are low in B, and crops with a high requirement, such as alfalfa, respond to B applications of 3 or more lb/a. Sandy soils with fine-textured subsoils generally do not respond to B the same way as those with coarse-textured subsoils. B added to soils remains soluble, and up to 85% can be leached in low-OM, sandy soils. Finer-textured soils retain added B for longer periods than coarse-textured soils because of greater B adsorption in clays. The fact that clays retain B more than sands does not imply that plants will absorb B in greater quantities from clays than from sands. Plants can take up much larger quantities of B from sandy soils than from fine-textured soils at equal concentrations of water-soluble B.

TABLE 8.12   B Uptake and Percentage Recovery of Added B by Five Harvests of Tall Fescue at Five Soil pH Levels

| | Amount of Added B (mg/pot) | | | | | | |
| | 0 | 4.5 | 8.9 | 17.8 | 4.5 | 8.9 | 17.8 |
Soil pH	B Uptake (mg/pot)				Percentage Recovery		
4.7	0.47	1.85	4.15	9.40	30.7	41.3	50.2
5.3	0.45	1.92	4.45	9.51	32.7	44.9	50.9
5.8	0.44	1.98	4.14	9.10	34.2	41.6	48.7
6.3	0.45	1.98	4.03	9.37	34.0	40.2	50.1
7.4	0.22	0.80	1.40	3.76	12.9	13.3	19.9

SOURCE: Peterson and Newman, *Soil Sci. Soc. Am. J.*, 40:280 (1976).

INTERACTIONS WITH OTHER ELEMENTS   Low tolerance to B occurs when plants have a low Ca supply. When Ca availability is high, there is a greater requirement for B. The occurrence of Ca^{2+} in alkaline and recently overlimed soils will restrict B availability; thus, high solution Ca^{2+} can protect crops from excess B. The Ca/B ratio in leaf tissues has been used to assess the B status of crops. B deficiency is indicated by ratios > 1,200:1 for most crops.

At low levels of B, increased rates of K may accentuate B-deficiency symptoms. B deficiency in alfalfa and other sensitive crops can be aggravated by K fertilization to the extent that B addition is needed to prevent yield loss. The effect of K may be related to its influence on Ca absorption.

SOIL MOISTURE   B deficiency is often associated with dry weather and low soil moisture conditions. This behavior is related to restricted release of B from OM and to reduced B uptake due to lack of moisture in the root zone. Although B levels in soil may be high, low soil moisture impairs B diffusion and mass flow to absorbing root surfaces.

PLANT FACTORS   Because of the narrow range between sufficient and toxic levels of available soil B, the sensitivity of crops to excess B is important (Table 8.13). Genetic variability contributes to differences in B uptake. Investigations with tomatoes revealed that susceptibility to B deficiency is controlled by a single recessive gene. Tomato variety T3238 is B inefficient, while the variety Rutgers is B efficient. Corn hybrids exhibit similar genetic variability related to B uptake.

B Sources

ORGANIC B   Only small quantities of B occur in animal wastes, ranging between 0.001 and 0.005% (0.02 and 0.1 lb/t). Thus, with most manures, average application rates will provide sufficient plant available B. Similar to the other micronutrients, increasing OM content and associated chelation properties with waste application will increase B concentration in soil solution and plant availability. B content in municipal waste is usually very low, with an average B content of 0.01%, or 0.2 lb/t.

TABLE 8.13   Relative Sensitivity of Selected Crops to B Deficiency

High Sensitivity		Moderate Sensitivity		Low Sensitivity	
Alfalfa	Peanut	Apple	Cotton	Asparagus	Pea
Cauliflower	Sugar beet	Broccoli	Lettuce	Barley	Peppermint
Celery	Table beet	Cabbage	Parsnip	Bean	Potato
Rapeseed	Turnip	Carrot	Radish	Blueberry	Rye
Conifers		Clovers	Spinach	Cucumber	Sorghum
Canola			Tomato	Corn	Spearmint
				Grasses	Soybean
				Oat	Sudangrass
				Onion	Sweet corn
					Wheat

SOURCE: Robertson et al., *Mich. Coop. Ext. Bull.* E-1037 (1976).

TABLE 8.14   Principal B Fertilizers and Their Formulas and B Percentages

Source	Formula	% of B
Borax	$Na_2B_4O_7 \cdot 10H_2O$	11
Boric acid	H_3BO_3	17
Colemanite	$Ca_2B_6O_{11} \cdot 5H_2O$	10–16
Sodium pentaborate	$Na_2B_{10}O_{16} \cdot 10H_2O$	18
Sodium tetraborate	$Na_2B_4O_7 \cdot 5H_2O$	14–15
Solubor	$Na_2B_4O_7 \cdot 5H_2O + Na_2B_{10}O_{16} \cdot 10H_2O$	20–21

INORGANIC B   B is one of the most widely applied micronutrients. Sodium tetraborate, $Na_2B_4O_7 \cdot 5H_2O$, is the most commonly used B source and contains about 15% B (Table 8.14). Solubor is a highly concentrated, completely soluble source of B that can be applied as a spray or dust directly to foliage. It is also used in liquid and suspension fertilizer formulations. Solubor is preferred to borax because it dissolves more readily. The Ca borate mineral, colemanite, is often used on sandy soils because it is less soluble and less subject to leaching than the sodium borates.

The most common methods of B application are broadcast, banded, or applied as a foliar spray or dust. In the first two methods, the B fertilizer source is usually mixed with N-P-K-S products and applied to soil. B salts can also be coated on dry fertilizer materials.

B fertilizers should be applied uniformly to soil because of the narrow range between deficiency and toxicity. Segregation of granular B sources in dry fertilizer blends must be avoided. Application of B with fluid fertilizers eliminates the segregation problem.

Foliar application of B is practiced for perennial tree-fruit crops, often in combination with pesticides other than those formulated in oils and emulsions. B may also be included in sprays of chelate, Mg, Mn, and urea. Foliar applications of B with insecticides are also used in cotton. B may be used with herbicides for peanuts.

Rates of B fertilization depend on plant species, soil cultural practices, rainfall, liming, soil OM, and other factors. Application rates of 0.5 to 3 lb/a are generally recommended. The amount of B recommended depends on the method of application. For example, the B rate for vegetable crops is 0.4 to 2.7 lb/a broadcast, 0.4 to 0.9 lb/a banded, and 0.09 to 0.4 lb/a foliar applied.

Chloride

Cl Cycle

Nearly all of the chloride (Cl^-) in soils exists in the soil solution (Fig. 8.28). The mineral and organic fractions contain only small quantities of Cl^-. Cl^- adsorption to clay or oxide surfaces is negligible. Because of its high solubility and mobility in soils, appreciable Cl^- leaching can occur under conditions of high water transport through soil.

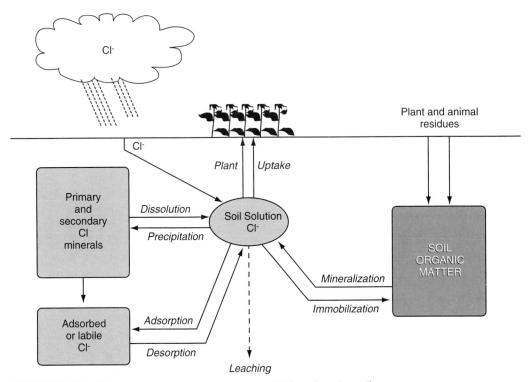

FIGURE 8.28   Diagrammatic representation of Cl cycling in soils.

CL IN PLANTS   Cl is absorbed by plants as the Cl^- ion through both roots and leaves. Cl can be readily transported in plant tissues. Its normal concentration in plants is about 0.2 to 2.0%, although levels as high as 10% are not uncommon. All of these values are much greater than the physiological requirement of most plants. Concentrations of 0.5 to 2.0% in the tissues of sensitive crops can lower yield and quality. Similar reductions in yield and quality can occur when Cl^- levels approach 4% in tolerant crops (Table 8.15).

Chloride has not been found in any true metabolite in higher plants. The essential role of Cl^- seems to lie in its biochemical inertness. This inertness enables it to fill osmotic and cation neutralization roles, which may have important bio-

TABLE 8.15   Sensitivity of Crops to Low Levels of Available Cl

High Sensitivity	Mild Sensitivity	Low Sensitivity
Avocado	Potatoes	Sugar beets
Peach	Wheat	Barley
Legumes	Oats	Corn
Tobacco	Soybean	Spinach
Lettuce	Cotton	Tomatoes

chemical and/or biophysical consequences. Cl^- is the counterion during rapid K fluxes, thus contributing to leaf turgor.

Observations that loss of leaf turgor is a symptom of Cl^- deficiency support the concept that Cl^- is an active osmotic agent. A high level of Cl^- nutrition will increase total leaf water potential and cell sap osmotic potential. Some of the favorable action of Cl^- fertilization is attributed to improved moisture relations.

Cl appears to have a definite role in the evolution of O_2 in photosystem II in photosynthesis. Concentrations of nearly 11% Cl^- have been detected in chloroplasts.

Chlorosis in younger leaves and an overall wilting of the plants are the two most common symptoms of Cl^- deficiency (see color plates inside book cover). Necrosis in some plant parts, leaf bronzing, and reduction in root growth may also be seen in Cl^--deficient plants. Tissue concentrations below 70 to 700 ppm are usually indicative of deficiency.

Excesses of Cl^- can be harmful, and crops vary widely in their tolerance to this condition (Table 8.15). Leaves become thickened and tend to roll with excessive amounts of Cl^-. The storage quality of tubers is adversely affected by excessive Cl^-. The principal effect of too much Cl^- is to increase the osmotic pressure of soil water and thereby lower the availability of water to plants.

CL IN SOIL

Mineral Cl. Cl concentration is 0.02 to 0.05% in the earth's crust, and it occurs primarily in igneous and metamorphic rocks. Most of the soil Cl^- commonly exists as soluble salts such as $NaCl$, $CaCl_2$, and $MgCl_2$. Cl^- is sometimes the principal anion in extracts of saline soils. The quantity of Cl^- in soil solutions may range from 0.5 ppm or less to more than 6,000 ppm.

The majority of Cl^- in soils originates from salts trapped in parent material, from marine aerosols, and from volcanic emissions. Nearly all of the soil Cl^- has been in the oceans at least once, being returned to the land surface either by uplift and subsequent leaching of marine sediments or by oceanic salt spray carried in rain or snow. Annual Cl^- depositions of 12 to 35 lb/a in precipitation are common and may increase to more than 100 lb/a in coastal areas.

Levels of 2 ppm Cl^- are typical in the precipitation near seacoasts. The actual quantities depend on the amount of sea spray, which is related to temperature; the foam formation on tops of waves; the strength and frequency of winds sweeping inland from the sea; the topography of the coastal region; and the amount, frequency, and intensity of precipitation. Salty droplets or dry salt dust may be whirled to great heights by strong air currents and carried over long distances. Concentration of Cl^- in precipitation drops off rapidly inland, with areas about 500 miles inland averaging approximately 0.2 ppm.

Solution Cl^-. The Cl^- anion is very soluble in most soils. Exchangeable Cl^- occurs in acidic soils that have significant pH-dependent positive charge (AEC). Because of the mobility of Cl^- it will accumulate where the internal drainage of soils is restricted and in shallow groundwater where Cl^- can be moved by capillarity into the root zone and deposited at or near the soil surface.

Problems of excess Cl^- occur in some irrigated areas and are usually the result of interactions of two or more of the following factors:

1. Significant amounts of Cl^- in the irrigation water.
2. Failure to apply sufficient water to leach out Cl^- accumulations adequately.

3. Unsatisfactory physical properties and drainage conditions for proper leaching.
4. High water table and capillary movement of Cl^- into the root zone.

Environmental damage in localized areas from high concentrations of Cl^- has resulted from road deicing, water softening, saltwater spills associated with the extraction of oil and natural gas deposits, and disposal of feedlot wastes and various industrial brines.

Plant Responses

Depression of Cl^- uptake by high concentrations of NO_3^- and SO_4^{2-} or vice versa has been observed in a number of plants (Fig. 8.29). Here potato yields increase as the Cl^- level in petioles increases from 1.1 to 6.9% and that of NO_3^- decreases. Although beneficial effects of Cl^- on plant growth are not fully understood, improved plant-water relationships and inhibition of plant diseases are two important factors. The negative interaction between Cl^- and NO_3^- has been attributed to competition for carrier sites at root surfaces.

The effect of Cl^- fertilization on root and leaf disease suppression has been observed on a number of crops (Table 8.16). Several mechanisms have been suggested and include (1) increased NH_4^+ uptake through inhibition of nitrification by Cl^-, which reduces take-all root disease by decreased rhizosphere pH or (2) competition between Cl^- and NO_3^- for uptake. Plants with a low NO_3^- level are less susceptible to root-rot diseases.

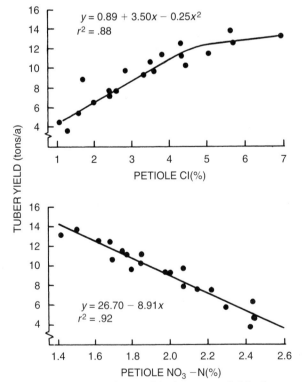

FIGURE 8.29   Relationship between yield of potatoes and Cl^- and NO_3^- concentrations in petiole samples.   *Jackson et al., unpublished data, Oregon State Univ., 1981.*

TABLE 8.16   Diseases Suppressed by Cl Fertilization

Location	Crop	Suppressed Disease
Oregon	Winter wheat	Take-all
	Winter wheat	Septoria
	Potatoes	Hollow heart
	Potatoes	Brown center
North Dakota	Winter wheat	Tanspot
	Spring wheat	Common root rot
	Barley	Common root rot
	Barley	Spot blotch
	Durum wheat	Common root rot
South Dakota	Spring wheat	Leaf rust
	Spring wheat	Tanspot
	Spring wheat	Septoria
New York	Corn	Stalk rot
California	Celery	Fusarium yellows
Saskatchewan	Spring wheat	Common root rot
	Barley	Common root rot
Manitoba	Spring wheat	Take-all
Alberta	Barley	Common root rot
	Barley	Net blotch
Germany	Winter wheat	Take-all
Great Britain	Winter wheat	Stripe rust
India	Pearl millet	Downy mildew
Indonesia	Rice	Stem rot
	Rice	Sheath blight
Philippines	Coconut palm	Gray leaf spot

SOURCE: Fixen, *2nd National Wheat Res. Conf.,* (1987).

In some regions, the Cl^- response in some crops has not been related to disease suppression. For example, in South Dakota, Cl^- deficiency is caused by low soil Cl^-, with the probability of a response to Cl^- fertilization increasing with decreasing water-extractable soil Cl^- (Table 8.17).

Cl Sources

ORGANIC CL   Because of the solubility and mobility of Cl^-, most animal and municipal wastes contain little or no Cl^-.

TABLE 8.17   Frequency of Spring Wheat
Response to Cl Fertilization as a Function of
Soil Cl Levels

Category	Soil Cl Content (lb/a–2 ft)	Yield Response Frequency (%)
Low	0–30	69
Medium	31–60	31
High	>60	0

SOURCE: Fixen, *J. Fert. Issues,* 4:95 (1979).

INORGANIC CL   When additional Cl^- is desirable, it can be supplied by the following sources:

Ammonium chloride (NH_4Cl)   66% Cl
Calcium chloride ($CaCl_2$)   65% Cl
Magnesium chloride ($MgCl_2$)   74% Cl
Potassium chloride (KCl)   47% Cl
Sodium chloride (NaCl)   60% Cl

Rates of Cl^- vary, depending on the crop, method of application, and purpose of addition (i.e., for correction of nutrient deficiency, disease suppression, or improved plant water status). Where take-all root rot of winter wheat is suspected, banding 35 to 40 lb/a of Cl^- with or near the seed at planting is recommended. Broadcasting 75 to 125 lb/a of Cl^- has effectively reduced crop stress from take-all and by leaf and head diseases (i.e., stripe rust and septoria). The Cl^- needs for high yields of most temperate-region crops are usually satisfied by only 4 to 10 kg/ha.

Molybdenum

Mo Cycle

The main forms of Mo in soil include nonexchangeable Mo in primary and secondary minerals, exchangeable Mo held by Fe/Al oxides, Mo in the soil solution, and organically bound Mo. Although Mo is an anion in solution, the relationships between these fractions are similar to those of other metal cations (Fig. 8.30).

MO IN PLANTS   Mo is a nonmetal anion absorbed as molybdate (MoO_4^{2-}). This is a weak acid and can form complex polyanions such as phosphomolybdate. Sequestering of Mo in this form may explain why it can be taken up in relatively large amounts without any apparent toxicity.

Normally, the Mo content of plant material is < 1 ppm, and deficient plants usually contain < 0.2 ppm. Mo concentrations in plants are frequently low because of the extremely small amounts of MoO_4^{2-} in the soil solution. In some cases, however, Mo levels in crops may exceed the range 1,000 to 2,000 ppm.

Mo is an essential component of the enzyme NO_3^- reductase, which catalyzes the conversion of NO_3^- to NO_2^-. Most of the Mo in plants is concentrated in this enzyme, which primarily occurs in chloroplasts in leaves. The Mo requirement of plants is influenced by inorganic N supplied to plants, with either NO_3^- or NH_4^+ effectively lowering its need. It is also a structural component of nitrogenase, the enzyme actively involved in N_2 fixation by root-nodule bacteria of leguminous crops, by some algae and actinomycetes, and by free-living, N_2-fixing organisms such as *Azotobacter*. Mo concentrations in the nodules of legume crops of up to 10 times higher than those in leaves have been observed. Mo is also reported to have an essential role in Fe absorption and translocation in plants (for an example of this deficiency symptom see color plates inside book cover).

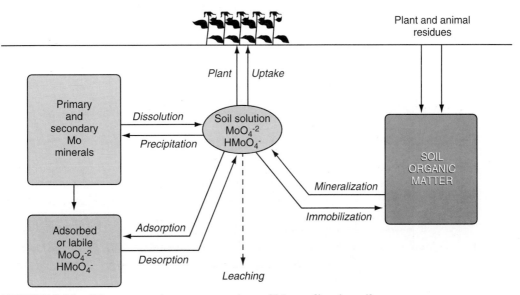

FIGURE 8.30   Diagrammatic representation of Mo cycling in soils.

Excessive amounts of Mo are toxic, especially to grazing cattle or sheep. High-Mo forage may occur on wet soils mostly neutral to alkaline in reaction, often with either a thick A_1 horizon or an A_1 horizon capped by a thin surface layer of peat or muck. Pockets of peats may also be present in these problem areas. Molybdenosis, a disease in cattle, is caused by an imbalance of Mo and Cu in the diet, when the Mo content of the forage is greater than 5 ppm. Mo toxicity causes stunted growth and bone deformation in the animal and can be corrected by oral feeding of Cu, injections of Cu, or the application of $CuSO_4$ to the soil. Other practices used to decrease Mo toxicity are application of S or Mn and improvement of soil drainage.

Mo in Soil

Mineral Mo. The average concentration of Mo in the earth's crust is about 2 ppm, and in soils it typically ranges from 0.2 to 5 ppm. The minerals in soil controlling MoO_4^{2-} concentration are $PbMoO_4$ and $CaMoO_4$ (Fig. 8.31). The Ca mineral predominates in both acidic and calcareous soils. The solubility of Mo in soils generally follows soil Mo, which is very close to the solubility of $PbMoO_4$ or wulfenite.

Soil Solution Mo. Mo in solution occurs predominantly as MoO_4^{2-}, $HMoO_4^-$, and $H_2MoO_4^0$. Concentration of MoO_4^{2-} and $HMoO_4^-$ increases dramatically with increasing soil pH (Fig. 8.32). The extremely low concentration of Mo in soil solution is reflected in the low Mo content of plant material ($\sim$1 ppm Mo). At concentrations above 4 ppb in the soil solution, Mo is transported to plant roots by mass flow, while Mo diffusion to plant roots occurs at levels <4 ppb.

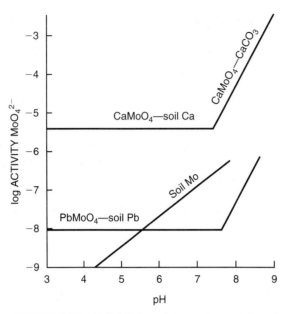

FIGURE 8.31 Solubilities of the common Mo minerals, which increase with increasing pH. *Lindsay,* Chemical Equilibria in Soils, *John Wiley & Sons, New York,*

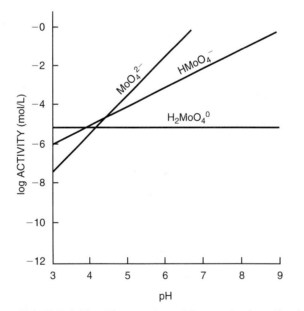

FIGURE 8.32 The common Mo species in soil solution as influenced by pH. *Lindsay,* Chemical Equilibria in Soils, *John Wiley & Sons, New York, 1979.*

Factors Affecting Mo Availability

SOIL PH MoO_4^{2-} availability, unlike that of other micronutrients, increases about 10-fold per unit increase in soil pH (Figs. 8.32 and 8.33). Liming to correct soil acidity will increase Mo availability and prevent Mo deficiency. Alternatively, Mo availability is decreased by application of acid-forming fertilizers such as $(NH_4)_2SO_4$ to a coarse-textured soil.

ABSORPTION TO FE/AL OXIDES   Mo is strongly adsorbed by Fe/Al oxides, a portion of which becomes unavailable to the plant. Soils that are high in amorphous Fe/Al oxides tend to be low in available Mo.

INTERACTIONS WITH OTHER NUTRIENTS   P enhances Mo absorption by plants, probably due to exchange of adsorbed MoO_4^{2-}. In contrast, high levels of SO_4^{2-} in solution depress Mo uptake by plants (Table 8.18). On soils with marginal Mo deficiencies, the application of SO_4^{2-}-containing fertilizers may induce a Mo deficiency in plants.

Both Cu and Mn can also reduce Mo uptake; however, Mg has the opposite effect and will encourage Mo absorption by plants.

Nitrate N encourages Mo uptake, while NH_4^+ sources reduce Mo uptake. This beneficial effect of NO_3^- nutrition is perhaps related to the release of OH^- ions and an accompanying increase in solubility of soil Mo.

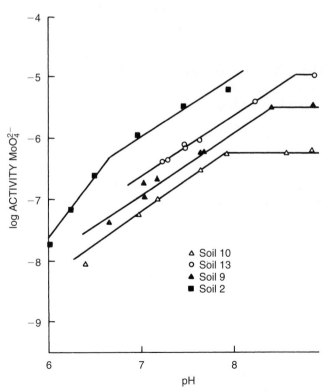

FIGURE 8.33   pH dependence of Mo solubility in four Colorado soils.   *Vlek and Lindsay,* Soil Sci. Soc. Am. J., *41:42, 1977.*

TABLE 8.18   Effect of S and Mo on the Yield and S and Mo Concentration of Brussels Sprouts

Treatment	Yield (g/pot)	Above-ground Tissue	
		Mo (ppm)	S (%)
Sulfur*			
No S added	12.6	5.09	0.25
50 ppm S	13.8	0.88	0.60
100 ppm S	13.8	0.50	0.70
Molybdenum			
No Mo added	12.7	0.08	0.53
Seed treated with Mo	13.6	0.16	0.49
2.5 ppm Mo	13.9	6.23	0.51

*S treatments did not alter soil pH or the exchangeable Mo content.
SOURCE: Gupta, *Sulphur Inst. J.,* 5(1):4 (1969).

CLIMATIC EFFECTS   Mo deficiency is more severe under dry soil conditions, probably due to reduced mass flow or diffusion under low soil moisture content.

PLANT FACTORS   Crops vary in their sensitivity to low solution Mo (Table 8.19). Mo-efficient and Mo-inefficient varieties of alfalfa, cauliflower, corn, and kale have been identified. The differential susceptibility of cauliflower varieties to Mo deficiency is related to their ability to extract soil Mo.

Mo Sources

ORGANIC MO   Only very small quantities of Mo occur in animal wastes, ranging between 0.0001 and 0.0005% (0.0002 and 0.01 lb/t). Thus, with most manures, average application rates will provide sufficient plant available Mo. Mo content in municipal waste is usually very low, with an average Mo content of <0.0001%, or 0.002 lb/t.

INORGANIC MO   Sources of Mo used for fertilizers are listed in Table 8.20. Rates of Mo application are very low, only 0.5 to 5 oz/a, and the solution may be applied to soil, sprayed on foliage, or put on seed before planting. The optimum

TABLE 8.19   Sensitivity of Crops to Low Levels of Available Mo

High Sensitivity	Mild Sensitivity	Low Sensitivity
Broccoli	Beet	Sugar beets
Brussels sprouts	Cotton	Barley
Legumes	Lettuce	Corn
Cauliflower	Spinach	Wheat
Rapeseed	Potatoes	Flax
Clover	Soybean	Oats
Alfalfa	Tomatoes	Ryegrass

TABLE 8.20   Sources of Mo Used for Fertilizer

Sources	Formula	% Mo
Ammonium molybdate	$(NH_4)_6Mo_7O_{24} \cdot 2H_2O$	54
Sodium molybdate	$Na_2MoO_4 \cdot 2H_2O$	39
Molybdenum trioxide	MoO_3	66
Molybdenum frits	Fritted glass	1–30

SOURCE: Mortvedt, *Farm Chem.*, 143:42 (1980).

Mo rate depends on the application method, with lower rates used in the latter two methods. Seeds treated with a solution of sodium molybdate before seeding are widely used because of the low application rates needed. Seed treatments with a slurry or dust are also effective. To obtain satisfactory distribution of the small quantities of Mo applied to soil, Mo sources are sometimes combined with N-P-K fertilizers. Foliar spray applications with NH_4 or Na molybdate are also effective in correcting deficiencies.

Application of Mo to clovers will in some cases increase the yield equivalent to that achieved with the addition of limestone. Since liming can be more expensive, Mo fertilization is often preferred.

Cobalt

CO IN PLANTS   Co is essential for microorganisms fixing N_2. Co is thus needed in the nodules of legumes and N_2-fixing algae. Co concentration in plant dry matter ranges from 0.02 to 0.5 ppm. Only 10 ppb of Co in nutrient solution was found to be adequate for N_2 fixation by alfalfa.

Co is essential for the growth of symbiotic microorganisms such as *Rhizobia*, free-living N_2-fixing bacteria, and blue-green algae. Co forms a complex with N important for synthesis of vitamin B_{12} coenzyme. Co is also important in the synthesis of vitamin B_{12} in ruminant animals; thus, soil is an important source of plant Co for animals. Because Co behaves similarly to Fe or Zn, excess Co produces visual symptoms similar to Fe and Mn deficiencies.

CO IN SOILS   The average total Co concentration in the earth's crust is 40 ppm. Acidic rocks, including granites, containing large amounts of Fe-rich ferromagnesian minerals are low in Co, with levels ranging from 1 to 10 ppm. Soils formed on granitic glacial drift are also generally low in total Co. Much higher levels (100 to 300 ppm Co) are present in Mg-rich ferromagnesian minerals. Sandstones and shales are normally low in Co, with concentrations frequently below 5 ppm.

Total Co content of soils typically ranges from 1 to 70 ppm and averages about 8 ppm. Co deficiencies in ruminants are often associated with forages produced on soils containing less than 5 ppm of total Co.

Soils in which Co deficiency can occur are (1) acidic, highly leached, sandy soils with low total Co; (2) some highly calcareous soils; and (3) some peaty soils.

Co is adsorbed on the exchange complex and occurs as clay-OM complexes similar to those of the other metal cations (Fig. 8.23). The order of adsorption is muscovite > hematite > bentonite = kaolinite. Co concentration in the soil solution is very low, often <0.5 ppm.

Among the several factors that influence Co availability is the presence of Fe/Al/Mn oxides. These minerals have a high adsorption capacity for Co and are capable of fixation of soil-applied Co fertilizer. Co appears to replace Mn in the surface layers of these minerals. Co availability is favored by increasing acidity and waterlogging conditions, which solubilize Mn oxide; therefore, liming and drainage reduce Co availability.

CO FERTILIZERS   Co deficiency of ruminants can be corrected by (1) adding it to feed, salt licks, or drinking water; (2) drenching; (3) using Co bullets; and (4) fertilizing forage crops with small amounts of Co. Co fertilization with 1.5 to 3 oz/a as $CoSO_4$ is recommended.

Soils too low in available Co for satisfactory nodulation and N_2 fixation by clover and alfalfa require applications of 0.5 to 2 oz/a of Co, as $CoSO_4$. Superphosphate, with small amounts of $CoSO_4$, has also been used to increase the Co concentration in subterranean clover.

Sodium

NA IN PLANTS   Na is essential for halophytic plant species that accumulate salts in vacuoles to maintain turgor and growth. The increased growth produced by salt in halophytes is believed to be due to increased turgor. The beneficial effects of Na on plant growth are often observed in low-K soils, because Na^+ can partially substitute for K^+. Crops have been categorized according to their potential for Na uptake (Table 8.21). Growth of those crops with high and medium ratings will be favorably influenced by Na^+.

Na is absorbed by plants as Na^+, and its concentration varies widely, from 0.01 to 10%, in leaf tissue. Sugar beet petioles frequently contain levels at the upper end of this range. Many plants that possess the C_4 dicarboxylic photosynthetic pathway require Na as an essential nutrient. It also has a role in inducing crassulacean acid metabolism, which is considered part of a general response to water stress.

Water economy in plants seems to be related to the C_4 dicarboxylic photosynthetic pathway of plants. Many plant species that have the extremely efficient C_4 CO_2-fixing

TABLE 8.21   Na Uptake Potential of Various Crops

High	Medium	Low	Very Low
Fodder beet	Cabbage	Barley	Buckwheat
Sugar beet	Coconut	Flax	Maize
Mangold	Cotton	Millet	Rye
Spinach	Lupins	Rape	Soybean
Swiss chard	Oats	Wheat	Swede
Table beet	Potato		
	Rubber		
	Turnip		

system occur naturally in arid, semiarid, and tropical conditions, where the closure of stomata to prevent wasteful water loss is essential for growth and survival. CO_2 entry must also be restricted when stomata tend to remain closed. The ratio of weights of CO_2 assimilated to water transpired by C_4 plants is often double that of C_3 plants. It is also noteworthy that C_4 plants are often found in saline habitats.

Sugar beets appear to be particularly responsive to Na. Na influences water relations in this crop and increases the resistance of sugar beets to drought. In low-Na soils the beet leaves are dark green, thin, and dull in hue. The plants wilt more rapidly and may grow horizontally from the crown. There may also be an interveinal necrosis similar to that resulting from K deficiency. Some of the effects ascribed to Na may also be due to Cl^- since the usual source of Na is NaCl.

NA IN SOILS   Na content in the earth's crust is about 2.8%, while soils contain 0.1 to 1%. Low Na in soils indicates weathering of Na from Na-containing minerals. Very little exchangeable and mineral Na occurs in humid-region soils, whereas Na is common in most arid- and semiarid-region soils. In arid and semiarid soils, Na exists in silicates, as well as NaCl, Na_2SO_4, and Na_2CO_3.

The soil solution contains between 0.5 and 5 ppm Na^+ in temperate-region soils. Solution and exchangeable Na^+ vary greatly among soils. In humid-region soils the proportion of exchangeable Na^+ to other cations is $Ca^{2+} > Mg^{2+} > K^+ = Na^+$. Exchangeable Na^+ can be utilized by crops. Sugar beets respond to fertilization when exchangeable Na^+ in soil is <0.05 meq/100 g.

In arid regions and if soils are irrigated with sodic waters, exchangeable Na^+ levels generally exceed those of K^+. Sodium salts accumulating in poorly drained soils of the arid and semiarid regions will be contributors to soil salinity and sodicity (see Chapter 2).

NA SOURCES   Responses to Na have been observed in crops with a high uptake potential (Table 8.21). The Na demand of these crops appears to be independent of, and perhaps even greater than, their K demand. The important Na-containing fertilizers are the following:

- K fertilizers with various NaCl contents.
- $NaNO_3$ (about 25% Na).
- Multiple-nutrient fertilizers with Na.

Silicon

SI IN PLANTS   Si is absorbed by plants as silicic acid, $H_4SiO_4^0$. Cereals and grasses contain 0.2 to 2.0% Si, while dicotyledons may accumulate only one-tenth of this amount. Concentrations of up to 10% occur in Si-rich plants.

Si contributes to the structure of cell walls. Grasses, sedges, nettles, and horsetails accumulate 2 to 20% of the foliage dry weight as Si. Si primarily impregnates the walls of epidermal and vascular tissues, where it appears to strengthen the tissues, reduce water loss, and retard fungal infection. Where large amounts of Si are accumulated, intracellular deposits known as *plant opals* can occur.

The involvement of Si in root functions is believed to be its contribution to the drought tolerance of crops such as sorghum. Although no biochemical role for Si in plant development has been positively identified, it has been proposed that enzyme-Si complexes that act as protectors or regulators of photosynthesis and enzyme activity form in sugarcane. Si can suppress the activity of invertase in sugarcane, resulting in greater sucrose production. A reduction in phosphatase activity is believed to provide a greater supply of essential high-energy precursors needed for optimum cane growth and sugar production.

The beneficial effects of Si have been attributed to correction of soil toxicities arising from high levels of available Mn^{2+}, Fe^{2+}, and Al^{3+}; plant disease resistance; greater stalk strength and resistance to lodging; increased availability of P; and reduced transpiration.

Freckling, a necrotic leaf spot condition, is a symptom of low Si in sugarcane receiving direct sunlight. Ultraviolet radiation seems to be the causative agent in sunlight since plants kept under plexiglass or glass do not freckle. There are suggestions that Si in the sugarcane plant filters out harmful ultraviolet radiation.

This element has also had a favorable influence on rice production. Si tends to maintain erectness of rice leaves, increases photosynthesis because of better light interception, and results in greater resistance to diseases and insect pests. The oxidizing power of rice roots and accompanying tolerance to high levels of Fe and Mn were found to be very dependent on Si nutrition. Supplemental Si was beneficial when the Si concentration in rice straw fell below 11%. Heavy applications of N render rice plants more susceptible to fungal attack because of decreases in Si concentration in the straw. To correct this problem, Si-bearing materials are added when high rates of N fertilizer are used.

SI IN SOILS   Si is the second most abundant element in the earth's crust, averaging 28%, while Si in soils ranges between 23 and 35%. Unweathered sandy soils can contain as much as 40% Si, compared with as little as 9% Si in highly weathered tropical soils. Major sources of Si include primary and secondary silicate minerals and quartz (SiO_2). Quartz is the most common mineral in soils, comprising 90 to 95% of all sand and silt fractions.

Low-Si soils exist in intensively weathered, high-rainfall regions. Properties of the Si-deficient soils include low total Si, high Al, low base saturation, and low pH. In addition, they all have extremely high P-fixing capacity due to their high AEC and Fe/Al oxide content. Plant available Fe^{2+} and Mn^{2+} may also be high in these soils.

Silicic acid ($H_4SiO_4^0$) is the principal Si species in solution. Si concentrations of less than 0.9 to 2 ppm in soil solution are insufficient for proper nutrition of sugarcane. By comparison, levels of 3 to 37 ppm Si in solution have been reported for a wide range of normal soils. Si levels adequate for rice production are >100 ppm.

The concentration of $H_4SiO_4^0$ in soil solutions is largely controlled by a pH-dependent adsorption reaction. Si is adsorbed on the surfaces of Fe/Al oxides. Si leaching in highly weathered soils will reduce solution Si and Si uptake.

SI SOURCES   Primary Si fertilizers include the following:

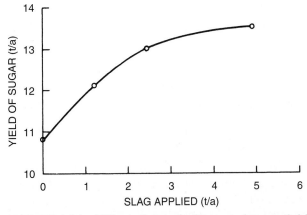

FIGURE 8.34   Effect of electric furnace slag on yield of sugar from sugarcane grown on an aluminous humic ferruginous latosol in Hawaii. Mean results for plant and ratoon crops combined.   *Ayres, Soil Sci., 101:216, 1966.*

Calcium silicate slag ($CaAl_2Si_2O_8$)	18 to 21% Si
Calcium silicate ($CaSiO_3$)	31% Si
Sodium metasilicate ($NaSiO_3$)	23% Si

Minimum rates of at least 5,000 lb/a of $CaSiO_3$ are broadcast applied and incorporated before planting sugarcane (Fig. 8.34). Annual $CaSiO_3$ applications of between 500 and 1,000 lb/a applied in the row have also improved sugarcane yields. Additions of lime that increase Ca levels and decrease soil acidity do not produce similar dramatic improvements in the growth of sugarcane. Rates of 1.5 to 2.0 t/ha of silicate slag usually provide sufficient Si for rice produced on low-Si soils.

Selenium

SE IN PLANTS   Se is not needed by plants, but it must be present in forage since it is essential for animals. A greater frequency of livestock nutritional disorders caused by low Se has been observed after cold, rainy summers than after hot, dry ones. High summer temperatures are amenable to increased Se concentration in feedstuffs.

Plant species differ in Se uptake. Certain species of *Astragalus* absorb many times more Se than do other plants growing in the same soil, because they utilize Se in an amino acid peculiar to the species. Plants such as the cruciferae (e.g., cabbage, mustard) and onions, which require large amounts of S, absorb intermediate amounts of Se, while grasses and grain crops absorb low to moderate amounts.

SE IN SOILS   Se occurs in very small amounts in nearly all materials of the earth's crust. It averages only 0.09 ppm in rocks and is found mainly in sedimentary minerals. Se is similar in behavior to S; however, it has five important oxidation states: −2, 0, +2, +4, and +6.

The total Se concentration in most soils is between 0.1 and 2 ppm and averages about 0.3 ppm. Extensive areas of high-Se soils in western North America and in other semiarid regions produce vegetation toxic to livestock. The parent materials of these soils are predominantly sedimentary shale deposits. High-pH, calcareous soils in regions of low rainfall (<20 in. total precipitation) are usually high in Se.

The forms of Se generally considered to be present in soil are selenides (Se^{2-}), elemental Se^0, selenites (Se^{4+}), selenates (Se^{6+}), and organic Se compounds (Fig. 8.35). Se species in soils and sediments are closely related to redox potential, pH, and solubility.

Selenides (Se^{2-}). Selenides are largely insoluble and are associated with S^{2-} in soils of semiarid regions where weathering is limited. They contribute little to Se uptake because of their insolubility.

Elemental Se (Se^0). Se^0 is present in small amounts in some soils. Significant amounts of Se^0 may be oxidized to selenites and selenates by microorganisms in neutral and basic soils.

Selenites (SeO_3^{2-}). A large fraction of Se in acidic soils may occur as stable complexes of selenites with hydrous iron oxides. The low solubility of Fe-selenite complexes is apparently responsible for the nontoxic levels of Se in plants growing on acidic soils having very high total Se contents. Plants absorb selenite but generally to a lesser extent than selenate.

Selenates (SeO_4^{2-}). Selenates are frequently associated with SO_4^{2-} in arid-region soils and are stable in many well-aerated, semiarid seleniferous soils. Other forms of Se will be oxidized to selenates under these conditions. Only limited quantities of selenate occur in acidic and neutral soils. Selenates are highly soluble and readily available to plants and thus are largely responsible for toxic accumulations in plants grown on high-pH soils. Most of the water-soluble Se in soils probably occurs as selenates.

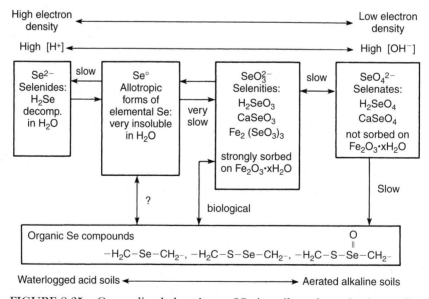

FIGURE 8.35   Generalized chemistry of Se in soils and weathering sediments. *Allaway, in D. E. Hemphill (Ed.),* Trace Substances in Environmental Health, *II. Univ. of Missouri, 1968.*

Organic Se. Organically complexed Se can be an important fraction, since up to 40% of the total Se in some soils is present in humus. Soluble organic Se compounds are liberated through the decay of seleniferous plants. Such substances derived from accumulator or indicator plants are readily taken up by other plants. Se in residue is stable in semiarid areas, and much of it remains available in soil. Organic Se is more soluble in basic than acidic soils, which would enhance availability to plants in semiarid-region soils.

Insufficient plant uptake of Se is usually caused by one or both of the following soil factors: low total Se in the soil parent material or low availability of Se in acidic and poorly drained soils.

Plant uptake of Se is generally greater in high-pH soil than in acidic soils. Se in the soil solution is lowest at slightly acidic to neutral pH and increases under both more acidic and basic soil pH. High soil pH facilitates the oxidation of selenites to the more readily available selenates.

Increased yields with N and S fertilization may lower Se concentrations in crops through dilution. There has been some concern about increased incidence and severity of Se deficiencies in cattle due to the negative interaction of SO_4^{2-} on SeO_4^{2-} uptake by crops.

SE SOURCES   Although Se deficiency disorders such as muscular dystrophy or white muscle disease in cattle and sheep can be corrected by therapeutic measures, there is interest in Se fertilization to produce forages adequate in Se for grazing animals rather than to satisfy any particular plant requirements. Se fertilization is acceptable if proper precautions are taken:

1. At no stage should herbage become toxic to grazing animals; topdressing of growing plants must be avoided.
2. High levels of Se in edible animal tissue should be prevented.
3. Protection against Se deficiency should be provided for at least one grazing season following application during the dormant season.

Fertilization with selenites is preferred because they are slower acting and less likely to produce excessive levels of Se in plants than the rapidly available selenates, which are effective if rapid Se uptake is desired. The addition of Na selenite at rates of 1 oz/a of Se is satisfactory for forages. Foliar application of Na selenite at 6 g/a of Se is an efficient way to increase Se in field corn.

Se is present in phosphate rocks and in superphosphate produced from them. Superphosphate containing 20 ppm or more Se may provide sufficient Se to the plants in Se-deficient areas to protect livestock from Se-deficiency disorders.

Nickel

Ni is the latest (1987) nutrient to be established as essential to higher plants, since Cl^- was established as an essential micronutrient in 1954. The Ni content of crop plants normally ranges from about 0.1 to 1.0 ppm dry weight. It is readily taken up by most species as Ni^{2+}.

TABLE 8.22   Effect of Ni Supply on Germination and Yield of Barley

Ni Supplied in Nutrient Solution (μM)	Germination (%)	Ni Concentration (ng/g dry wt)	Total Grain Wt (g dry wt)
0	11.6	7.0	7.3
0.6	56.6	63.8	7.5
1.0	94.0	129.2	8.4

SOURCE: Brown et al., *Plant Physiol.* 85:801 (1987).

Ni is the metal component of urease that catalyzes the reaction $CO(NH_2)_2 + H_2O \rightarrow 2\,NH_3 + CO_2$. Apparently Ni is essential for plants supplied with urea and for those in which ureides are important in N metabolism. Nodule weight and seed yield of soybeans have been stimulated by Ni.

Results clearly demonstrate the beneficial role of Ni for legumes, with their particular type of N metabolism. Ni-deficient plants accumulate toxic levels of urea in leaf tips because of reduced urease activity. Ni-deficient plants may develop chlorosis in the youngest leaves that progresses to necrosis of the meristem. Ni may also be involved in plant disease resistance, again due to changes in N metabolism.

Ni has been demonstrated as essential to small-grain crops (Table 8.22). These data show increasing barley germination and grain yield with increasing solution Ni concentration.

High levels of Ni may induce Zn or Fe deficiency because of cation competition. Application of some sewage sludge may result in elevated levels of Ni in crop plants.

Vanadium

Low concentrations of V are beneficial for the growth of microorganisms, animals, and higher plants. Although it is considered to be essential for the green alga *Scenedesmus*, there is still no decisive evidence that V is necessary for higher plants. Some workers suggest that V may partially substitute for Mo in N_2 fixation by microorganisms such as the *Rhizobia*. It has also been speculated that it may function in biological oxidation-reduction reactions. Increases in growth attributable to V have been reported for asparagus, rice, lettuce, barley, and corn. The V requirement of plants is said to be less than 2 ppb dry weight, whereas the normal concentration in plant material averages about 1 ppm.

Selected References

BARBER, S. A. 1984. *Soil Nutrient Bioavailability: A Mechanistic Approach.* John Wiley & Sons, New York.

EPSTEIN, E. 1972. *Mineral Nutrition of Plants: Principles and Perspectives.* John Wiley & Sons, New York.

FOLLETT, R. H., L. S. MURPHY, and R. L. DONAHUE. 1981. *Fertilizers and Soil Amendments.* Prentice-Hall, Englewood Cliffs, N.J.

MENGEL, K., and E. A. KIRKBY. 1987. *Principles of Plant Nutrition.* International Potash Institute, Bern, Switzerland.

MORTVEDT, J. J., and F. R. COX. 1985. Production, marketing, and use of calcium, magnesium, and micronutrient fertilizers, pp. 455–82. In O. P. ENGELSTAD (Ed.), *Fertilizer Technology and Use*. Soil Science Society of America, Madison, Wisc.

MORTVEDT, J. J., et al. (Eds.) 1991. *Micronutrients in Agriculture*, No. 4. Soil Science Society of America, Madison, Wisc.

RÖMHELD, V., and H. MARSCHER. 1991. Function of micronutrients in plants. In J. J. MORTVEDT et al. (Eds.), *Micronutrients in Agriculture*, No. 4. Soil Science Society of America, Madison, Wisc.

CHAPTER 9

Soil Fertility Evaluation

Optimum productivity of any cropping system depends on an adequate supply of plant nutrients. The quantity of nutrients required by plants depends on many interacting factors including (1) plant species and variety, (2) yield level, (3) soil type, (4) environment (i.e., water, temperature, and sunlight), and (5) management. The next several chapters detail the influences of these primary factors on nutrient demand and utilization by plants. Plants vary greatly in their nutrient requirements, which, depending on the other factors listed earlier, greatly influence the quantity of additional nutrient supplied to optimize yield. Table 9.1 illustrates the wide range in average nutrient use by selected crops. Continued removal of nutrients, with little or no replacement, increases the potential for future nutrient-related plant stress and yield loss.

When the soil does not supply sufficient nutrients for normal plant development and optimum productivity, supplemental nutrients must be applied. The proper rate of plant nutrients is determined by knowing the nutrient requirement of the crop (Table 9.1) and the nutrient-supplying power of the soil. Diagnostic techniques, including identification of deficiency symptoms and soil and plant tests, are helpful in determining specific nutrient stresses and the quantity of nutrients needed to optimize yield. By the time a plant has shown deficiency symptoms, a considerable reduction in yield potential will already have occurred; thus, the analysis of the nutrient-supplying power or capacity of the soil is essential for quantifying the probability of a crop response to nutrient additions.

The value of soil and plant analysis in quantifying nutrient requirements depends on careful sampling and analysis and using tests that are calibrated or correlated with plant response. Knowledge of the relationship between test results and crop response is essential for providing the most appropriate nutrient recommendation. Several techniques are commonly employed to assess the nutrient status of a soil:

1. Nutrient-deficiency symptoms of plants.
2. Analysis of tissue from plants growing on the soil.
3. Biological tests in which the growth of either higher plants or certain microorganisms is used as a measure of soil fertility.
4. Soil analysis.

TABLE 9.1   Nutrient Removal or Uptake Values for Selected Agricultural Crops.

Crop	Yield/a	N	P	K	Ca	Mg	S	Cu	Mn	Zn
					lbs/a					
				Grains						
Barley (grain)	60 bu	65	14	24	2	6	8	0.04	0.034	0.08
Barley (straw)	2 ton	30	10	80	8	2	4	0.01	0.32	0.05
Canola	45 bu	145	32	100	—	—	28	—	—	—
Corn (grain)	200 bu	150	40	40	6	18	15	0.08	0.10	0.18
Corn (stover)	6 tons	110	12	160	16	36	16	0.05	1.50	0.30
Flax	25 bu	65	8	29	—	—	12	—	—	—
Oats (grain)	80 bu	60	10	15	2	4	6	0.03	0.12	0.05
Oats (straw)	2 tons	35	8	90	8	12	9	0.03	—	0.29
Peanuts (nuts)	4,000 lb	140	22	35	6	5	10	0.04	0.30	0.25
Peanuts (vines)	5,000 lb	100	17	150	88	20	11	0.12	0.15	—
Rye (grain)	30 bu	35	10	10	2	3	7	0.02	0.22	0.03
Rye (straw)	1.5 tons	15	8	25	8	2	3	0.01	0.14	0.07
Sorghum (grain)	80 bu	65	30	22	4	7	10	0.02	0.06	0.05
Sorghum (stover)	4 tons	80	25	115	32	22	—	—	—	—
Soybean (beans)	50 bu	188	41	74	19	10	23	0.05	0.06	0.05
Soybean (stover)	6,100 lb	89	16	74	30	9	12	—	—	—
Sunflower	50 bu	70	13	30	—	—	12	—	—	—
Wheat (grain)	60 bu	70	20	25	2	10	4	0.04	0.10	0.16
Wheat (straw)	2.5 tons	45	5	65	8	12	15	0.01	0.16	0.05
				Forages and Turf						
Alfalfa	6 tons	350	40	300	160	40	44	0.10	0.64	0.62
Bentgrass	2 tons	230	22	100	—	—	—	—	—	—
Bluegrass	2 tons	60	12	55	16	7	5	0.02	0.30	0.08
Bromegrass	4 tons	140	22	180	—	15	15	—	—	—
Clover	6 tons	320	40	260	—	—	—	—	—	—
Coastal Bermuda	8 tons	400	45	310	48	32	32	0.02	0.64	0.48
Cowpea	2 tons	120	25	80	55	15	13	—	0.65	—
Fescue	3.5 tons	135	18	160	—	13	20	—	—	—
Orchardgrass	6 tons	300	50	320	—	25	35	—	—	—
Red Clover	2.5 tons	100	13	90	69	17	7	0.04	0.54	0.36
Ryegrass	5 tons	215	44	200	—	40	—	—	—	—
Sorghum-Sudan	8 tons	320	55	400	—	47	—	—	—	—
Soybean	2 tons	90	12	40	40	18	10	0.04	0.46	0.15
Timothy	4 tons	150	24	190	18	6	5	0.03	0.31	0.20
Vetch	6 tons	360	38	250	—	—	—	—	—	—
				Fruits and Vegetables						
Apples	500 bu	30	10	45	8	5	10	0.03	0.03	0.03
Bean, Dry	30 bu	75	25	25	2	2	5	0.02	0.03	0.06
Bell Peppers	180 cwt	137	52	217	—	43	—	—	—	—
Cabbage	20 tons	130	35	130	20	8	44	0.04	0.10	0.08

TABLE 9.1   Continued

Crop	Yield/a	N	P	K	Ca	Mg	S	Cu	Mn	Zn
					lbs/a					
				Fruits and Vegetables						
Onions	7.5 tons	45	20	40	11	2	18	0.03	0.08	0.31
Peaches	600 bu	35	20	65	4	8	2	—	—	0.01
Peas	25 cwt	164	35	105	—	18	10	—	—	—
Potatoes (white)	30,000 lb	90	48	158	5	7	7	0.06	0.14	0.08
Potatoes (vines)	—	61	20	54	—	12	7	—	—	—
Potatoes (sweet)	300 bu	40	18	96	4	4	6	0.02	0.06	0.03
Potatoes (vines)	—	30	4	24	—	5	—	—	—	—
Snap Beans	4 tons	138	33	163	—	17	—	—	—	—
Spinach	5 tons	50	15	30	12	5	4	0.02	0.10	0.10
Sweet Corn	90 cwt	140	47	136	—	20	11	—	—	—
Tomatoes	20 tons	120	40	160	7	11	14	0.07	0.13	0.16
Turnips	10 tons	45	20	90	12	6	—	—	—	—
				Other Crops						
Cotton (seed & lint)	2,600 lb	63	25	31	4	7	5	0.18	0.33	0.96
Cotton (stalks, leaves, & burs)	3,000 lb	57	16	72	56	16	15	0.05	0.06	0.75
Sugar beets	20 tons	200	20	320	—	50	25	—	—	—
Sugarcane	40 tons	180	40	250	—	25	22	—	—	—
Tobacco, flue-cured (leaves)	3,000 lb	85	15	155	75	15	12	0.03	0.55	0.07
Tobacco, flue-cured (stalks)	3,600 lb	41	11	102	—	9	7	—	—	—
Tobacco, burley (leaves)	4,000 lb	145	14	150	—	18	24	—	—	—

Nutrient-Deficiency Symptoms of Plants

Growing plants act as integrators of all growth factors (Fig. 9.1). Therefore, careful inspection of the growing plant can help identify a specific nutrient stress (see color plates inside book cover). If a plant is lacking a particular nutrient, characteristic symptoms may appear. Deficiency of a nutrient does not directly produce symptoms. Rather, the normal plant processes are thrown out of balance, with an accumulation of certain intermediate organic compounds and a shortage of others. This leads to the abnormal conditions recognized as symptoms. Visual evaluation of nutrient stress should be used only as a supplement to other diagnostic techniques (i.e., soil and plant analysis).

In addition, nutrient deficiencies have a marked effect on the extent and type of root growth (Fig. 9.2). Plant roots have not received much attention because of the difficulty of observing them; however, considering that roots absorb most

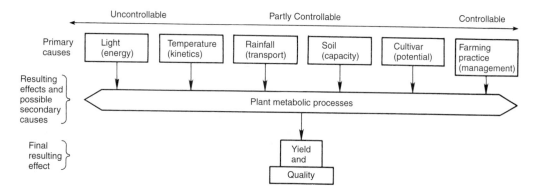

FIGURE 9.1 Schematic representation of the interrelationships between crop yield and quality, metabolic process, and external and genetic factors. *Beaufils,* Soil Sci. Bull. 1, *Univ. of Natal, Pietermaritzburg, South Africa, 1973.*

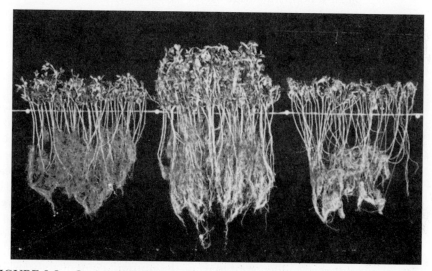

FIGURE 9.2 Omitting P (*left*) or K (*right*) reduced the growth of alfalfa roots as well as tops the spring after seeding in soil deficient in P and K. *Courtesy of the Potash & Phosphate Institute, Atlanta, Ga.*

of the nutrients needed by the plant, inspection of root growth can be an important diagnostic tool.

Each symptom must be related to some function of the nutrient in the plant (see Chapters 4–8). A given nutrient may have several functions, which makes it difficult to explain the physiological reason for a particular deficiency symptom. For example, when N is deficient, the leaves of most plants become pale green or light yellow. When the quantity of N is limiting, chlorophyll production is reduced, and the yellow pigments, carotene and xanthophyll, show through. A number of nutrient deficiencies produce pale-green or yellow leaves, and the difficulty must be further related to a particular leaf pattern or location on the plant.

Apparent visual deficiency symptoms can be caused by many factors other than a specific nutrient stress. Precautions in interpreting nutrient-deficiency symptoms include the following:

1. *The visual symptom may be caused by more than one nutrient.* For example, N-deficiency symptoms may be identified, although S may also be deficient and its symptoms may not be readily apparent.
2. *Deficiency of one nutrient may be related to an excessive quantity of another.* For example, Mn deficiency may be induced by adding large quantities of Fe, provided that soil Mn is marginally deficient. Also, at a low level of P supply, the plant may not require as much N compared with normal or adequate P. In other words, once the first limiting factor is eliminated, the second limiting factor will appear (Liebig's law of the minimum).
3. *It is difficult to distinguish among the deficiency symptoms in the field, because disease, insect, or herbicide damage can resemble certain micronutrient deficiencies.* For example, leaf hopper damage can be confused with B deficiency in alfalfa.
4. *A visual symptom may be caused by more than one factor.* For example, sugars in corn combine with flavones to form anthocyanins (purple, red, and yellow pigments), and their accumulation may be caused by an insufficient supply of P, low soil temperature, insect damage to the roots, or N deficiency.

Nutrient-deficiency symptoms appear only after the nutrient supply is so low that the plant can no longer function properly. In such cases, supplemental nutrients were needed long before the symptoms appeared. If the symptom is observed early, it might be corrected during the growing season. Since the objective is to get the limiting nutrient into the plant as quickly as possible, with some nutrients and under some conditions this may be accomplished with foliar applications or sidedressings. Usually the yield is reduced below the quantity that would have been obtained if adequate nutrients had been available at planting. However, if the problem is properly diagnosed, the deficiency can be corrected the following year.

With most nutrients on most crops, significant responses can be obtained even though no recognizable symptom have appeared. The question, then, is how to identify this "hidden symptom" or the nutrient content above which deficiency symptoms are visible but still considerably below that needed for optimum crop production. Testing of plants and soils is essential for planning or modifying nutrient management programs to avoid yield loss from nutrient stress.

Nutrients may be present in sufficient quantities when conditions are ideal, but in drought, excessive moisture, or unusual temperature conditions the plant may not be able to obtain an adequate supply. For example, with cooler temperatures, nutrient uptake is generally reduced because

1. Mass flow of nutrients is reduced by decreased growth rate and transpiration.
2. Nutrient diffusion rate decreases with declining temperature and a lower concentration gradient.
3. Mineralization of nutrients complexed with OM is reduced.

Nutrient-deficiency symptoms appearing during early growth may disappear as the growing season progresses, or there may be no measurable yield benefit from supplemental additions of the nutrient(s) in question. For example, fertilizer P may improve the early growth of crops, but at harvest there may be no measurable yield response. Such occurrences are probably related to seasonal effects or to penetration of roots into areas of the soil having higher fertility levels.

To eliminate plant nutrients as a limiting factor, the nutrient content of the plant must be raised to a level that takes advantage of a good season and prevents climate-related nutrient stress. Plant nutrient analysis can be an invaluable tool in identifying *hidden hunger* or verifying the nutrient stress suspected from the visual deficiency symptom.

Plant Analyses

Two general types of plant analysis have been used: (1) tests on fresh tissue in the field, and (2) tissue analysis performed in the laboratory. Plant analyses are based on the premise that the amount of a given nutrient in a plant is related to the nutrient availability in the soil. Since a shortage of a nutrient will limit growth, other nutrients may accumulate, regardless of their supply. For example, if corn is low in NO_3^-, the P content may be high. This is no indication, however, that if adequate N were supplied the supply of P would be adequate.

Tissue tests and plant analyses are made for the following reasons:

1. To help identify deficiency symptoms and to determine nutrient shortages before they appear as symptoms.
2. To aid in determining the nutrient-supplying capacity of the soil. They are employed in conjunction with soil tests and management history.
3. To aid in determining the effect of fertility treatment on the nutrient supply in the plant.
4. To study the relationship between the nutrient status of the plant and crop performance.

Tissue Tests

Rapid tests for the determination of nutrients in fresh tissue are important in diagnosing the nutrient needs of growing plants. The concentration of nutrients in the cell sap is usually a good indication of how well the plant is supplied *at the time of testing*. Through the proper application of tissue testing, it is possible to anticipate or forecast certain production problems while the crop is still in the field.

GENERAL METHODS   In one test the plant parts may be chopped up and extracted with reagents. The intensity of the color developed is compared with standards and used as a measure of the supply of the nutrient. In another more rapid test, plant tissue is squeezed with pliers to transfer plant sap to filter paper and color-developing reagents are added. The resulting color is compared with a standard chart that indicates very low, low, medium, or high nutrient content. Semi-quantitative values for the N, P, and K status of a plant can be obtained in about a minute.

Tissue tests are easy to conduct and interpret, and many tests can be made in a few minutes. Because laboratory tests take longer, there is a tendency to guess rather than send samples to the laboratory. It is important to recognize that application of nutrients to correct a nutrient stress identified with a tissue test may not be feasible because (1) the deficiency may have already caused yield loss, (2) the crop may not respond to the applied nutrient at the specific growth stage tested, (3) the crop may be too large to apply nutrients, and (4) climatic conditions may be unfavorable for fertilization and/or for the crop to benefit from nutrient additions.

PLANT PARTS TO BE TESTED   It is essential to test the part of the plant that will give the best indication of the nutritional status (Fig. 9.3). In general, the conductive tissue of the latest mature leaf is used for testing, while immature leaves at the top of the plant are avoided.

TIME OF TESTING   The plant growth stage is important in tissue testing, because nutrient status and demand changes during the season. Although the most critical growth stage for tissue testing is at bloom or from bloom to the early fruiting stage, recent information shows that in high yielding corn there are two distinct peak periods of dry matter and plant nutrient accumulation. Tissue testing during these periods of potential maximum utilization of nutrients will be useful to monitor adequacy of nutrient supply.

The first peak occurs during vegetative growth stages V12 to V18 when ear size and number of kernels are being established at the same time as photosynthetic reserves are accumulating in the stalk and leaves. The second peak is during grain-fill when final kernel numbers and sizes are determined.

For corn, it has been recommended to sample the leaf opposite and just below the uppermost ear at silking. At this growth stage, however, it may be difficult to apply nutrients to correct any deficiencies.

Measurement of NO_3^- concentration in the lower cornstalk at physiological maturity has been used to assess N sufficiency during the growing season (Fig. 9.4). Concentrations below 0.05 to 0.15% stalk NO_3^- indicate that additional N available during the growing season likely would have increased grain yield. Also, stalk NO_3^- has been positively correlated with soil NO_3^- concentration.

The time of day can influence the NO_3^- level in plants; this nutrient is usually higher in the morning than in the afternoon if the supply is short. It accumulates at night and is utilized during the day as carbohydrates are synthesized. Therefore, tests should not be made in early morning or late afternoon.

The value of tissue tests for in-season adjustments of N in cotton is illustrated in Figure 9.5. Petiole samples of the newest mature cotton leaf are collected at or near first bloom stage and either sent to a laboratory for NO_3^- analysis or quick-tested using in-field colorimetric analysis of the petiole tissue sap. For example, if first bloom petiole samples tested 0.2% NO_3-N (see Fig. 9.5), then a foliar application of 5 lb/a of N would be recommended. Usually three foliar applications are required to optimize cotton yield. As the cotton plant matures, the petiole NO_3^- concentration declines, similar to the data shown in Figure 9.5.

A few points relative to tissue tests are as follows:

1. Follow the uptake of nutrients through the season by testing five or six times. Nutrient levels should be higher in the early season when the plant is not under stress.
2. There can be two peak periods of nutrient demand. The first is during maximum vegetative growth, and the second is during the reproductive stage. To determine the adequacy of the fertilization program, these are the optimum times for tissue testing; however, at the later peak period, it is generally too late for corrective action.

Sampling Chart

Plant	Test	Part to sample	(To avoid hidden hunger) Minimum level
Corn			
Under 15 in.	NO$_3$	Midrib, basal leaf	High
	PO$_4$	Midrib, basal leaf	Medium
	K	Midrib, basal leaf	High
15 in. to ear showing	NO$_3$	Base of stalk	High
	PO$_4$	Midrib, first matue leaf*	Medium
	K	Midrib, first mature leaf*	High
Ear to very early dent	NO$_3$	Base of stalk	High
	PO$_4$	Midrib, leaf below ear	Medium
	K	Midrib, leaf below ear	Medium
Soybeans			
Early growth to midseason	NO$_3$	Not tested	
	PO$_4$	Pulvinus (swollen base of (petiole), first mature leaf*	High High
	K	Petiole, first mature leaf	High
Midseason to good pod development	PO$_4$	Pulvinus, first mature leaf	Medium
	K	Petiole, first mature leaf	Medium
Cotton			
To early bloom	NO$_3$	Petiole, basal leaf*	High
	PO$_4$	Petiole, basal leaf*	High
	K	Petiole, basal leaf*	High
Boil setting to 2/3 maturity	NO$_3$	Petiole, first mature leaf*	High
	PO$_4$	Petiole, first mature leaf*	High
	K	Petiole, first mature leaf*	High
2/3 maturity to maturity	NO$_3$	Petiole, first mature leaf*	Medium
	PO$_4$	Petiole, first mature leaf*	Medium
	K	Petiole, first mature leaf*	Medium
Alfalfa			
Before first cutting	PO$_4$	Middle 1/3 of stem	High
	K	Middle 1/3 of stem	High
Before other cuttings	PO$_4$	Middle 1/3 of stem	Medium
	K	Middle 1/3 of stem	Medium
Small Grains Shoot stage to milk stage	NO$_3$	Lower stem	High
	PO$_4$	Lower stem	Medium
	K	Lower stem	Mediium

*First Mature Leaf—Avoid the immature leaves at the top of the plant. Take the most recently fully matured leaf near the top of the plant.

FIGURE 9.3   Part of the plant used for tissue tests.   *Wickstrom et al.,* Better Crops Plant Food, *47(3):18, 1964.*

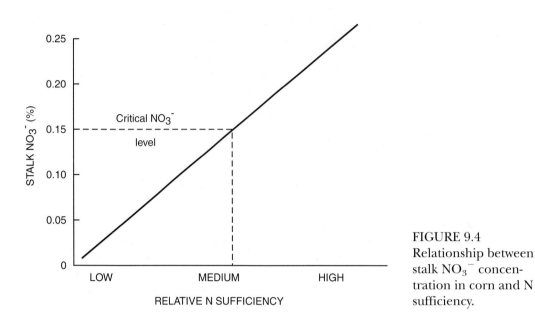

FIGURE 9.4
Relationship between
stalk NO_3^- concen-
tration in corn and N
sufficiency.

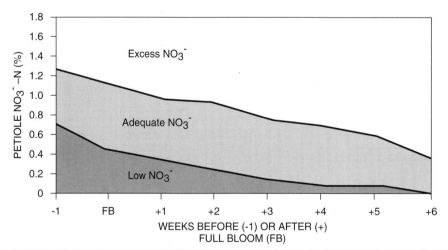

FIGURE 9.5   Cotton petiole NO_3^- concentration as influenced by sampling
time. FB represents first bloom and the minus or plus sign represents weeks before or after first bloom. Thus, FB+4 is 4 weeks after first bloom. Foliar applications of N would be needed to optimize cotton yield when a low petiole NO_3^-
concentration was measured.   *Courtesy of Dr. Steve Hodges, Dept. Soil Science,
NCSU, 1997.*

3. Comparison of plants in a field is helpful. Test plants from deficient areas and compare them with plants from normal areas.
4. Plants vary; thus, test 10 to 15 plants and average the results.

INTERPRETATION   Important factors in sampling and interpreting results from tissue tests are

1. General performance and vigor of the plant.
2. Levels of other nutrients in the plant.
3. Incidence of insects or disease.
4. Soil conditions, such as moisture, aeration, structure, and so on.
5. Climatic conditions.

If a plant appears to be discolored or stunted and the tissue tests high for N, P, and K, then some other factor is limiting growth. Only after this condition has been corrected can tissue tests reveal which plant nutrients may be limiting growth. Generally, low to medium test results for N, P, or K in the early part of the growing season mean that the yield will be considerably less than optimum. At bloom stage, a medium to high test result is adequate in most crops.

Total Analysis

Total analysis is performed on the whole plant or on specific plant parts. After sampling, the plant material is dried, ground, and the nutrient content determined following digesting or ashing of the plant material. With total analysis, the content of all elements, essential and nonessential, can be determined. As in tissue tests, the plant part selected is important, with the most recently matured leaf preferred (Fig. 9.6). Plant sampling guidelines for selected crops are shown in Table 9.2.

INTERPRETATION   Plants that are severely deficient in an essential nutrient exhibit a visual deficiency symptom (Fig. 9.7). Plants that are moderately deficient usually exhibit no visual symptoms, although yield potential is reduced. Added nutrients will maximize yield potential and increase nutrient concentration in the plant. The term *luxury consumption* means that plants continue to absorb a nutrient in excess of that required for optimum growth. This extra consumption results in an accumulation of the plant nutrient without a corresponding increase in growth. However, with higher crop yields, a greater concentration of nutrients is required. When nutrient *toxicity* occurs plant growth and yield potential decrease, increasing the nutrient concentration in the plant.

The critical nutrient concentration (CNC) is commonly used in interpreting plant analysis results and diagnosing nutritional problems (Fig. 9.7; also see Fig. 1.9). The CNC is located in that portion of the curve where the plant-nutrient concentration changes from deficient to adequate; therefore, the CNC is the level of a nutrient below which crop yield, quality, or performance is unsatisfactory. For example, CNCs in corn are about 3% N, 0.3% P, and 2% K in the leaf opposite and below the uppermost ear at silking time (Table 9.3). For crops such as sugar beets or malting barley, in which excessive con-

TABLE 9.2   Plant Sampling Guidelines for Selected Crops

Crop	When to Sample	Part of Plant to Sample	Number of Plants to Sample
Field Crops			
Alfalfa	At one-tenth bloom stage or before	Mature leaf blades about one-third of the way down the plant	45–55
Cereal grains (including rice)	Seedling stage or	Entire above-ground portion	50–75
	Before heading	Four uppermost blades from the top of the plant	30–40
Clover	Before bloom	Mature leaf blades about one-third of the way down the plant	50–60
Corn	Seedling stage or	Entire above-ground portion	25–30
	Before tasseling or	First fully developed leaves from the top	15–20
	From tasseling to silking	Leaves below and opposite the ear	15–20
Cotton	Before or at first bloom or when first squares appear	Youngest fully mature leaves on the main stem	30–35
Hay, forage, or pasture grasses	Before seed head emergence or at the stage for best quality	Four uppermost leaf blades	50–60
Milo sorghum	Before or at heading	Second leaf from the top of the plant	20–25
Peanuts	Before or at bloom stage	Fully developed leaves from the top of the plant	45–50
Soybeans	Seedling stage or	Entire above-ground portion	20–30
	Before or during initial flowering	First fully developed leaves from the top	20–30
Sugar beets	Midseason	Fully mature leaves midway between the younger center leaves and the oldest leaf whorl on the outside	30–35
Sugarcane	Up to 4 months old	Fourth fully developed leaf from the top	25–30
Tobacco	Before bloom	Top fully developed leaf	8–12
Ornamentals and Flowers			
Carnations	Unpinched plants	Fourth or fifth leaf pair from base of plant	20–30
	Pinched plants	Fifth or sixth leaf pair from top of primary laterals	20–30
Chrysanthemums	Before or during early flowering	Top leaves on flowering stem	20–30
Ornamental trees and shrubs	Current year's growth	Fully mature leaves	30–75
Poinsettias	Before or during early flowering	Most recently mature, fully expanded leaf	15–20
Roses	During flowering	Upper leaves on the flowering stem	25–30
Turf	During growing season	Leaf blades; avoid soil contamination	2 cups of material

TABLE 9.2   Continued

Crop	When to Sample	Part of Plant to Sample	Number of Plants to Sample
Vegetable Crops			
Beans	Seedling stage or	Entire above-ground portion	20–30
	Before or during initial flowering	Two or three mature leaves at the top of the plant	25–30
Cabbage and the like (head crops)	Before heading	First mature leaves from center of whorl	10–20
Celery	Midgrowth	Petiole of youngest mature leaf	20–30
Cucumber	Before fruit set	Mature leaves near the base of the main stem	20–25
Leaf crops (lettuce, spinach, and the like)	Midgrowth	Youngest mature leaf	30–50
Melons	Before fruit set	Mature leaves near the base of the main stem	20–30
Peas	Before or during initial flowering	Leaves from the third node down from the top of the plant	30–50
Potato	Before or during early bloom	Third to sixth leaf from growing tip	20–30
Root crops (carrots, beets, onions, and the like)	Before root or bulb enlargement	Center mature leaves	25–35
Sweet corn	Before tasseling or	Entire fully mature leaf below the whorl	20–25
	At tasseling	Entire leaf at the ear node	20–25
Tomato (field)	Before or during early bloom stage	Third or fourth leaf from growing tip	20–25
Tomato (greenhouse)	Before or during fruit set	Young plants: leaves from second and third clusters	20–25
		Older plants: leaves from fourth to sixth clusters	20–25
Fruit and Nut Crops			
Apple, apricot, almond, cherry, peach, pear, plum	Midseason	Leaves near base of current year's growth	75–100
Grapes	End of bloom period	Petioles from leaves adjacent to fruit clusters	75–100
Lemon, lime	Midseason	Mature leaves from last flush of growth on nonfruiting terminals	30–40
Orange	Midseason	Spring cycle leaves, 4 to 7 months old, from nonfruiting terminals	25–30
Pecan	6–8 weeks after bloom	Leaves from terminal shoots,	30–45
Raspberry	Midseason	Youngest mature leaves on laterals of primo canes	25–40
Strawberry	Midseason	Youngest fully expanded mature leaves	50–70
Walnut	6–8 weeks after bloom	Middle leaflet pairs from mature shoots	30–40

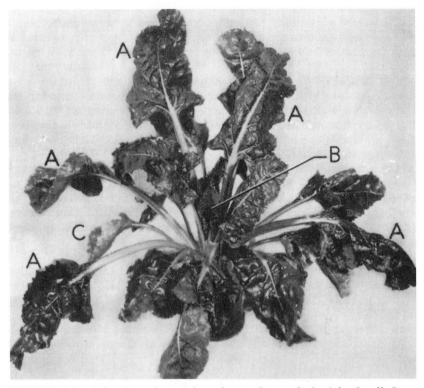

FIGURE 9.6   Selection of sugar beet leaves for analysis. A leaf stalk from any one of the recently matured, fully expanded leaves marked *A* may be included in the sample. The small leaves in the center and the old leaves should be avoided.   *Courtesy of the Potash & Phosphate Institute, Atlanta, Ga.*

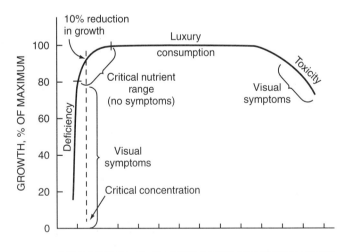

FIGURE 9.7   Relationship between nutrient concentration in the plant and crop yield. The critical nutrient range (CNR) represents an economic loss in yield without visual deficiency symptoms.

TABLE 9.3   Critical Nutrient Range for Macro- and Micronutrients

| Nutrient | Corn | | | |
	Whole Plant 24–45 Days*	Third Leaf, 45–80 Days[†]	Earleaf Green Silks[‡]	Earleaf Brown Silks[§]
N, %	4.0–5.0	3.5–4.5	3.0–4.0	2.8–3.5
P, %	0.40–0.60	0.35–0.50	0.30–0.45	0.25–0.40
K, %	3.0–5.0	2.0–3.5	2.0–3.0	1.8–2.5
Ca, %	0.51–1.6	0.20–0.80	0.20–1.0	0.20–1.2
Mg, %	0.30–0.60	0.20–0.60	0.20–0.80	0.20–0.80
S, %	0.18–0.40	0.18–0.40	0.18–0.40	0.18–0.35
B, ppm	6–25	6–25	5–25	5–25
Cu, ppm	6–20	6–20	5–20	5–20
Fe, ppm	40–500	25–250	30–250	30–250
Mn, ppm	40–160	20–150	20–150	20–150
Zn, ppm	25–60	20–60	20–70	20–70

*Seedlings 6 to 16 in. tall; 24 to 45 days after planting.
[†]Third leaf from top; plants over 12 in. tall; before silking.
[‡]70 to 90 days after planting.
[§]Grain in developing stage up to "roasting ear."
SOURCE: Schulte and Kelling, *National Corn Handbook, NCH-46,* Purdue Univ. Coop. Ext. Service.

| Soybean | | Small Grain | |
Nutrient	Sufficiency Range*	Nutrient	Sufficiency Range
N, %	4.3–5.50	N (winter grains)	1.75–3.00
P, %	0.3–0.50	(spring grains)	2.00–3.00
K, %	1.7–2.50	P	0.20–0.50
Ca, %	0.4–2.00	K	1.50–3.00
Mg, %	0.3–1.00	Ca (except barley)	0.20–0.50
		(barley)	0.30–1.20
Mn, ppm	21–100	Mg	0.15–0.50
Fe, ppm	51–350	S	0.15–0.40
B, ppm	21–55		
Cu, ppm	10–30	Mn	25–100
Zn, ppm	21–50	Zn	15–70
Mo, ppm	1.0–5.0	Cu	5–25

*Upper fully developed trifoliate leaves sampled before pod set.

Alfalfa

Plant Part	N	P	K	S	Ca	Mg
	%					
Top 6 in.	4.0–5.0	0.20–0.30	1.8–2.4	0.18–0.30	0.8–1.5	0.2–0.3
Upper one-third	—	0.18–0.22	1.7–2.0	0.20–0.30	—	—
Whole tops	—	0.20–0.25	1.5–2.2	0.20–0.24	1.4–2.0	0.28–0.32
N/S tops*	—	—	—	12–17	—	—

*N/S = N to S ratio

Grasses

	Fescue	Kentucky Bluegrass	Ryegrass	Coastal Bermuda
	%			
N	2.8–3.4	2.4–2.8	3.2–3.6	1.8–2.2
P	0.26–0.32	0.24–0.30	0.28–0.34	0.20–0.26
K	2.5–2.8	1.6–2.0	2.6–3.0	1.8–2.1

centrations of N seriously affect quality, the CNC for N is a maximum rather than a minimum.

In addition, it is difficult to determine an exact CNC since considerable variation exists in the transition zone between deficient and adequate nutrient concentrations. Consequently, it is more realistic to use the critical nutrient range (CNR), which is defined as that range of nutrient concentration at a specified growth stage above which the crop is amply supplied and below which the crop is deficient. Critical nutrient ranges have been developed for most of the essential nutrients in many crops (Table 9.3).

INCREASE IN YIELD WITH INCREASE IN NUTRIENT CONTENT   When a nutrient is deficient, increasing its availability will increase its content in the plant and the crop yield until the CNR is exceeded (Fig. 9.7). For example, applied N increased the % N in wheat (Fig. 9.8) and in corn (Fig. 9.12). Above the CNR, % N increases with no yield advantage.

Other examples of the relationship between nutrient concentration in the plant and plant growth or yield are shown in Figure 9.9. Plant analysis interpretations based on the CNR and sufficiency range concepts have limitations. Unless the crop sample is taken at the proper growth stage, the analytical results will have little value (Fig. 9.10). Also, considerable skill on the part of the diagnostician is needed to interpret the crop analysis results in terms of the overall production conditions.

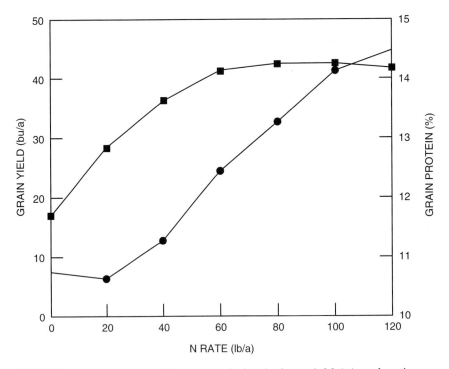

FIGURE 9.8   Influence of N rate on dryland wheat yield (■) and grain protein content (•). *Halvorson et al., N.D. Farm Res., 33:3–9, 1976.*

BALANCE OF NUTRIENTS   One of the problems in interpreting plant analyses is nutrient balance. Ratios of nutrients in plant tissue can be used to study nutrient balance in crops. For example, N/S, K/Mg, K/Ca, Ca + Mg/K, N/P, and other ratios are commonly used.

When a nutrient ratio is optimal, optimum yield occurs unless some other limiting factor reduces the yield. When a ratio is too low, a response to the nutrient in the numerator will be obtained if it is limiting. If the nutrient in the denominator is excessive, a yield response may or may not occur. When the ratio is too high, the reverse is true. These conclusions are supported by the following examples based on the assumption of an optimum range for N/S in a particular plant part, where crop yield is maximized.

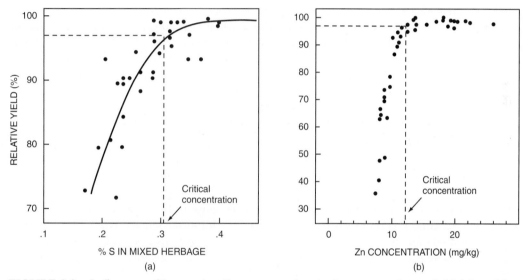

FIGURE 9.9   Influence of increasing S concentration in forage on clover yield (a) and increasing Zn concentration in leaves on wheat growth (b).

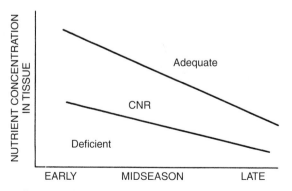

FIGURE 9.10   Generalized interpretive guide based on the concept of critical nutrient range (CNR) for tissue sampled at different times throughout the season.   *Dow and Roberts,* Agron. J., *74:401, 1982.*

When N/S is optimum or balanced, it is identified by a horizontal arrow ($\rightarrow$). Ratios above the optimum are recognized by an upward vertical arrow ($\uparrow$), and those below it are assigned a downward vertical arrow ($\downarrow$).

In situations with N/S = $\rightarrow$ (optimal range), three possibilities exist:

$$\frac{N\rightarrow}{S\rightarrow} \quad \text{or} \quad \frac{N\uparrow}{S\uparrow} \quad \text{or} \quad \frac{N\downarrow}{S\downarrow}$$

| Both numerator and denominator optimal | Both numerator and denominator excessive | Both numerator and denominator insufficient |

It is not possible to determine from the ratio alone which of the three situations is occurring in the plant. All that can be said is that the two nutrients are in relative balance.

Where the N/S ratio is either above or below the optimal range, two possibilities exist in each case:

$$\frac{N}{S} = \uparrow \frac{N\rightarrow}{S\downarrow} \quad \text{or} \quad \frac{N\uparrow}{S\rightarrow}$$
$$\text{S insufficiency} \qquad\qquad \text{N excess}$$

$$\frac{N}{S} = \downarrow \frac{N\rightarrow}{S\uparrow} \quad \text{or} \quad \frac{N\downarrow}{S\rightarrow}$$
$$\text{S excess} \qquad\qquad \text{N insufficiency}$$

With N/S above the optimal range, a response to S will be obtained only if S is lacking. If N is excessive and S is normal, additional S may not improve the yield. The same is true with respect to N when the N/S ratio is below the optimal range. This analysis demonstrates why, when a ratio has a given value outside the optimum range, a yield response is not always obtained.

Optimum nutrient ratios are established in the same way as sufficiency levels for individual nutrients are established (Fig. 9.11). These data illustrate that when S concentration < 0.12% or N/S < 17, wheat grain yield is likely to respond to S fertilization.

Chlorophyll Meters for Plant N Status

Evaluation of the nutrient status of the plant may not always involve destructive sampling of the plant or leaf. Use of a handheld chlorophyll meter may provide an indication of the leaf N status of the crop. Leaf chlorophyll content is related to N nutrition (in addition to other nutrients): therefore, measuring the relative chlorophyll content can provide an indication or index of the N status in the plant. Chlorophyll meter readings do not directly indicate the chlorophyll content, but the value recorded can be related to % N in leaf and grain yield as influenced by N rate (Fig. 9.12). Increasing the rate of fertilizer N increases grain yield and both % N in the leaf and leaf chlorophyll meter reading at V8 and silking growth stages in irrigated corn. For N management purposes, chlorophyll readings have greater value at the V8 stage, because a greater corn yield response

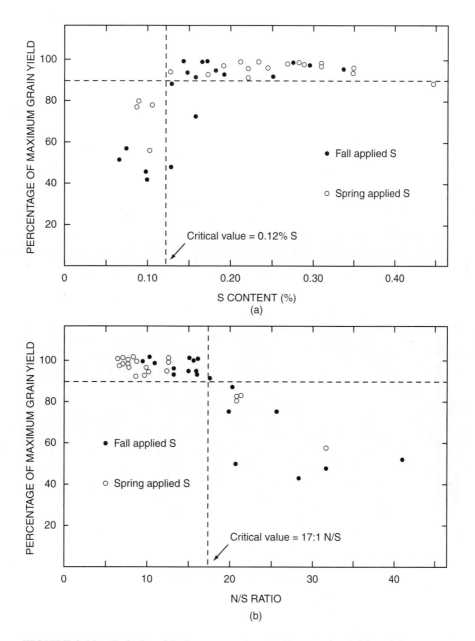

FIGURE 9.11 Relationship between winter wheat grain yield and S concentration (a) and N/S ratio (b). N and S were determined in whole plant samples collected at boot stage.

to additional fertilizer N at this growth stage would occur than with N applied later at silking.

The chlorophyll meter can also be used to identify S deficiency (Fig. 9.13). In this example, chlorophyll meter values <45 in the flag leaf indicate S-deficient wheat; the values are similar to those for N in corn leaves at V8 growth stage.

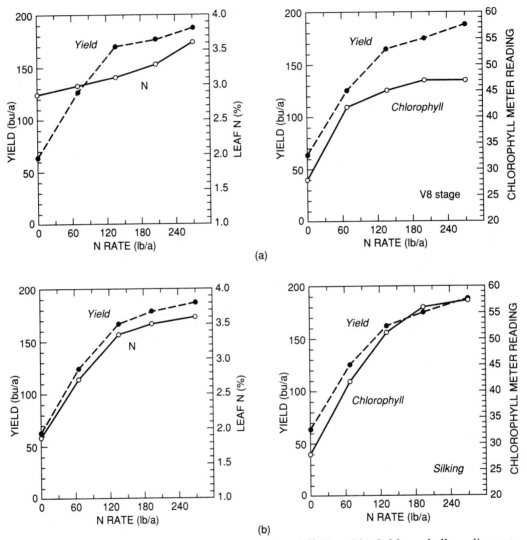

FIGURE 9.12   Relationship between corn yield and % N and leaf chlorophyll readings at V8 growth stage (a) and at silking (b) for irrigated corn.   *Schepers et al.,* Proc. Great Plains Soil Fert. Conf., *p. 42, 1992.*

Grain Analysis for N Sufficiency

Grain samples are often collected to provide nutritional information to the grower. For example, wheat grain protein can be used to indicate whether sufficient N was available for optimum yield or if insufficient N was applied to the crop (Fig. 9.14). These data show that when grain protein was <11.5%, the wheat crop most likely would have responded in grain yield to additional fertilizer N. Alternatively, > 11.5% grain protein indicated sufficient N availability for maximum grain yield. Although grain analysis can be very helpful in N management, it is a postmortem analysis. However, monitoring grain protein for several consecutive years will help growers to identify more accurately the appropriate N rate for a specific crop.

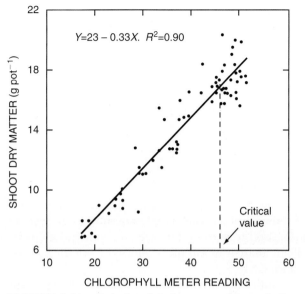

$Y = 23 - 0.33X.$ $R^2 = 0.90$

FIGURE 9.13   Relationship between chlorophyll content in flag leaves and shoot dry matter.

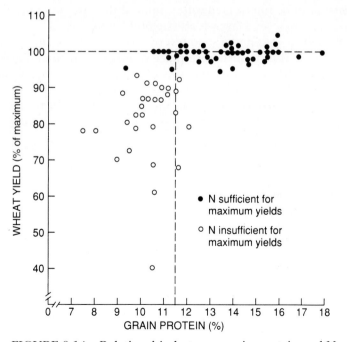

FIGURE 9.14   Relationship between grain protein and N sufficiency in winter wheat.   *Goos et al.*, Agron J., *74:130–133, 1982.*

Biological Tests

Field Tests

The field-plot method is essential for measuring the crop response to nutrients. After the specific treatments are selected, they are randomly assigned to an area of land. The treatments are replicated several times to obtain more reliable results and to account for variations in soil.

For example, when various rates of N are applied, the yield results are helpful in determining N recommendations. When many tests are conducted on well-characterized soils, recommendations can be extrapolated to other soils with similar characteristics.

Field tests are used in conjunction with laboratory and greenhouse studies in the calibration of soil and plant tests. Field experiments are essential in establishing the equation used to provide fertilizer recommendations that will optimize crop yield. Plant analysis of samples collected from the various treatments can also help establish CNR.

There can be difficulties with this technique when results from small plots exhibiting little spatial variability in soil are applied to large farm fields subject to much greater variability.

Strip Tests on Farmers' Fields

Narrow field strips undergoing selected nutrient treatments can help verify the accuracy of recommendations based on soil or plant tests (Fig. 9.15). The results of these tests must be interpreted with caution if they are unreplicated. Repetition of strip tests on several farms is also helpful.

Figure 9.16 illustrates how a strip test of several nutrient rates might be located in the field. It is important to place treatments in as similarly uniform areas as possible. If several soil types or conditions occur in the same field, locate treatments so that each soil type occurs equally in each treatment. The use of a yield-monitoring combine to measure and record treatment yields makes strip tests a valuable tool in assessing the accuracy of management recommendations.

Laboratory and Greenhouse Tests

Simpler and more rapid laboratory/greenhouse techniques utilize small amounts of soil to quantify nutrient availability. Generally, soils are collected to represent a wide range of soil chemical and physical properties that contribute to the variation in availability for a specific nutrient. Selected treatments are applied to the soils, and a crop is planted that is sensitive to the specific nutrient being evaluated. Crop response to the treatments can then be determined by measuring total plant yield and nutrient content. Figure 9.16 illustrates the use of a greenhouse test to separate Fe-deficient and Fe-sufficient soils. Soils were selected to represent a range in DTPA-extractable Fe (see Chap. 8). Sorghum plants show decreasing Fe deficiency as DTPA-extractable Fe increases.

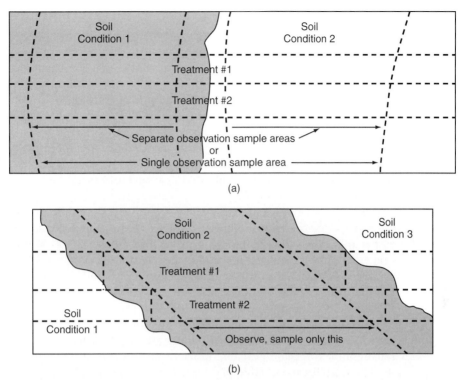

(a)

(b)

FIGURE 9.15   Example of strip tests located in a field with two (a) or three (b) soil types or conditions in the field.

FIGURE 9.16   Greenhouse test used to evaluate the ability of DTPA to separate Fe-deficient and Fe-sufficient soils. Sorghum was used as an indicator crop. Fe stress in sorghum decreased with increasing DTPA-extractable Fe (ppm).

Soil Testing

Although plant analyses are extremely valuable in diagnosing nutrient stress, analysis of the soil is essential in determining the supplemental nutrient requirement of a crop. A soil test is a chemical method for estimating the nutrient-supplying power of a soil. Compared with plant analysis, the primary advantage of soil testing is its ability to determine the nutrient status of the soil *before* the crop is planted.

A soil test measures part of the total nutrient supply in the soil and represents only an *index* of nutrient availability. Soil tests do not measure the exact quantity of a nutrient potentially taken up by a crop. To predict the nutrient needs of crops, the soil test must be calibrated against nutrient rate experiments in the field and in the greenhouse.

The quantity of nutrient extracted by the soil test should be closely related but not equal to the quantity of nutrient absorbed by the crop. For example, the Bray-1 extractable P in a 6-a (2.5-ha) field varies from about 20 to more than 80 ppm (Fig. 9.17). If the P soil test represents the P-supplying power of the soil, then the variability in % P in the crop should reflect the variability in Bray-1 extractable P. Comparison of the two graphs in Figure 9.17 shows that the spatial distribution of P concentration in wheat grain reflects the distribution of plant available P as measured by soil test P. Specifically, high and low soil test P results in high and low grain P, respectively. These data demonstrate the ability of a soil test to provide a reliable index of plant available nutrient, in this case P. They also illustrate the spatial variability in available plant nutrient levels in a field.

Objectives of Soil Tests

Information gained from soil testing is used in many ways:

1. *To provide an index of nutrient availability or supply in a given soil.* The soil test or extractant is designed to extract a portion of the nutrient from the same "pool" (i.e., solution, exchange, organic, or mineral) used by the plant.
2. *To predict the probability of obtaining a profitable response to lime and fertilizer.* Although a response to applied nutrients will not always be obtained on low-testing soils because of other limiting factors, the probability of a response is greater than on high-testing soils.
3. *To provide a basis for recommendations on the amount of lime and fertilizer to apply.* These basic relations are obtained by careful laboratory, greenhouse, and field studies.
4. *To evaluate the fertility status of soils on a county, soil area, or statewide basis by the use of soil test summaries.* Such summaries are helpful in developing both farm-level and regional nutrient management programs.

Expressed simply, the objective of soil testing is to obtain a value that will help to predict the amount of nutrients needed to supplement the supply in the soil. For example, a soil testing high will require little or no addition of nutrients, in contrast to soil with a low test value (Fig. 9.18).

Sufficiency levels are also commonly utilized in soil testing, where a high soil test might represent 90 to 100% sufficiency in supplying adequate plant available nutrients from the soil. Sufficiency levels decrease with decreasing soil test levels.

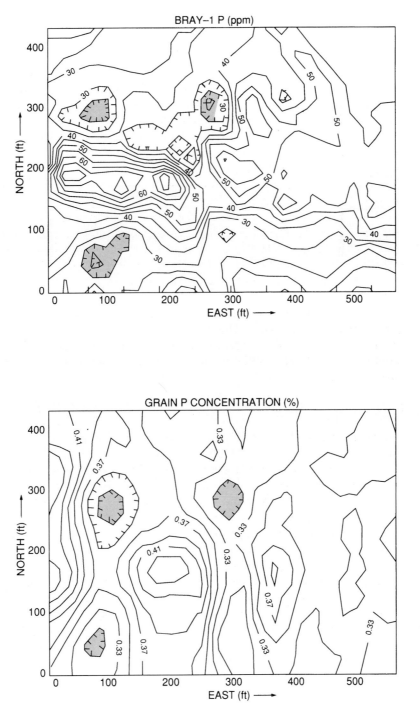

FIGURE 9.17   Spatial distribution of Bray-1 P and wheat grain P concentration over a 6-a field. Areas of high soil test P correspond to areas of high % P in the grain (left center region of both figures). Low Bray-1 P levels result in low % P in the grain (see two shaded areas above and one below high soil test P area). *Havlin and Sisson*, Proc. Dryland Farming Conf. *p. 406–408.,1990.*

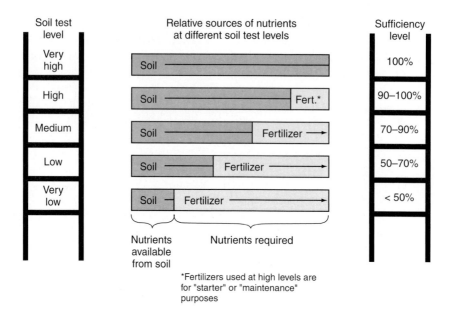

FIGURE 9.18   As soil test levels decrease, nutrient rates applied for maximizing yield potentials increase. High soil test levels represent a 90 to 100% sufficiency of the soil in supplying nutrients at needed levels. A low soil test level would represent a 50 to 70% sufficiency in supplying nutrients at needed levels.

The soil testing–nutrient recommendation system is comprised of four consecutive steps:

1. Collect a representative soil sample from the field.
2. Determine the quantity of plant available nutrient in the soil sample (soil test).
3. Interpret the soil test results (soil test calibration).
4. Estimate the quantity of nutrient required by the crop (nutrient recommendation).

Soil Sampling

The most critical aspect of soil testing is obtaining a soil sample that is representative of the field. Usually, a composite sample of only 1 pint of soil (about 1 lb) is taken from a field or sampling area, which represents, for example, a 10-a field or about 20 million lb of surface soil. There is considerable opportunity for sampling error; thus, it is essential that the field be sampled correctly. If the sample does not represent the field, it is impossible to provide a reliable fertilizer recommendation. The sampling error in a field is much greater than the error in laboratory analyses.

There are two approaches to soil sample collection: (1) sampling whole fields or parts of fields to provide "average" soil test value(s), or (2) describing spatial variability in soil test values. Currently, soil sampling to obtain the field average soil test is the most common method, but site-specific nutrient management in which the spatial distribution of soil test values is quantified will likely become a more common practice.

FIELD AVERAGE SAMPLING   The size of the sampling area for one composite sample varies greatly but usually ranges from 10 to 40 a or more. Areas that vary in appearance, slope, drainage, soil types, or past treatment should be sampled separately (Fig. 9.19), and small areas might be omitted from the sample. Each soil sample is a composite consisting of the soil from cores taken at several places in the field. The purpose of this procedure is to minimize the influence of any local nonuniformity in the soil. For example, in fields in which lime or nutrient has recently been applied, plant nutrients may be incompletely mixed with the soil. There may be areas where materials were unevenly distributed or where crop residue from harvesting was concentrated. A sample taken entirely from such an area would be completely misleading. Consequently, most recommendations call for sampling 15 to 40 locations over the field for each composite sample (Fig. 9.19). The soil samples from each area are mixed well, and a subsample is sent to the laboratory for analysis.

In the quantification of soil variability through intensive sampling, even on apparently uniform fields, plant nutrients can be highly variable and the values may not fall within a "normal" population distribution or "bell-shaped" curve. When

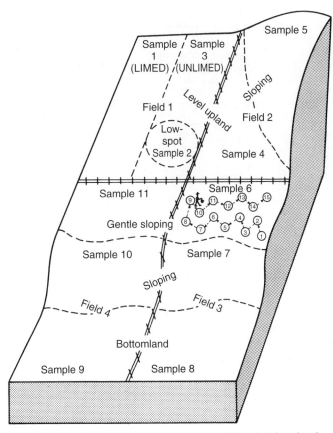

FIGURE 9.19   Samples that are representative of the field to be fertilized are important. The sampling pattern recommended by the various agricultural agencies should be followed.   *Courtesy of the Nebraska Agricultural Extension Service.*

nutrient levels are normally distributed the average soil test value is the one that occurs most frequently. If soil test results do not follow a bell shaped normal distribution, the data is skewed and the average does not represent the most frequently occurring value or "mode." The results in Table 9.4 show how the average soil test level in all cases was consistently greater than the mode which represents the largest percentage of the field. Fertilizer recommendations based on average soil test values would have been too low because the mode was much lower.

Soil tests can overestimate nutrient availability because of a few outliers in the data. This is clearly evident in Figure 9.20 which plots frequency distribution of soil test K data from an intensively sampled site. According to official recommendations, this field would not receive fertilizer K, yet 30% of the field tested low enough to require K and another 33% was marginal in K status. The small area testing high inflated the entire field average.

The problem of a few outliers can be particularly critical for a nutrient like S, where there is often extreme variability. It only takes one or two sample cores testing very high to distort the field average by so much that the results become useless.

SITE-SPECIFIC SAMPLING   Soils are heterogeneous, and wide variability can occur even in fields that appear uniform. Intensive soil sampling is the most effective way to quantify variability.

TABLE 9.4   Statistical Characteristics of Soil Tests from a 220 ft × 220 ft sampling grid (55 samples) at Stettler, Alberta

Nutrient	Range	Average	Mode
		ppm	
NO_3-N	2–24	11	8
P	0–104	15	9
K	127–598	276	155
SO_4-S	7–9440	480	10

SOURCE: Penney et al., *Proc. Great Plains Soil Fert. Conf.* 6:126 (1996).

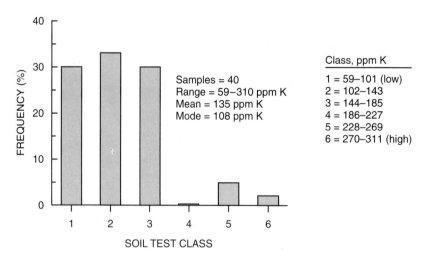

FIGURE 9.20   Frequency distribution of soil test K from a 220′ × 220′ sampling grid in Mundare, Alberta *(Penney et al. 1996).*

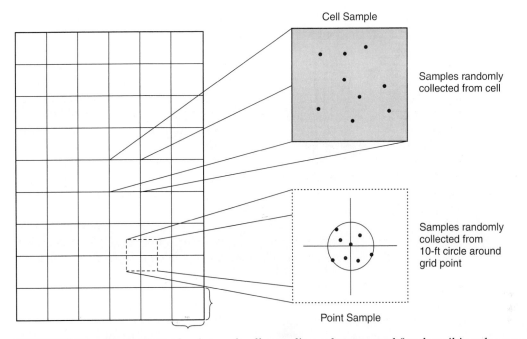

FIGURE 9.21   Illustration of point and cell sampling schemes used for describing the spatial distribution of soil properties.

Use of application equipment capable of variably applying nutrients according to variability in soil test levels for a specific nutrient requires intensive, georeferenced field sampling. Georeferenced soil sampling refers to the use of the Global Positioning System (GPS) to record the latitude and longitude of each sampling point.

GRID SAMPLING   Grid sampling consists of collecting equally spaced soil samples throughout the field and analyzing each sample separately (Fig. 9.21). Grid sizes range from 1 to 5 a, depending on the field. Typically 2- to 3-a grids are used, representing 300- to 360-ft square grids, respectively. Decreasing the grid size increases the number of samples collected in each field and thus associated costs. Figure 9.22 illustrates the effect of increasing grid size from 0.75 a (180-ft grid) to 7.5 a (570-ft grid) on the distribution Bray-1 P soil test level. These data show that 3.0-a grid size adequately describes the spatial variability.

Grid samples can be collected as *cell* or *point* samples (Fig. 9.21). With *cell* sampling, random samples are collected within each grid and composited. With large grids (e.g., 2 a), compositing samples within a cell will mask variability within the grid. To avoid the averaging that occurs with cell sampling, point samples can be collected (Fig. 9.21). With *point* sampling five to ten individual soil samples are composited from a 10-ft-diameter circle centered over each intersection of vertical and horizontal grid lines. Thus, with point sampling more of the within-field variability is quantified.

DIRECTED SAMPLING   To reduce the cost of grid sampling, sampling locations can be identified by using other spatially variable parameters in the field. Figure 9.23 illustrates that soil color, OM, and elevation could be used to identify

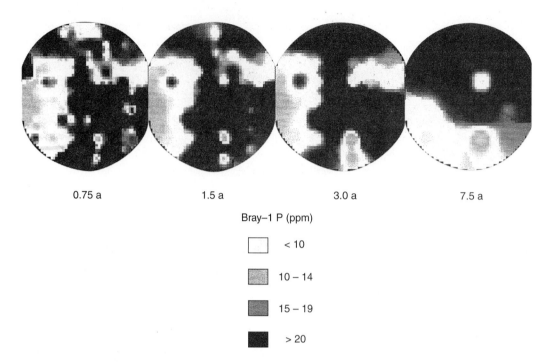

0.75 a   1.5 a   3.0 a   7.5 a

Bray–1 P (ppm)

☐	< 10
▨	10 – 14
▦	15 – 19
■	> 20

FIGURE 9.22   Influence of grid sampling size on the spatial distribution of Bray-1 P soil test levels; 3 a or less adequately described the variability.   *Havlin et al., Argon. Abstracts, p. 184., 1996.*

Soil Color   Elevation   Soil OM   Bray-1 P

FIGURE 9.23   The spatial distribution of soil color, using aerial black-and-white photography, is related to soil OM and soil test P. Dark soil color regions correspond to regions of high soil OM and Bray-1 P. Georeferenced soil sampling could be guided by the soil color map to reduce the samples required to grid sample the field.   *Courtesy of R. Ferguson and J. Schepers, Univ. of Nebraska, 1995.*

the spatial distribution of Bray-1 P. Yield-monitored data and other remote sensing information could also help direct specific sampling locations.

Other Soil Sampling Considerations

BANDED NUTRIENTS   Band application of immobile nutrients in the soil (i.e., P and K) often results in residual available nutrient in the old fertilizer bands for several years after application. Residual availability depends on the rate of application, soil chemical and physical properties, quantity of nutrient removed by the crop, crop rotation and intensity, and time after application. For example, the variation in soil test P level with P rate and method of placement is shown in Figure 9.24. Increasing the broadcast P rate increased soil test P. Similarly, increasing band-applied P increased soil test P in the band. Band-applied P increased soil test P more than the same rate of broadcast P. Thus, if only the bands are sampled, the soil test P is much higher than if none of the bands are sampled (i.e., if only the between-band areas are sampled). Few guidelines have been established for soil sampling fields in which immobile nutrients have been band applied. In wheat-fallow-wheat systems the following recommendation has been developed:

$$S = \frac{8 \,(\text{row spacing})}{30}$$

where S = ratio of off-band to on-band samples. Thus, for 30-cm (12-in.) band spacing, eight samples between the bands are required for every sample taken on the band. If similar recommendations do not exist in other regions, then increasing the sampling intensity should provide an adequate estimate of the average soil test level in the field.

DEPTH OF SAMPLING   For cultivated crops samples are ordinarily taken to the depth of tillage, which can vary from 6 to 12 in. (Fig. 9.25). Tillage generally mixes previous lime and nutrient applications within the tillage layer. When lime

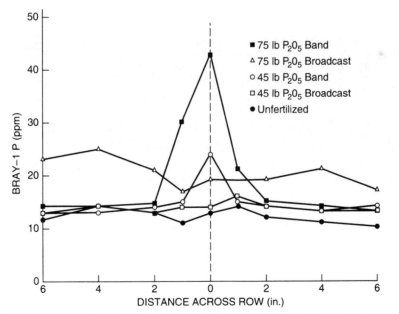

FIGURE 9.24   Effect of P rate and placement on Bray-1 P levels measured in 1-in. increments across the wheat row. P_2O_5 fertilizer was broadcast and band applied (2 in. below the seed) at 15, 45, and 75 lb/a at planting in September 1986. Soils were sampled in August 1988. The soil test levels for the 15 lb/a rate of P_2O_5 are not shown because they were similar to the unfertilized "check" treatment.   *Havlin et al.,* Proc. Fluid Fert. Found. *p. 193., 1989.*

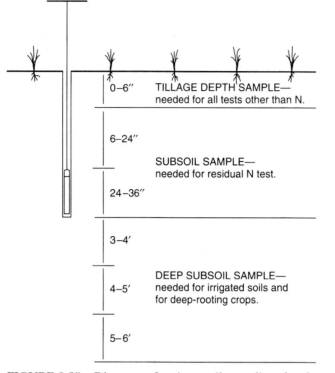

FIGURE 9.25   Diagram of various soil sampling depths used to collect samples for nutrient analyses.

and nutrients are broadcast on the surface for established pastures and lawns, a sample from the upper 2 in. is satisfactory. With no-till or minimum-till operations, it is best to take a sample from the surface 2 in. and another sample from the 2- to 8-in. layer. Considerable nutrient stratification occurs with reduced and no-till systems, as shown in Tables 9.5 and 9.6. Although plants obtain nutrients from below the tillage layer, little information is available for interpreting the analyses of subsoil samples.

Soil sampling of 2 to 6 ft in lower-rainfall areas is done to measure NO_3^- and moisture in the profile. The soil profile NO_3^- content is generally correlated with the crop response to N fertilization.

TABLE 9.5   Influence of Tillage on Stratification of Soil Test P and K

Depth (in.)	Plow	Chisel	No-Till	Plow	Chisel	No-Till
		Bray P_1 (ppm)			Exch. K (ppm)	
0–3	37	85	90	150	230	285
3–6	47	35	27	165	105	100
6–9	30	15	18	140	100	100
9–12	8	8	8	100	100	100

SOURCE: Mengel, *Agron. Guide AY-268,* Purdue Univ. (1990).

TABLE 9.6   Influence of Tillage and N Rate on Soil pH

Yearly N Rate (lb/a)	No-Till		Plow Tillage	
	0–2 in.	2–6 in.	0–2 in.	2–6 in.
0	5.75	6.05	6.45	6.45
75	5.20	5.90	6.40	6.35
150	4.82	5.63	5.85	5.83
300	4.45	4.88	5.58	5.43

SOURCE: Blevins et al., *Soil Tillage Res.* 3:136–46 (1983).

TIME OF SAMPLING   Ideally, samples should be taken just before seeding or when the crop is growing. However, these times are largely impractical because of constraints in taking samples, obtaining test results, and supplying the needed lime and fertilizer. Consequently, samples are customarily taken any time soil conditions permit. Samples from spring-planted crops are often taken in the fall after harvesting. In drier regions where NO_3^- levels are used to assess the N status of soil, sampling in the fall to diagnose the needs of annual spring-seeded crops is often postponed until the surface soil temperature drops to 5°C.

Most recommendations call for testing each field about every 3 years, with more frequent testing on coarse soils. In most instances, this frequency is sufficient to check soil pH and to determine whether the fertilization program is adequate for the crop rotation. For instance, if the P level is decreasing, the rate of application can be increased. If it has risen to a satisfactory level, application may be reduced to maintenance rates.

Soil Tests

Many chemical extractants have been developed for use in soil testing (Table 9.7). The ability of an extractant to extract a plant nutrient in quantities related to plant requirements depends on the reactions that control nutrient supply and availability (Fig. 2.1). An effective soil test will, to some extent, simulate plant removal of the specific nutrient, with subsequent resupply to solution from the nutrient pools (i.e., exchange, OM, mineral, and so forth) that control availability.

TABLE 9.7   Common Soil Test Extractants and the Nutrient Source, or Pool, Extracted in the Soil

Plant Nutrient	Common Extractants	Nutrient Source
NO_3^-	KCl, $CaCl_2$	Solution
NH_4^+	KCl	Solution/CEC
$H_2PO_4^-/HPO_4^{2-}$	NH_4F/HCl (Bray-P)	Fe/Al mineral solubility
	$NaHCO_3$ (Olsen-P)	Ca mineral solubility
	NH_4F/CH_3COOH (Mehlich-P)	Fe/Al and Ca mineral solubility
K^+	NH_4OAc	CEC
SO_4^{2-}	$Ca(H_2PO_4)_2$, $CaCl_2$	Solution/AEC
$Zn^{2+}, Fe^{3+}, Mn^{2+}, Cu^{2+}$	DTPA	Chelation
$H_3BO_3^0$	Hot water	Solution
Cl^-	Water	Solution

N SOIL TESTS   Generally, in regions where a crop's evapotranspiration demand exceeds annual precipitation, measuring the preplant profile NO_3^- content is valuable in predicting N requirements. However, in humid regions where annual precipitation normally exceeds evapotranspiration, leaching can occur; thus, profile N measurements have not been reliable in predicting fertilizer N requirements. Laboratories that utilize residual profile NO_3^- in making N recommendations are generally in regions where little or no water percolates below the root zone (Fig. 9.26).

Since NO_3^- is found predominantly in the soil solution, a simple water extract of the soil sample would recover a large part of soluble N. However, salt solution (i.e., 2 M KCl) is commonly used. The Cl^- would likely exchange for the small amount of exchangeable NO_3^- absorbed to positively charged sites on OM or minerals. The K^+ would remove NH_4^+ as well. Usually a 2- to 3-ft profile sample is collected before planting to provide plant available NO_3^-. Relationships between extractable NO_3^- and relative crop yield are shown in Figure 9.27. Typical calibration of the profile NO_3^- soil test for corn is shown in Figure 9.28.

Most soil-testing laboratories in regions where annual percolation below the root zone <17.5 cm use NO_3^- soil tests for predicting N requirements.

PRESIDEDRESS NO_3^- TEST (PSNT)   The recent development of a NO_3^- soil test for corn will improve the accuracy of N recommendations. In this test, 12-in. surface soil samples are collected between corn rows when corn is about 12 in. tall and analyzed for NO_3^- (Fig. 9.29). The PSNT has been calibrated in several

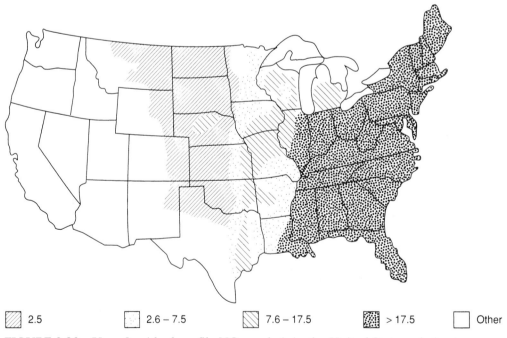

2.5   2.6 – 7.5   7.6 – 17.5   > 17.5   Other

FIGURE 9.26   Use of residual profile NO_3 analysis in the United States relative to average annual potential percolation (shading indicates values in cm) below the root zone. *Hergert*, SSSA Spec. Publ. No. 21, *1987*.

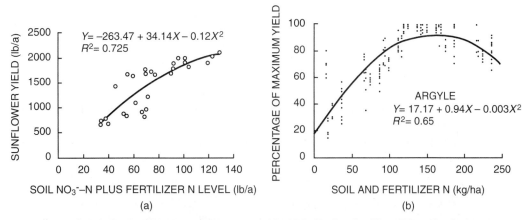

FIGURE 9.27 Relationship between extractable NO_3^- plus fertilizer N and relative crop yield of sunflower (a) and Kentucky bluegrass (b). *Black and Bauer, 1992.* Proc. Great Plains Soil Fert. Conf.

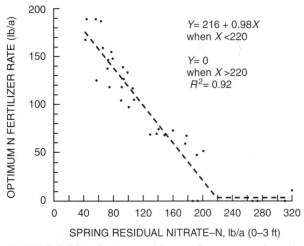

FIGURE 9.28 Optimum N rates for corn and spring soil nitrate content in Wisconsin soils.

states throughout the Northeast and Midwest (Fig. 9.30). These data suggest that when $NO_3^- > 20$ to 25 ppm, additional sidedress N applications are unnecessary.

The theoretical basis of this test is that the soil sample is collected at or near maximum N mineralization; thus, the contribution of organic N is more accurately quantified (Fig. 9.31). Applying a portion of the fertilizer N requirement at an early crop growth stage could increase fertilizer N efficiency through synchrony of fertilization with the period of maximum crop N uptake.

Sidedress N recommendations are commonly based on the critical PSNT level of 25 ppm NO_3^-. For the Iowa example (Fig. 9.32) maximum sidedress N recommendations for <10 ppm NO_3^- would be between 110 and 160 lb/a of N, depending on yield level. N recommendations decrease with increasing PSNT level.

New technologies may allow description of the spatial variation in N availability while reducing the dependence on soil sampling and analysis. Figure 9.33

shows how a grain protein sensor mounted on a yield-monitoring combine recorded the distribution of grain protein in winter wheat. As an indicator of N availability (Fig. 9.33), grain protein content was highly correlated with soil profile N content. These new technologies will ultimately enhance our ability to more accurately predict crop N requirements.

P SOIL TESTS   The soil tests commonly used for P are based on chemical principles that relate to inorganic P minerals (see Chapter 5). When solution P decreases with plant uptake, P minerals can dissolve or adsorbed P can be

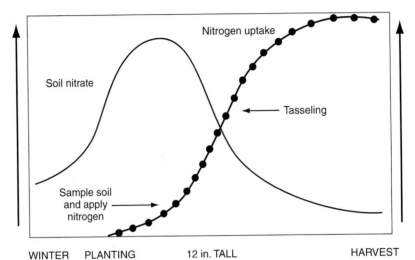

FIGURE 9.29   Relationship between soil NO_3^- and N uptake by corn through the growing season.

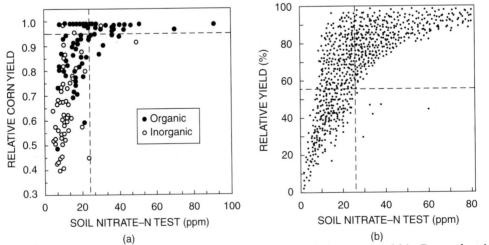

FIGURE 9.30   Relationship between soil NO_3^- and relative corn yield in Pennsylvania (a) and Iowa (b). *Organic* represents soils with a history of manure or legumes, whereas *inorganic* indicates soils without this history.   *Beegle*, Proc. Indiana Ag. Chem. Conf., *1982*.

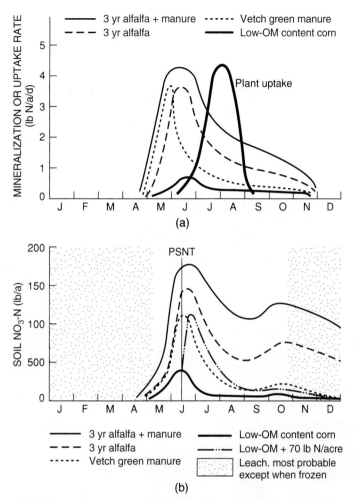

FIGURE 9.31 Synchronization of N mineralization and crop N uptake (*a*) and subsequent accumulation of NO_3 (*b*) as influenced by previous crop. Theoretically, the presidedress NO_3 test (PSNT) is determined from a soil sample collected at maximum N mineralization/NO_3 production, which corresponds to 12-in.-high corn.   *Magdoff,* J. Prod. Ag. *4:297–305, 1991.*

released to resupply soil solution P (Fig. 5.1). The chemical extractants for P simulate this process, because they reduce solution Al or Ca through precipitation as Al-P or Ca-P minerals. As solution Al or Ca decreases during extraction, native Al-P or Ca-P minerals dissolve to resupply solution Al or Ca. Solution P then concurrently increases, which provides a measure of the soil's ability to supply or buffer plant available P.

The *Bray-1* extractant was developed for use in acidic soils and contains 0.025 $M\,HCl + 0.03\,M\,NH_4F$. From Figure 5.14 we see that $AlPO_4$ is the primary P mineral controlling solution P concentration. The F complexes Al^{3+} in solution, and as the Al^{3+} concentration in solution decreases, $AlPO_4$ dissolves to buffer or resupply solution Al and release P into solution. The subsequent increase in solution P is measured, which represents an estimate of the soil's capacity to supply plant available P. The HCl in the extractant also dissolves Ca-P minerals present in slightly acidic and neutral soils.

The *Olsen* (Bicarb-P) extractant was developed for use in neutral and calcareous soils and contains 0.5 M NaHCO$_3$ buffered at pH 8.5. In these soils, Ca-P minerals control the solution P concentration (Fig. 5.15). The HCO$_3^-$ ion causes CaCO$_3$ to precipitate during extraction, which reduces the Ca^{2+} concentration

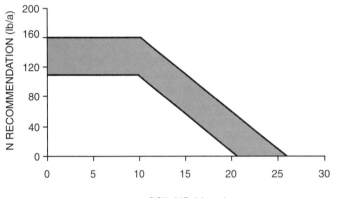

FIGURE 9.32   Sidedress N recommendations for corn in Iowa based on PSNT. For example, at 10 ppm NO$_3^-$ the range in N rates is 110 to 160 lb/a, whereas at 20 ppm NO$_3^-$, N rates of 10 to 60 lb/a would be recommended for sidedress application.   *Iowa State Univ. Extension, p. 1381, 1991.*

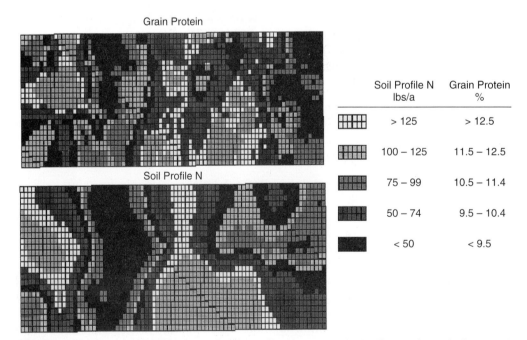

Soil Profile N lbs/a	Grain Protein %
> 125	> 12.5
100 – 125	11.5 – 12.5
75 – 99	10.5 – 11.4
50 – 74	9.5 – 10.4
< 50	< 9.5

FIGURE 9.33   Spatial distribution of soil profile N content is similar to that of wheat grain protein content determined with a grain protein sensor mounted on a yield-monitoring combine. Grain protein content is high where soil NO$_3^-$ is high.

in solution. Consequently, Ca-P minerals dissolve to buffer solution Ca^{2+} and release P into solution. As with the Bray-1 soil test, the increase in solution P provides a measure of the soil's ability to supply plant available P.

Although the Bray-1 and Olsen soil tests were developed for acidic and calcareous soils, respectively, both have been used to quantify plant available P in both soil types. For example, the Bray-1 soil test extracts P in quantities that are not equal to, but highly correlated with, Olsen-extractable P in calcareous soils. Provided that a given soil test has been carefully calibrated with the crop response to fertilizer P, either soil test can be used.

The *Mehlich* test, containing NH_4F + CH_3COOH + NH_4NO_3/HNO_3 or NH_4Cl/HCl, extracts P in the same manner as the Bray-1 test.

A less known method known as the Kelowna ($0.015N$ NH_4F + $0.25N$ HOAc) and modified Kelowna ($0.015N$ NH_4F + $0.5N$ HOAc + $1N$ NH_4Oac) is recommended for plant available P in western Canada, especially on high pH calcareous soils. It was found to be more accurate than older testing procedures including the Olsen method.

Soil tests for P reflect the relative responsiveness of crops to P fertilization on soils with varying extractable P levels. As soil test P increases, the response to P fertilization decreases, as measured by percentage of yield (Fig. 9.34).

Although soil test calibrations differ among regions and crops, general sufficiency levels for the common P soil tests are shown in Table 9.8. These categories show that the Bray-1 and Mehlich P tests extract similar quantities of P, while the Olsen P test extracts about half as much P.

K SOIL TESTS   The exchangeable plus soil solution K is usually extracted with 1 *M* NH_4OAc (Table 9.7). K^+ moves to the plant root through diffusion and mass flow, and the mechanisms of K release and adsorption complicate the measurement.

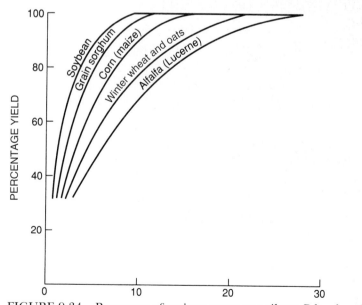

FIGURE 9.34   Response of various crops to soil test P level.   *Olsen et al.,* Nat. Corn Handbook., *1984.*

Laboratories vary in analytical approaches. Some adjust for CEC, removal in harvested crop, and/or yield goal. Extraction of some K^+ not available to crops can occur in soils, especially those that exhibit high K release or fixation. In general, the NH_4OAc soil test extracts K in concentrations related to K availability to plants (Fig. 9.35). A typical calibration of the NH_4OAc soil test for corn is shown in Table 9.9.

S Soil Tests   Like NO_3^-, SO_4^{2-} is very mobile in the soil; thus, in humid regions, extractable SO_4^{2-} has not been a reliable measure of the S status of the soil. However, a number of laboratories determine the water-, $Ca(H_2PO_4)_2$, or $CaCl_2$ extractable SO_4^{2-}. In some soils, especially in regions of low rainfall, these soil tests have been somewhat reliable. As previously discussed, most of the plant available SO_4^{2-} is supplied by mineralization of organic S during the growing season. Since organic S represents about 90% of total S in most soils, S soil tests that estimate mineralizable S might be more accurate in identifying S-deficient soils. Analysis

TABLE 9.8   Calibrations for the Bray-1, Mehlich III, and Olsen Soil Tests

P Sufficiency Level	Bray-1	Mehlich III	Olsen	Fertilizer P Recommendation	
		ppm		*lb P₂O₅/a*	*kg P/ha*
Very low	<5	<7	<3	50	25
Low	6–12	8–14	4–7	30	15
Medium	13–25	15–28	8–11	15	8
High	>25	>28	>12	0	0

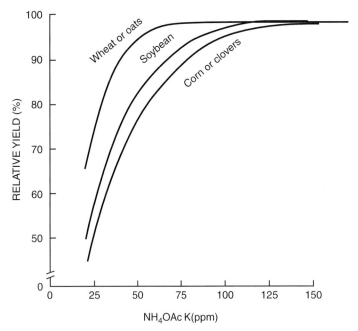

FIGURE 9.35   Relationship between NH_4OAc extractable K and relative yield of several crops.   *McLean, 1976.*

TABLE 9.9   Relationship between Soil Test K, Sufficiency Level, and K
Recommendations for Corn

NH_4OAc-K level (ppm)	Sufficiency Level	Sufficiency (%)	K Recommendation (lb/a)
<40	Very low	<50	120–160
41–80	Low	50–70	80–120
81–120	Medium	70–90	40–80
121–160	High	90–100	0–40
>160	Very high	100	0

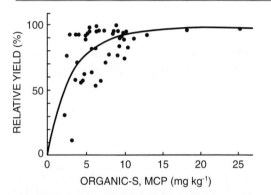

FIGURE 9.36   Relationship between relative annual yield of pasture and
monocalcium phosphate (MCP) extractable organic S.

of $Ca(H_2PO_4)_2$ or KH_2PO_4 extracts for total S has shown promise in estimating
extractable organic S. Total S in these extracts represents organic S + SO_4^{2-} and
should be a better indicator of S availability (Fig. 9.36). Extraction with 0.2M
KH_2PO_4 is the recommended S test in New Zealand, while 0.25 M KCl is a reliable S test in Australia. A typical calibration of the S soil test for mixed grass pasture is shown in Table 9.10.

Another way to assess the potential for S deficiency is to evaluate soil texture
and OM content. Crop response to S is more likely on coarse-textured, low-OM
soils (Chapter 7). Other factors to consider are (1) the crop requirement for S,
(2) the crop history, (3) the use of manures, (4) the proximity to industrial S
emissions, and (5) the S content of irrigation water.

FE, ZN, MN, CU SOIL TESTS   Chelate-micronutrient relationships and stability in soils are utilized in soil testing for micronutrients. Figure 8.10 shows that
when EDDHA is added to soil, it is 100% complexed with Fe over the pH range in
soil. Therefore, EDDHA might make a good extractant for Fe; however, Fe-EDDHA

TABLE 9.10   Interpretation of SO_4 and Organic S Soil Test Values

Soil S Status	SO_4 Test (ppm)	Organic S Test (ppm)	Interpretation
Deficient	0–6	0–10	10–20 lb/a of S
Adequate	7–12	10–20	0
Above optimum	>12	≥20	0

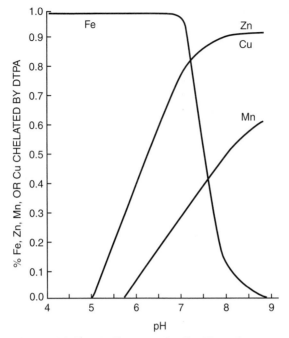

FIGURE 9.37   Influence of soil pH on the amount of Fe, Zn, Mn, or Cu chelated with DTPA. Over the pH range of most soils (pH 5 to 8), DTPA complexes micronutrients in quantities related to crop response.

is so stable that very few other micronutrient cations would be complexed with ED-DHA. Although Fe-DTPA is not as stable at high pH as Fe-EDDHA (Fig. 8.10), the other micronutrients (i.e., Zn and Cu) exhibit considerable stability with DTPA, especially at pH > 7 (Fig. 9.37).

Knowledge of chelate stability in soil provides the basis for developing the DTPA soil test for Fe, Zn, Cu, and Mn, which is used in most soil-testing laboratories. Before chelate relationships were developed, the most common micronutrient soil test was based on an acid extraction, usually HCl. Although some laboratories still use acid-extractable micronutrient soil tests, the DTPA test is preferred. An example of the correlation between DTPA-extractable Cu and relative barley yields is shown in Figure 9.38.

An example of calibration of the soil test for Zn with crop response is shown in Figure 9.39. About 90% of the soils testing below 0.65 ppm Zn responded to Zn, whereas 100% above this level did not respond. The DTPA soil test has been calibrated for most crops, and the general interpretation for DTPA-extractable micronutrients is shown in Table 9.11.

B, CL, MO SOIL TESTS   Extraction with hot water is the most common soil test for B. Critical levels for most crops are 0.5 ppm B or less. When hot water–extractable B is greater than 4 to 5 ppm, B toxicity may be evident.

Since Cl^- is soluble, a simple water extract is used as a Cl^- soil test. Like NO_3^-, soil samples should be taken to at least a 2-ft depth. The critical water-extractable Cl^- level is 7 to 8 ppm Cl^- for most crops.

No reliable Mo soil test has been developed. Mo deficiency is uncommon in the United States, although both water and NH_4 oxalate extracts have been used.

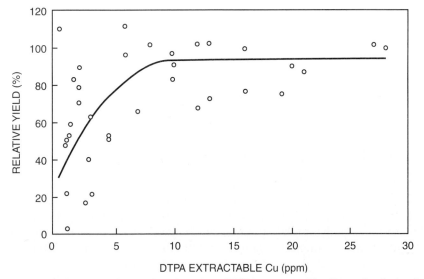

FIGURE 9.38   Relationship between DTPA-extractable Cu and relative barley yield.

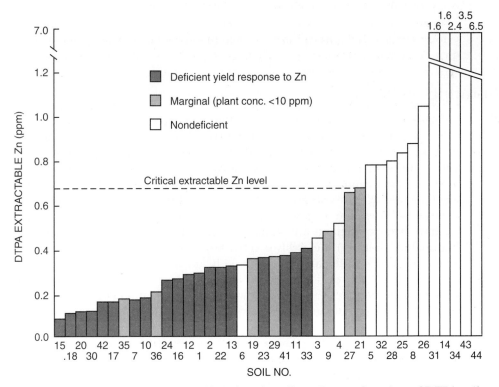

FIGURE 9.39   Corn response of 40 Colorado soils to Zn as a function of DTPA soil test levels.   *Havlin and Soltanpour, SSSAJ, 45:70–75, 1981.*

TABLE 9.11   DTPA-Extractable Fe, Zn, Cu, and Mn for Deficient, Marginal, and Sufficient Soils

Category	Fe	Zn	Mn	Cu
		ppm		
Low (deficient)	0–2.5	0–0.5	<1.0	0–0.4
Marginal	2.6–4.5	0.6–1.0	—	0.4–0.6
High (sufficient)	>4.5	>1.0	>1.0	>0.6

Ion Exchange Membranes in Soil Testing

Simultaneous extraction of plant nutrient ions from soils with ion exchange resins or membranes placed in intimate contact with soil provides a relatively under-utilized but reliable alternative to conventional soil testing. Ion exchange membranes closely simulate the nutrient removal action of plant roots over time and their measurements of plant nutrient availability are expected to be more biologically meaningful than those obtained with "extraction" by a particular chemical solution. They also have the advantage of elimination of collection, preparation and treatment of soil samples. Available P extracted from soil with anion exchange resin has long been considered the best measure of biologically available P.

Ion exchange resins in loose bead form or sown up in nylon net mesh bags and strips or sheets of membrane were used in the early testing of this method. A significant advance was made in the technology of the procedure by the introduction of ion exchange membranes encased in a plastic soil probe. Such probes are highly adaptable and are readily used either in-field or in the greenhouse or laboratory.

Ion exchange resin procedure can be used in routine soil testing and is being used in many laboratories.

Calibrating Soil Tests

Perhaps the greatest challenge in a soil testing program is the calibration of tests. It is essential that soil tests be calibrated against crop responses to applied nutrients in field experiments conducted over a wide range of soils. Yield responses from various rates of applied nutrients can then be related to the quantity of available nutrients in the soil indicated by the soil test. An accurately calibrated soil test (1) correctly identifies the degree of deficiency or sufficiency of the nutrient and (2) provides an estimate of the amount of nutrient required to eliminate the deficiency.

Controlled experiments are initially conducted in the greenhouse to provide information about (1) the ability of a soil test extractant to extract a nutrient in quantities related to the amount removed by the plant (i.e., to identify the best extractants), (2) the relationship between soil test level and relative yield and the CNR for various crops (see examples under Soil Tests), and (3) the range in soil test levels and crop responsiveness for the major soil types in the region.

After greenhouse studies have been completed, field calibration experiments are conducted on the major soil series and crops in the region. For example, if a P soil test is being calibrated, four to six rates of P will be applied and the crop response quantified by measuring yield (e.g., forage, grain, and fruit) and P content in the whole plant or plant part. The yield response data can be plotted in terms of percentage of yield or yield increase (Fig. 9.40). In Figure 9.40a,

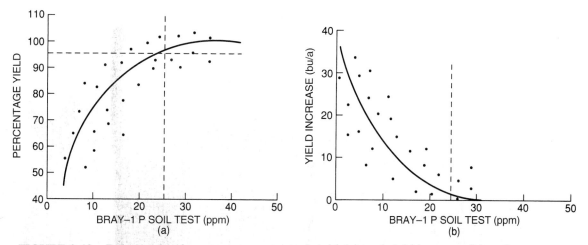

FIGURE 9.40   Relationship between percentage of yield (a) and yield increase (b) and Bray-1 P soil test level.

percentage of yield represents the ratio of the yield in the unfertilized soil to the yield obtained where P is nonlimiting (fertilized soil). For example, 70% yield means that the crop yield with the unfertilized soil is 70% of the yield obtained at the optimum level of P. Similarly, yield increase represents the increase in yield obtained with optimum fertilization (Fig. 9.40b). Thus, as soil test P increases, percentage of yield increases to 100%, which represents the soil test level at which there is no difference in yield between fertilized and unfertilized soil. Alternatively, as soil test P level increases, the yield increase to P fertilization decreases to 0 (Fig. 9.40b).

Generally, when percentage yield reaches 95 to 100% or when yield increase reaches 0 to 5%, the critical soil test level (CL) is obtained. The CL represents the soil test level above which no yield response to fertilization will be obtained. Soil test CLs can vary among crops, climatic regions, and extractants. For example, the CLs for the Bray-1 P, Olsen-P, and Mehlich-P tests are approximately 25, 13, and 28 ppm, respectively (Table 9.8).

Soil test calibration studies also provide the data to establish fertilizer recommendations. For example, at each location the P rate required for optimum yield can be determined and the results displayed similarly to Figure 9.41. Increasing soil test level corresponds to decreasing P rate required for optimum yield. These diagrams can be used to establish the fertilizer rates associated with very low, low, medium, and high soil test levels; however, most laboratories use an equation that describes the relationship in Figure 9.41.

Interpretation of Soil Tests

Many of the testing laboratories classify the fertility level of soils as very low, low, medium, high, or very high, based on the quantity of nutrient extracted, although the exact quantities are also reported. The probability of a response to fertilization increases with decreasing soil test level (Fig. 9.42). Soil fertility is only one of the factors influencing plant growth, but in general there is a greater chance of obtaining a response from a given nutrient with a low soil test result. For example, more than 85% of the fields testing very low may give a profitable increase; in the low range 60 to 85% may give increases, whereas in the very high

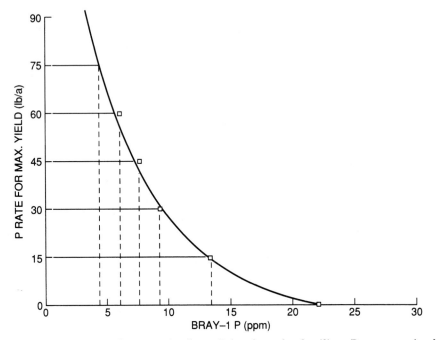

FIGURE 9.41   Influence of soil test P level on the fertilizer P rate required for maximum yield.

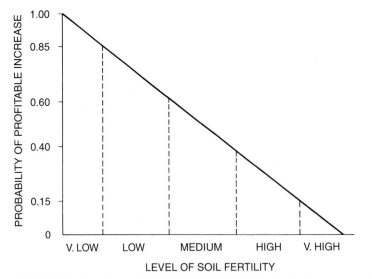

FIGURE 9.42   There is a greater probability of obtaining a profitable response from fertilization on soils testing low in an element than from soils testing high in that element.

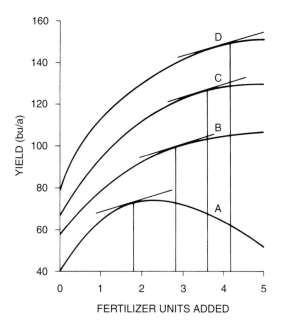

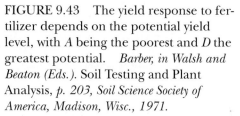

FIGURE 9.43   The yield response to fertilizer depends on the potential yield level, with *A* being the poorest and *D* the greatest potential.   *Barber, in Walsh and Beaton (Eds.).* Soil Testing and Plant Analysis, *p. 203, Soil Science Society of America, Madison, Wisc., 1971.*

range less than 15% will respond. These values are arbitrary, but they illustrate the concept of the probability of a response.

Soil test interpretation involves an economic evaluation of the relation between the soil test value and nutrient response. However, the response may vary due to several factors, including soil, crop, expected yield, level of management, and weather (Fig. 9.43). Factors *A–D* represent increasing yield potential or yield goal where *A* and *D* are low and high yield levels, respectively. However, *A–D* also represent other factors, including climate and crop/variety. For example, increasing rainfall from drought (*A*) to optimum rainfall (*D*) increases the yield potential and therefore the nutrient requirement. Different crops often respond differently to applied nutrients, where *A* might be a relatively unresponsive crop (e.g., soybeans) while *D* is extremely responsive (e.g., alfalfa or wheat) (Fig. 9.34).

Many laboratories make one recommendation, assuming best production practices for the region, and the grower may make adjustments as necessary. As technology and management practices improve or as economic incentives increase, yield potential and recommendations increase. *For the grower, the goal is to maintain plant nutrients at a level for sustained productivity and profitability, which means that nutrients should not be a limiting factor at any stage, from plant emergence to maturity.*

PERCENTAGE OF SUFFICIENCY AND MOBILITY   The interpretation of soil tests for purposes of making nutrient recommendations is influenced by the mobility of the nutrient in question. With mobile nutrients, crop yield is proportional to the total quantity of nutrient present in the root zone, because of minimal interaction with soil constituents (e.g., CEC and OM) (Fig. 9.44). Recall that for NO_3^-, SO_4^{2-}, and Cl^-, a 2- to 3-ft soil profile sample is important for accurately assessing plant availability. In contrast, yield response to immobile nutrients (i.e., $H_2PO_4^-$, K^+, Zn^{2+}) is proportional to the concentration of nutrients near the root surface, because these nutrients strongly interact with or are buffered by soil constituents (Fig. 9.44).

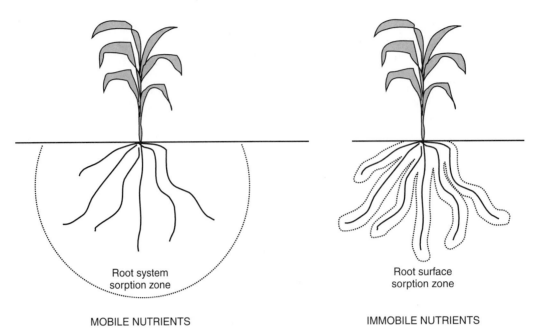

MOBILE NUTRIENTS   IMMOBILE NUTRIENTS

FIGURE 9.44   Difference in nutrient extraction zones between mobile and immobile nutrients in the soil. For mobile nutrients, the soil test must estimate the total nutrient available in the root zone. For immobile nutrients, the soil test is an index of the quantity available to the plant. *Courtesy B. Raun and G. Johnson, Oklahoma State Univ.*

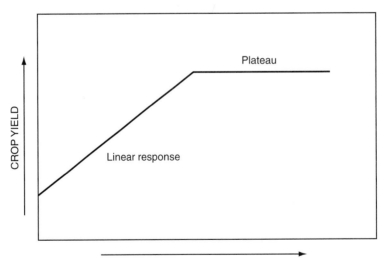

FIGURE 9.45   Illustration of the linear crop response to mobile nutrients in the soil. Yield potential is limited by the total mobile nutrient available in the root zone.

The *mobility concept* refers to the situation in which crop response to an increasing concentration of mobile nutrients is linear until the yield potential for the given environment is achieved or is limited by the depletion of the mobile nutrient or other yield limiting factors (Fig. 9.45). For example, if the profile N soil test indicated sufficient N available for only 100 bu/a, additional N would be recommended if the yield goal was 150 bu/a. N recommendations are thus usually based on yield goal, where the amount of N required to produce each unit of yield is known (i.e., 2.4 lb/bu of N for winter wheat, see page 350). This concept is also evident when sidedress or topdress N recommendations are made during the growing season because better-than-average growing conditions increased yield potential above initial estimates provided before planting.

With N, some evidence exists that yield goal may not be important in developing N recommendations. Long-term N response data show that the optimum N rate was similar for both high- and low-yielding years (Figs. 9.46 and 9.47). Simi-

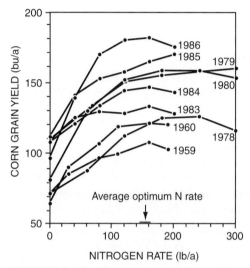

FIGURE 9.46   Typical corn yield response to applied N on Plano silt loam soils. Bar represents ± standard error of the mean of economically optimum N rates. *Vanotti and Bundy,* J. Prod. Agric. *7:243.,1994.*

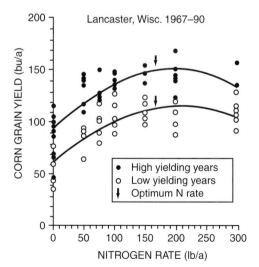

FIGURE 9.47   Corn yield response to applied N and economic optimum N rates in high- and low-yielding years.   *Vanotti and Bundy,* J. Prod. Agric. *7:243.,1994.*

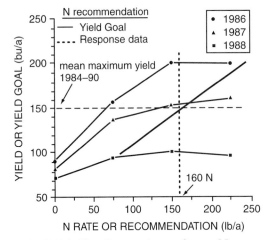

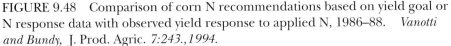

FIGURE 9.48   Comparison of corn N recommendations based on yield goal or N response data with observed yield response to applied N, 1986–88.   *Vanotti and Bundy,* J. Prod. Agric. *7:243.,1994.*

lar N requirements for different yield goals may be due to greater recovery of available N in the soil profile observed in years with optimum moisture and temperature conditions.

Comparison of the two systems of making N recommendations (yield goal and response data) shows errors in N rates with very low or high yield goals (Fig. 9.48). Although N recommendations based on N response data may be more accurate, substantially more field calibration data for the major soils and regions of interest are required.

With immobile nutrients, crop yield potential is limited by the quantity of nutrient available at the soil-root interface (Fig. 9.44). Generally, soil solution concentrations of immobile nutrients are low, and although some replenishment occurs through exchange and mineralization reactions, the solution concentration in the soil-root zone can be depleted to levels that limit yield. Plant growth and yield will be limited to the extent the immobile nutrient is deficient. Alternatively, the yield potential is governed by the *sufficiency* of immobile nutrient supply. For example, with <100% sufficiency, crop yield potential is less than that possible under the given environment. Thus, with 80% sufficiency, actual crop yield is only 80% of yield potential, regardless of yield level. Soil tests for immobile nutrients provide an index of nutrient availability that is independent of environment. In a good year more roots will explore soil with similar soil test levels, and if these levels are <100% sufficient to support the growth, yield will be reduced by this percentage. Examples of percentage of sufficiency in relation to soil test levels are provided in Figure 9.18 and Table 9.9.

In contrast, some recommendation models for immobile nutrients account for yield potential similarly to N recommendations. Incorporation of yield potential into the P or K recommendation, for example, is based on "replacement" of nutrients removed as a function of yield level. For example, 70 bu/a wheat production would ultimately deplete soil test P if nutrients were applied for 40 bu/a yield potential. Alternatively, P rates applied for 70 bu/a production are inappropriate for 40 bu/a yields. Significant soil test P buildup would occur with overfertilization. As a result, some laboratories provide recommendations for immobile nutrients based on soil test level and yield potential (Table 9.12).

TABLE 9.12   P and N Recommended for Corn

Soil Test Bray P_1 (lb/a of P)	Yield Goals (bu/a)		
	80	120	160
	Annual Application of P_2O_5 (lb/a)		
5	55	70	85
15	45	60	75
25	35	50	65
30–60	30	45	60
75	20	30	45
90	20	20	30
	Annual Application of N (lb/a)		
Continuous corn	40	115	200

SOURCE: *Ohio Agronomy Guide 1983–1984.* Cooperative Extension Service, Ohio State Univ., Columbus (1983).

Whether the recommendation for immobile nutrients includes yield goal is not the important issue. A regular soil-testing program that allows accurate assessment of the change in soil test level with time as nutrients are added and removed will guide the producer to adjust rates depending on whether soil test levels are increasing or decreasing. Buildup of P or K with overfertilization is unnecessary; however, underfertilization that results in decreasing soil test levels below the critical level is more costly and should be avoided.

NUTRIENT RESPONSE FUNCTIONS   The most common models used in fitting nutrient response data are shown in Figure 9.49 and are given by the following:
Linear-plateau model:

$$Y = mx + b \qquad \text{linear portion with slope} = m$$
$$Y = b \qquad \text{plateau portion with slope} = 0$$

Exponential model:

$$Y = e^x$$

Quadratic model:

$$Y = a + bx + cx^2$$

where Y = yield, x = nutrient rate, and the letters a, b, c, and m represent constants or coefficients.

All of the preceding equations have been used to describe the yield response to both immobile and mobile nutrients. Regardless of the equation used, the response function will vary with the crop (Fig. 9.34), yield potential (Fig. 9.43), soil test level (Figs. 9.50 and 9.51), year (Fig. 9.48), previous crop (Fig. 9.53), and other factors discussed in Chapter 10.

N RECOMMENDATION MODEL   With mobile nutrients such as NO_3^-, SO_4^{2-}, and Cl^-, buildup and/or maintenance programs are not practical, because these

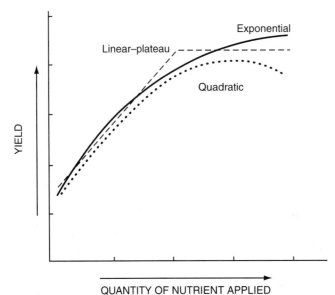

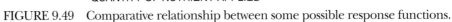

FIGURE 9.49   Comparative relationship between some possible response functions.

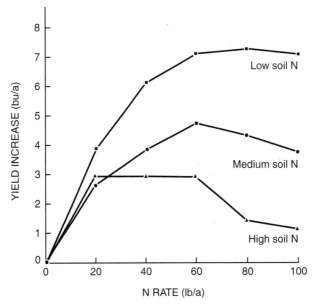

FIGURE 9.50   Interaction between wheat grain yield response to fertilizer N and soil profile NO_3. *Thompson*, Kansas Agric. Exp. Sta. Bull. 590, *1976*.

nutrients can readily leach below the root zone in many soils. Preventing potential NO_3^- contamination of groundwater by leaching while providing sufficient N for profitable crop production requires accurate N recommendations. N recommendations require knowledge of the quantity of N needed by the crop and supplied by the soil. In general, N recommendations are based on the following formula:

$$N_{fert} = N_{crop} - N_{soil} - (N_{OM} + N_{prev\ crop} + N_{manure}) \tag{1}$$

where

N_{fert}	= fertilizer N recommendation
N_{crop}	= yield goal × N yield
N_{soil}	= preplant soil profile NO_3^- content
N_{OM}	= organic N mineralization
$N_{prev\ crop}$	= legume N availability
N_{manure}	= manure N availability

The N_{crop} represents the N required by the crop and involves predicting the crop yield and the N needed to produce that yield. Underestimating the yield goal can cause considerable yield loss due to underfertilization. Alternatively, overestimating the yield goal results in overfertilization, which can greatly increase the profile N content after harvest and increase the potential for groundwater contamination if N leaches below the root zone.

Many growers commonly overestimate their yield goals and thus apply N in excess of the crop requirement (Fig. 9.52). These data illustrate that only 10% of growers attained their desired yield goals, and only 50% of growers reached 80% of their yield goal. More important, overestimating the yield goal by these producers resulted in N recommendations that were 40 lb/a greater than those required for a more realistic yield goal.

The quantity of N needed to produce the yield goal also varies among crops, regions or climates, and laboratories making the N recommendation (Table 9.1). For example, 1 bu of corn contains 0.7 lb N (N yield in eq. 1), assuming that all of the N applied enters the corn grain. Unfortunately, N recovered by the grain

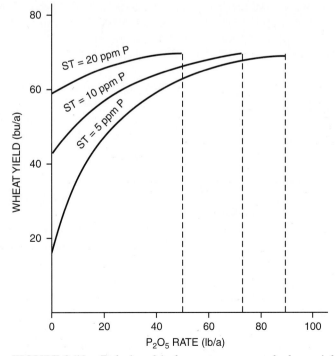

FIGURE 9.51   Relationship between expected wheat yield and P_2O_5 rate at three soil test (ST) levels.

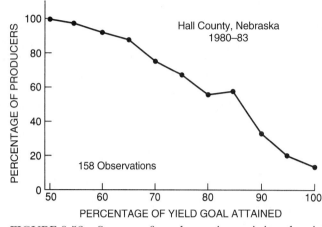

FIGURE 9.52   Success of producers in attaining the yield goal for irrigated corn.   *Schepers and Martin,* Proc. Agric. Impacts on Groundwater. *Well Water Journal Publ. Co., Dublin, Ohio, 1986.*

can vary between 40 and 75%; thus, usually 0.9 to 1.7 lb N/bu grain is used for "N yield" to determine N_{crop}. With winter wheat, for example, a 1.8- to 2.4-lb N/bu yield goal is used to estimate the N required. Thus, if the yield goal is 60 bu/a, the total N required by the crop would be 120 lb/a (using 2.0 lb/bu).

Once N_{crop} is estimated, this value is reduced by the N credits or potential N available in the soil. Profile N soil testing was discussed previously (p. 332). After adjustments for soil profile NO_3^- content, N_{crop} is adjusted for potential N mineralization from soil OM, legume N, and manure N. The N credit from manure varies with the rate applied and is approximately 5 lb N/ton in the first year following manure application. Generally, 50% of the manure N is available in the first year, 25% in the second year, and none in the third year.

The N credit for previous legume crops depends on the legume crop, the productivity of the legume (i.e., yield), and the length of time after the legume crop was rotated to the nonlegume crop (see Chapters 4 and 13). N credits from forage legumes are generally much greater than those from grain legumes, although low-yielding forage legumes can fix less N_2 than high-yielding grain legumes (Table 4.6).

The generalized relationship between N required by corn grown continuously or in rotation with alfalfa and mixed hay is shown in Figure 9.53. When nonlegume crops are grown on soils previously cropped to a forage legume, potential legume N mineralization decreases with time (Fig. 4.10). Thus, very little fertilizer N is required in the first year following the legume compared with subsequent years. The number of years a legume N credit is used in adjusting the fertilizer N rate also depends on the quantities of N fixed, which depends on the specific legume and its productivity. The N credit for soybeans, a grain legume, is usually 1 lb/bu soybean yield.

Few laboratories use a term for N_{OM} in their N recommendation models because of the difficulty in accurately estimating the quantity of N mineralization under variable climate (moisture and temperature) conditions from year to year. Although many tests have been evaluated, estimates of potential N mineralization from soil OM have not always been highly correlated with N availability. Some laboratories reduce the coefficient multiplied by yield goal (1.8 lb/bu of N instead of a higher value, for example) to account for some N mineralization. Alternatively, some laboratories use a measure of soil OM as an indicator of potential mineralizable N. Credits for N_{OM} generally range from 20 to 80 lb/a of N and are primarily based on the % OM content in the soil. For example, as % OM increases in soils, N_{OM} increases; thus, the N recommendation will decrease (Table 9.13).

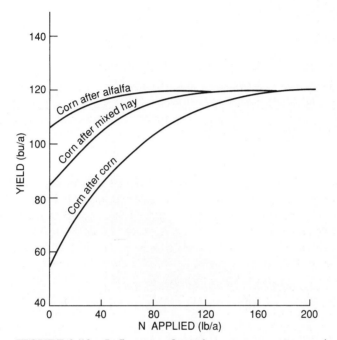

FIGURE 9.53   Influence of previous crop on corn grain yield response to N fertilization.   *Barber, SSSA Spec. Publ. No. 2., 1967.*

TABLE 9.13   Influence of Soil Profile NO_3^- and OM Content on Fertilizer N Recommendations for Dryland Winter Wheat in Colorado

NO_3^- * Soil Test (ppm)	Soil OM % (Surface Sample)		
	0–1.0	*1.1–2.0*	*>2.0*
	Fertilizer N (lb/a)		
0–4	75	75	65
5–8	75	75	50
9–12	75	60	35
13–16	70	45	20
17–20	60	30	5
21–24	45	15	0
25–28	30	0	0
29–32	15	0	0
>32	0	0	0

*0- to 2-ft sample.

SOURCE: *Guide to Fert. Recommendations,* XCM-37, Colorado State Univ. (1985).

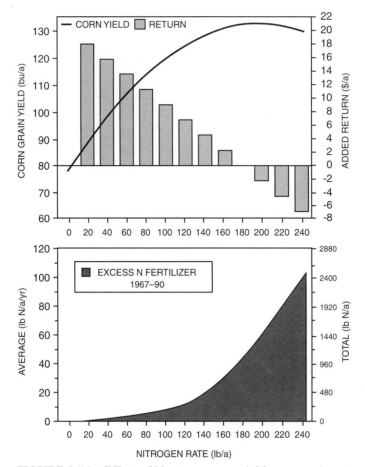

FIGURE 9.54   Effect of N rate on corn yield, economic return from fertilizer N, and amounts of excess N applied, Lancaster, Wisc., 1967–90.   *Vanotti and Bandy,* J. Prod. Agric. *7:243.,1994.*

In general, when the fertilizer N rate exceeds the N requirement for optimum crop yield, considerable quantities of residual fertilizer N exist after harvesting, which represents potentially leachable NO_3^- (Fig. 9.54). However, when the fertilizer N recommendation is estimated by accurately quantifying N_{crop} and the N credits, maximum recovery of fertilizer N by the crop as well as net return occurs. The data in Figure 9.54 illustrate that when fertilizer N exceeded the N rate needed for maximum yield (160 lb/a), residual profile NO_3^- after harvest increased greatly. Since rooting depth is only about 5 to 6 ft for corn, a portion of the profile NO_3^- moves below the root zone and thus is unavailable to the next crop. Eventual contamination of the groundwater could occur as this N continues to leach down through the profile. These data demonstrate the importance of accurately quantifying fertilizer N requirements for optimum production and minimum environmental risk.

IMMOBILE NUTRIENT RECOMMENDATIONS   Figure 9.55 represents several management options for immobile nutrients.

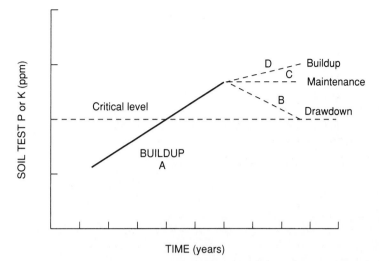

FIGURE 9.55   Diagram representing several management options for immobile nutrients. When the soil test level is increased above the CL, a grower could utilize residual fertilizer P reserves, which would cause the soil test level to decline (drawdown—B). Sufficient fertilizer P could be applied to maintain the soil test level (maintenance—C) or the soil test could be further increased with higher nutrient rates (buildup—D). The latter program would be advisable only if the grower knew an economic crop response to the higher rate was probable or if a crop with a higher P requirement followed in the cropping sequence.

Buildup.   When the soil test is below the CL, it may be desirable to apply rates of a nutrient to increase the soil test to the CL or above (*A* in Fig. 9.55). Generally, applications of 10 to 20 lb/a of P_2O_5 are required to increase the soil test P level 1 ppm, depending on the soil. Similarly, to increase the soil test K level 1 ppm, 5 to 10 lb/a of K_2O are needed.

When the soil test is above the CL (*D* in Figure 9.55), continued applications to increase it further are advisable only if the grower or consultant believes that an economic yield response is probable. In contrast, if the soil test is above the CL, it may be possible, depending on the crop, to "draw down" (*B*) the soil test level by not fertilizing the crop. Although this can be a viable option, soil test levels can decline rapidly; thus, annual soil testing would be required.

Maintenance.   When the soil test is at or above the CL, it can be maintained by fertilizer rates that replace losses by crop removal, erosion, and fixation. This approach demands well-calibrated tests, and there is a considerably greater chance of error. When capital is limited, the area is new, or the land is being rented, the maintenance program is probably the preferred method.

In some rotations, such as corn-soybeans, the corn is usually fertilized at sufficient rates for both crops. In double cropping, sufficient P and K are also recommended for both crops. Again, the removals, losses, and any buildup must be considered in making the recommendations.

With any fertilizer management program for an immobile nutrient, the soil must be monitored periodically to determine whether the soil fertility level is decreasing or increasing. The data in Figure 9.56 show that soil test P declined below the critical level when the irrigated alfalfa was not fertilized or was fertilized at a low P rate (50 lb/a of P_2O_5). In contrast, the P soil test was maintained at or slightly above the initial level with annual applications of 100 lb/a of P_2O_5. Notice how soil test level increased with a buildup rate of P (150 lb/a initially and every 3 years) and subsequently decreased with crop removal.

Decisions to maintain, draw down, or build up a soil test level dramatically affect yield potential, nutrient use efficiency, water quality, and profitability. Immobile nutrient management decisions must be based on the information provided from a periodic or annual soil-testing program. Figure 9.57 illustrates the relationship between soil test level and the relative nutrient rate needed for each management decision.

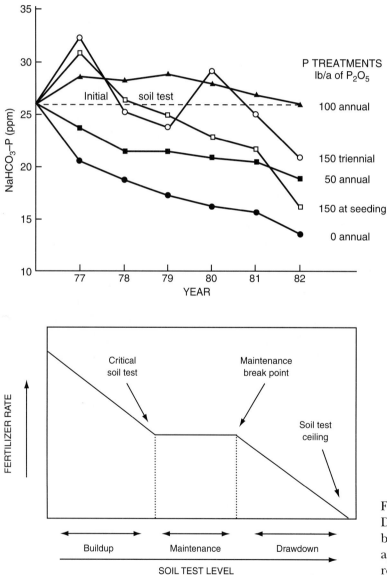

FIGURE 9.56 Influence of fertilizer P on the buildup, maintenance, and decline of soil test P in irrigated alfalfa. *Havlin et al.,* SSSAJ, *46:331–36, 1984.*

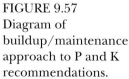

FIGURE 9.57 Diagram of buildup/maintenance approach to P and K recommendations.

Selected References

BROWN, J. R. (Ed.). 1987. *Soil Testing: Sampling, Correlation, Calibration, and Interpretation.* Special Publication No. 21. Soil Science Society of America. Madison, Wisc.

HAVLIN, J. L., and J. S. JACOBSON (Eds.). 1994. *New Directions in Soil Testing and Nutrient Recommendations.* No. 40. Soil Science Society of America, Madison, Wisc.

WALSH, L. M., and J. D. BEATON. 1973. *Soil Testing and Plant Analysis.* Soil Science Society of America, Madison, Wisc.

WESTERMANN, R. L. (Ed.). 1990. *Soil Testing and Plant Analysis.* No. 3. Soil Science Society of America, Madison, Wisc.

WHITNEY, D. A., J. T. COPE, and L. F. WELCH. 1985. Prescribing soil and crop nutrient needs. In O. P. ENGELSTAD (Ed.), *Fertilizer Technology and Use.* Soil Science Society of America, Madison, Wisc.

CHAPTER 10

Fundamentals of Nutrient Management

Efficient nutrient management programs supply adequate quantities of plant nutrients required to sustain maximum crop productivity and profitability while minimizing the environmental impact from nutrient use. Substantial economic and environmental consequences can occur when nutrients limit plant productivity. Ensuring optimum nutrient availability through effective nutrient management practices requires knowledge of the interactions between the soil, plant, and environment.

Crop Characteristics

Nutrient Utilization

The approximate quantity of nutrients required by crops (Table 9.1) varies depending on crop characteristics (crop, yield level, and variety or hybrid), environmental conditions (moisture and temperature), soil characteristics (soil type, soil fertility, and landscape position), and soil and crop management. Although these interacting factors affect the nutrient content in plants and recovery of applied nutrients, the accumulation of nutrients during the growing season generally follows a plant growth pattern (Fig. 10.1). The shape of the curve varies among plants, but nearly all plants exhibit a rapid or exponential increase in growth and nutrient accumulation rate up to a maximum, followed by a period of decline. Some plants exhibit rapid nutrient uptake and growth early in the growing season, while other plants exhibit maximum growth rate much later (Fig. 4.12). Soybeans, for example, exhibit changes in the nutrient uptake pattern through the season (Table 10.1). Regardless of the shape of the growth curve, nutrients are needed in the greatest quantities during periods of maximum growth rate. Thus, nutrient management plans are designed to ensure adequate nutrient supply before the exponential growth period.

Figure 10.2 illustrates the N uptake pattern for winter wheat as it relates to wheat growth stage. In this case, N should be applied preplant or split applied before the stem extension phase. All of the immobile nutrients, including P and K, should be applied before planting.

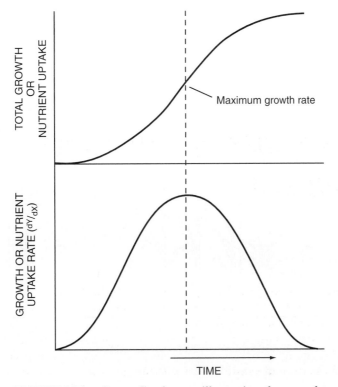

FIGURE 10.1   Generalized curve illustrating the growth pattern of an annual plant.

TABLE 10.1   Percentage of the Total Nutrient Requirement Taken Up at Different Growth Stages

| | Corn Growth Periods (days) | | | | |
	0–25	26–50	51–75	76–100	101–115
N	8	35	31	20	6
P	4	27	36	25	8
K	9	44	31	14	2
	Soybean Growth Periods (days)				
	0–40	41–80	81–100	101–120	121–140
N	3	46	3	24	24
P	2	41	7	25	24
K	3	53	3	21	20
	Sorghum Growth Periods (days)				
	0–20	21–40	41–60	61–85	86–95
N	5	33	32	15	15
P	3	23	34	26	14
K	7	40	33	15	5

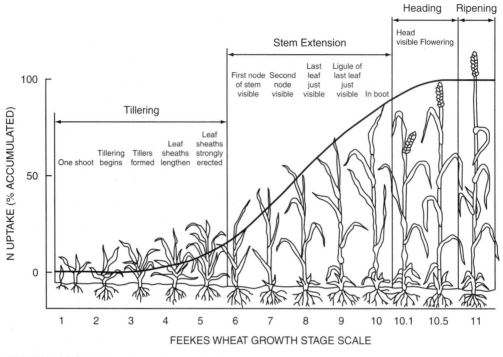

FIGURE 10.2   N uptake pattern of wheat.   *Mark Alley, Virginia Tech.*

Roots

Since the majority of plant nutrients are absorbed by roots, understanding rooting characteristics is important in developing efficient fertilization practices. Root systems are usually either fibrous or tap, and both occur with annuals, biennials, or perennials. The roots' ability to exploit the soil for nutrients and water depends on their morphological and physiological characteristics. Root radius, root length, root surface/shoot weight ratio, and root hair density are the main morphological features. The presence of mycorrhizae is also important.

SPECIES AND VARIETY DIFFERENCES   Knowledge of early rooting characteristics is helpful in determining the most effective method of nutrient management. If a vigorous taproot is produced early, applications should be placed directly under the seed. If many lateral roots are formed early, side placement is recommended.

A diagram of the extent of root development of several crops after planting shows that the corn root system is more extensive than that of cotton (Fig. 10.3). The root development of cotton suggests that, at least for early absorption of nutrients, the presence of nutrients under the plant is important. These differences tend to persist as long as 3 months after planting.

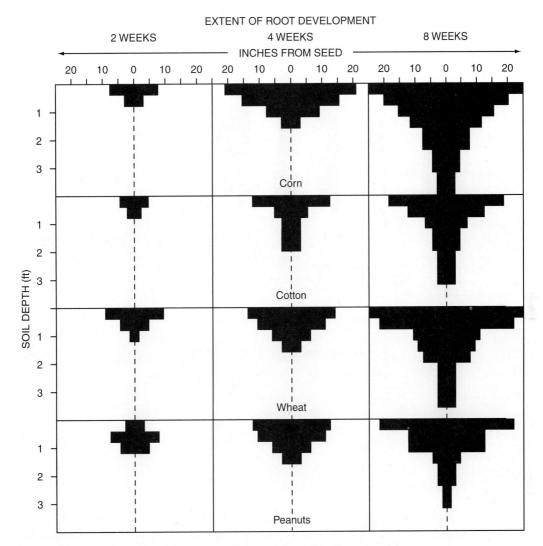

FIGURE 10.3   Root development at 2, 4, and 8 weeks after planting.

Corn develops an extensive root system and has a great capacity for utilizing the nutrients distributed throughout a large soil zone. The root systems of corn and soybean exploit the soil thoroughly, in contrast to those of cotton, potatoes, and other shallow-rooted crops (Fig. 10.4). Potatoes have a limited root system, often being confined by the hilled row, where the roots may penetrate only 10 to 20 in. below level ground.

Carrots show considerable root activity at a 33-in. soil depth (Fig. 10.5), much more than that of onions, peppers, and snap beans. The reduction in activity at 13 in. marks the beginning of a compact subsoil, although root activity increased below the compact layer compared with the peat soil.

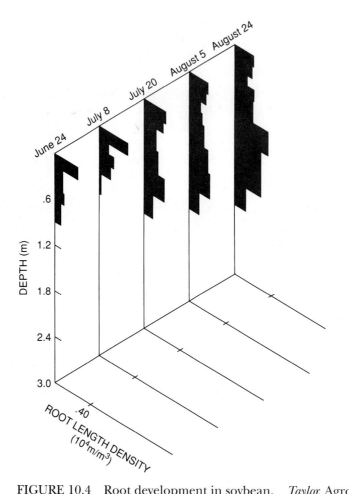

FIGURE 10.4   Root development in soybean.   *Taylor,* Agron. J., *72:543–47,*

Early root growth of most crops occurs mainly in the topsoil (Fig. 10.6). Both root surface per plant and root density in the subsoil increase considerably with plant age, and by harvest they are equal to or greater than those in the topsoil. On sandy soils, corn roots can reach a depth of 8 ft or more and completely extract available soil moisture down to 6 ft. Root weights can be as great as the above-ground biomass.

Small grains have extensive root systems (Figs. 10.7 and 10.8). The development of wheat roots from tillering to grain filling shows that, at maturity, most of the wheat roots are concentrated in the surface soil. An early response of small grains to P placed near the seed even on medium- to high-P soils is commonly observed. Alfalfa roots may penetrate 25 ft if soil conditions are favorable (Fig. 10.7). Depths of 8 to 10 ft are common even on compact soils. One of the advantages of deep-taprooted crops such as alfalfa and sweet clover is that they loosen compact subsoils by root penetration and subsequent decomposition. Also, legumes in pastures provide more animal feed during drought periods than do shallow-rooted grasses.

Root systems of the same species tend not to interpenetrate, which suggests an antagonistic or toxic effect (Fig. 10.9). Thus, with narrow row spacing and high populations, the characteristic root pattern is altered and there may be deeper rooting if soil conditions permit.

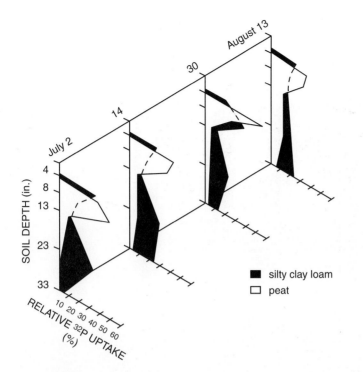

FIGURE 10.5 Root activity of carrots on a silty clay loam and a peat soil. *Hammes et al., Agron. J., 55:329, 1963.*

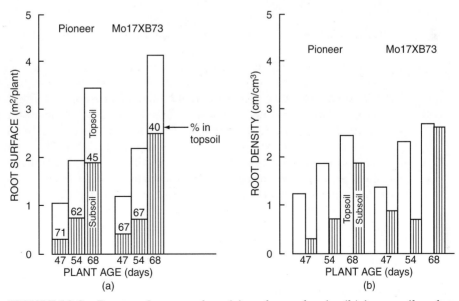

FIGURE 10.6 Root surface per plant (a) and root density (b) in topsoil and subsoil for two corn genotypes at three harvest dates. *Schenk and Barber*, Plant Soil, *54:65, 1980.*

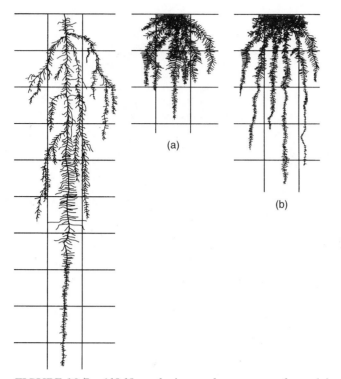

FIGURE 10.7   Alfalfa and winter wheat roots, where (a) and (b) are wheat roots under dryland and irrigated conditions, respectively. Grid lines are 30 cm apart.   *Russel,* Plant Root Systems, *McGraw-Hill, New York, 1977; taken from Weaver,* Root Development of Field Crops, *McGraw-Hill, New York, 1926.*

NUTRIENT-EXTRACTING POWER   Since roots occupy about 1% of the topsoil volume and much less in the subsoil, nutrient absorption characteristics and root-soil interactions can influence nutrient uptake. The exchange capacity of roots of dicotyledonous plants is much higher than that of monocotyledonous plants. Nonlegumes have a lower requirement for divalent cations and take up more monovalent cations. The relative absorption of cations and anions by the root is related to the release of H^+ or HCO_3^- by the root. Acidity develops from the release of H^+ from the root in response to absorption of NH_4^+, whereas pH increases with the release of HCO_3^- and/or OH^- following uptake of NO_3^-. Changes in rhizosphere pH affect the solubility and availability of many plant nutrients.

Mycorrhizal fungi are often associated with plant roots grown under low soil fertility conditions and can increase the ability of plants to absorb nutrients (Fig. 10.10). Mycorrhizal-infected roots exhibit greater uptake of P and other immobile nutrients compared with uninfected roots (Fig. 10.11). Excessive N and/or P fertilization and soil tillage can reduce the contribution of mycorrhiza-related nutrient uptake.

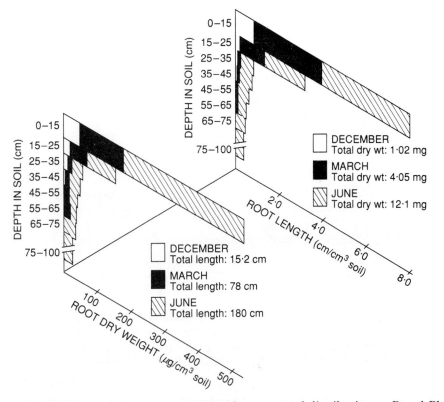

FIGURE 10.8 Winter wheat root development and distribution. *Russel,* Plant Root Systems, *McGraw-Hill, New York, 1977.*

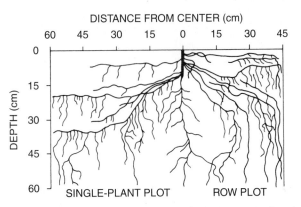

FIGURE 10.9 Contrasting rooting patterns of soybeans in single-plant plots and row plots. *Raper and Barber,* Agron. J., *62:581, 1970.*

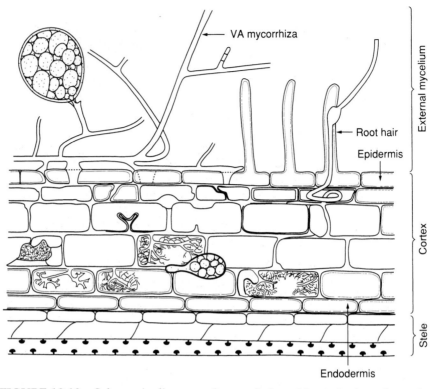

FIGURE 10.10   Schematic diagram of a root infected by vesicular arbuscular mycorrhizal fungi.   *Russel,* Plant Root Systems, *McGraw-Hill, New York, 1977.*

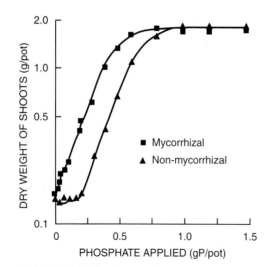

FIGURE 10.11   Example of increased P uptake by clover with or without inoculation with mycorrhizae.   *Bolan, 1983.*

Soil Characteristics

Basically, the soil is a rooting medium and a storehouse for nutrients and water. Hence, it is essential that the roots fully exploit the soil to obtain nutrients and reduce water stress. The yield of a crop is often directly related to the availability of stored soil water (Chapter 11).

The tillage system affects root distribution with depth (Table 10.2). When soil is cultivated annually, corn roots develop more extensively below 10 cm than with no-till systems, while intermediate root distribution occurs with rototill and chisel. When residues are removed, there is greater root growth in the surface 15 cm. Thus, residue decomposition products may inhibit root growth.

In some soils, no-till management can cause some restricted root growth because of increased bulk density compared with full tillage management (Fig. 10.12). Attempts to loosen plowpans or heavy subsoils have not been entirely successful. Subsoiling is most effective when the subsoil is dry so that shattering of the soil occurs; however, in most cases, there is a rapid resealing of the subsoil. One cultivation with a disk or another such implement may almost eliminate any effect of subsoiling. Vertical mulching, in which chopped plant residues are blown into the slit behind the subsoiler, serves to keep the channel open and improve water uptake.

In drought-prone areas with root-restrictive soil layers, subsoiling can increase rooting depth and plant available water, especially in crops such as soybeans that have limited ability to penetrate even moderately compacted soil layers. Both soybeans and corn have responded to in-row subsoiling.

Plant-Nutrient Effects

Adequate fertilization of surface soil encourages not only greater top growth but also a more vigorous and extensive root system. The proliferation of roots resulting from contact with localized zones of high nutrient concentration in infertile soil was illustrated in Figure 5.3. This stimulation of root development is related to the buildup of N and P in the cells, which hastens division and elongation.

Plants absorb nutrients only from areas in the soil in which roots are active. Plants cannot absorb nutrients from a dry soil; thus, root systems modified by shallow applications of fertilizer may be less effective in time of drought. In general, fertilizer should be placed in the portion of the root zone where stimulation of root growth is wanted; therefore, deep placement may be necessary in frequently droughty soils.

TABLE 10.2   Effect of Tillage Treatment on the Corn Root-Weight Distribution $(mg/100 \, cm^3$ of Soil) with Depth

Depth (cm)	Tillage Treatment				
	Conventional	Conventional, No Residues	Chisel	Rototill	No-Till
0–5	29	49	26	69	137
5–10	38	52	104	136	100
10–15	96	218	110	137	74
15–30	85	111	68	88	73
30–45	73	61	47	52	28
45–60	61	59	35	52	37

SOURCE: Barber, *Agron. J.*, 63:724 (1971).

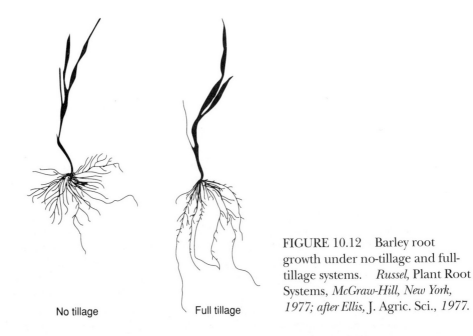

No tillage   Full tillage

FIGURE 10.12   Barley root growth under no-tillage and full-tillage systems.   *Russel,* Plant Root Systems, *McGraw-Hill, New York, 1977; after Ellis,* J. Agric. Sci., *1977.*

An example of the effect of proper management on corn root growth is shown in Figure 10.13. This soil has a claypan and is low in native fertility. The use of adequate plant nutrients with proper cropping systems, including legumes, was effective in the development of a much deeper root system.

The influence of plant nutrition on resistance to winter killing is important. With the addition of adequate plant nutrients to the soil, alfalfa is now being grown on many soils in which growth was once thought to be impossible. Simi-

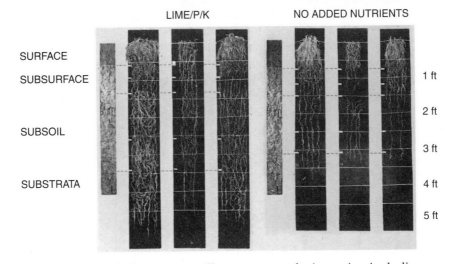

FIGURE 10.13   Soil treatment affects root growth. A rotation including corn, small grains, and legumes was followed. The corn roots on the left, however, were grown in soil receiving adequate lime, P, and K. Those on the right were grown in soil receiving no fertilizer or lime.   *Fehrenbacher et al.,* Soil Sci., *17:281. Copyright 1954 by The Williams & Wilkins Co., Baltimore.*

lar effects of plant nutrients on extending the root growth and winter survival of wheat have been observed (Fig. 10.14).

The previous discussion indicates that numerous soil physical and chemical factors influence the growth of roots and their ability to absorb water and nutrients in quantities sufficient for optimum productivity. Any management factor that improves the soil environment for healthy root growth will help ensure maximum yields. The data in Figure 10.15 illustrate the relationship between root length and soybean grain yield.

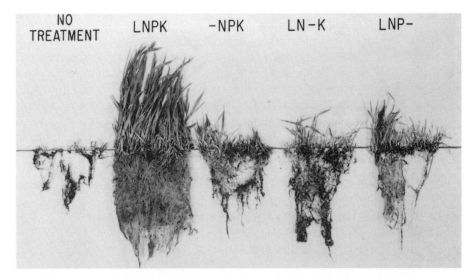

FIGURE 10.14   Balanced fertility aids winter survival of wheat. Early spring vigor means more stooling and more yield.

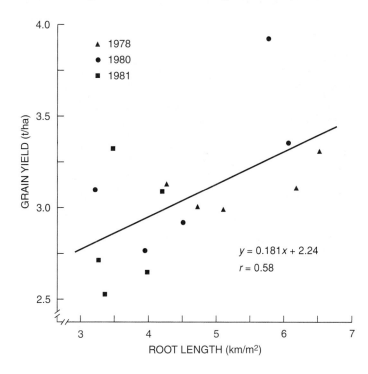

$y = 0.181x + 2.24$

$r = 0.58$

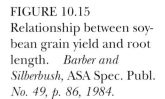

FIGURE 10.15 Relationship between soybean grain yield and root length. *Barber and Silberbush,* ASA Spec. Publ. *No. 49, p. 86, 1984.*

Nutrient Placement

Determining the proper zone in the soil to apply nutrients is just as important as choosing the correct amount of plant nutrients. Placement decisions involve knowledge of crop growth and soil characteristics, whose interactions determine nutrient availability. Numerous placement methods have been developed, and the following factors should be considered with fertilizer placement decisions:

1. *Efficient use of nutrients from plant emergence to maturity.* Vigorous seedling growth (i.e., no early growth stress) is essential for obtaining the desired yield potential and maximizing profitability. Merely applying nutrients does not ensure that they will be taken up by the plant.

2. *Prevention of salt injury to the seedling.* Soluble N, P, K, or other salts close to the seed may be harmful, although the potential for salt injury depends on the source and the crop sensitivity to salts. In general, there should be some fertilizer-free soil between the seed and the fertilizer band, especially for sensitive crops.

3. *Convenience to the grower.* Timeliness of all crop management factors is essential for obtaining the desired yield potential and maximum profit. In many areas, delay in planting after the optimum date often reduces yield potential. Consequently, growers often reject nutrient placement options, even when they may increase yield, to avoid delays in planting. However, placement decisions also influence yield potential; thus, planting date and nutrient placement effects on yield must be carefully evaluated.

Methods of Placement

Fertilizer placement options generally involve surface or subsurface applications before, at, or after planting (Fig. 10.16). Placement practices depend on the crop and crop rotation, degree of deficiency or soil test level, mobility of the nutrient in the soil, and equipment availability.

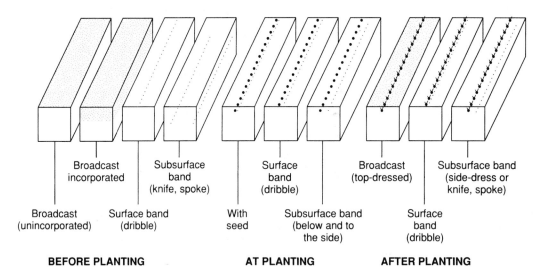

Broadcast incorporated		Subsurface band (knife, spoke)		Surface band (dribble)		Broadcast (top-dressed)		Subsurface band (side-dress or knife, spoke)
Broadcast (unincorporated)	Surface band (dribble)		With seed		Subsurface band (below and to the side)		Surface band (dribble)	

BEFORE PLANTING   **AT PLANTING**   **AFTER PLANTING**

FIGURE 10.16   Cross section of soil profile showing fertilizer placement.   *Adapted from Robertson, Agdex 542-5, Alberta Agriculture, August 1982.*

PREPLANT

1. *Broadcast.* The nutrient is applied uniformly over the field before planting the crop, and it is incorporated by tilling or cultivating. Where there is no opportunity for incorporation, such as on perennial forage or turf crops and in no-till cropping systems, materials may be broadcast on the surface. However, broadcast applications of N in no-till systems can greatly reduce fertilizer N recovery by the crop due to immobilization, denitrification, and volatilization losses (Table 4.18).

2. *Subsurface Band.* Crop recovery of nutrients can be increased with subsurface banding. The depth of placement varies between 2 and 8 in., depending on the crop and the nutrient source. Subsurface point or spoke injection of fluid fertilizers can be effective, especially with application of immobile nutrients (Fig. 10.17). Point injection of N in no-till systems is also more efficient than broadcast N.

3. *Surface Band.* Surface band- or "dribble"-applied fertilizers can be effective before planting. However, if not incorporated, dry surface soil conditions can reduce nutrient uptake, especially with immobile nutrients. Surface band applications of N can also improve N availability compared with broadcast application in some soils and cropping systems.

AT PLANTING

1. *Subsurface Band.* Fertilizer placement can occur at numerous locations near the seed, depending on the equipment and crop. Solid and fluid fertilizer sources can be used. At planting, the fertilizer is applied 1 to 2 in. directly below the seed or 1 to 3 in. to the side and below the seed, depending on the equipment (Fig. 10.18).

2. *Seed Band.* Fertilizer application with the seed is also a subsurface band but is commonly used as starter, or "pop-up," applications. These applications are generally used to enhance early seedling vigor, especially in cold, wet soils. Starter fertilization can also be placed near the seed instead of with the seed. Usually low rates of fertilizer are applied to avoid germination or seedling damage. Fluid or solid sources can be used.

FIGURE 10.17   Point or spoke injection application for fluid fertilizers.

FIGURE 10.18   Band application of fertilizer to the side and somewhat below the seed helps to avoid injury to the plant and permits more efficient use of fertilizer.   *Courtesy of the Potash & Phosphate Institute, Atlanta, Ga.*

3. *Surface Band.* Fertilizers can be surface applied or dribbled at planting in bands directly over the row or several inches to the side of the row (Fig. 10.19). Application over the row can be effective for placement of immobile nutrients with a hoe opener because soil can slough off over time and cover up the fertilizer band. Thus, the surface-applied band becomes a subsurface band placed slightly above the seed (Fig. 10.20).

AFTER PLANTING
1. *Topdressing.* Topdress applications of N are common on small grains and pastures; however, N immobilization in high surface residue systems can reduce the efficiency or recovery of topdress N. Topdressed P and K are not as effective as preplant applications. Both solid and liquid sources can be used.

2. *Sidedressing.* Sidedress application of N is very common with corn, sorghum, cotton, and other crops and is done with a standard knife or point injector applicator. Anhydrous NH_3 and fluid N sources are most common. Fluid fertilizer can also be surface band applied or dribbled beside the row after planting. Sidedressing allows a grower more flexibility in application time since sidedress applications can be made almost anytime the equipment can be operated without damage to the crop. Subsurface sidedress applications with a knife too close to the plant can cause damage by either root pruning or fertilizer toxicity (i.e., anhydrous NH_3). Sidedress application of immobile nutrients (i.e., P and K) is not recommended because most crops need P and K early in the season and during the reproductive growth stage.

FIGURE 10.19   Surface band or dribble applicator for fluid fertilizer. In this case fertilizer is applied over the row after the press wheel.

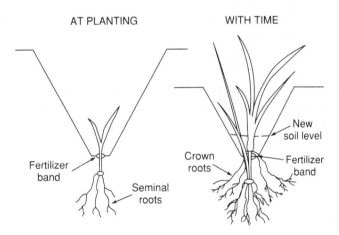

FIGURE 10.20   With surface band application of nutrients, soil will slough into the furrow and bury the fertilizer.   *Westfall et al.,* J. Fert. Issues, *4:114–21, 1987.*

General Considerations

Band Applications

Early stimulation of the seedlings is usually advantageous, and it is desirable to have N-P-K near the plant roots. Since photosynthesis occurs in the leaves, early leaf production will influence yield. It is important to have a small amount of nutrients near young plants to promote early growth and the formation of large, healthy leaves (Fig. 10.21).

FIGURE 10.21   A readily available supply of nutrients near the young plant helps to ensure rapid early growth and the formation of large leaves essential in photosynthesis.

Starter response to N-P-K can be independent of fertility level. With cool temperatures, the early available nutrient supplies may be inadequate because of slow mineralization of N, P, S, and micronutrients from the soil OM; restricted release of nutrients from soil minerals; reduced diffusion of P and K; or limited absorption of nutrients by the plant. The advantage of early stimulation depends on the crop and seasonal conditions. Some factors that might be considered are the following:

1. *Resistance to pests.* Under adverse conditions, a fast-growing young plant is usually more likely to resist insect and disease attacks.
2. *Competition with weeds.* Vigorous early growth of crops is important in reducing weed competition. Reduced weed pressure can improve herbicide effectiveness or reduce the number of cultivations.
3. *Early maturity.* Particularly with vegetables, an early crop is generally very important. A delay of only 3 or 4 days may make the difference between a good price in an early market and a break-even situation. Early maturity can be important in northern climates, where adverse fall weather may interrupt and delay the harvest.

Salt Index

Excessive concentrations of soluble salts in contact with roots or germinating seeds cause injurious effects through plasmolysis, restriction of moisture availability, or actual toxicity. The term *fertilizer burn* is often used. The plant dessicates and exhibits symptoms similar to those of drought.

Some N fertilizers contribute more to germination and seedling damage than is explained by the osmotic effects. Free NH_3 is toxic and can move freely through the cell wall, whereas NH_4^+ cannot. Urea, DAP, $(NH_4)_2CO_3$, and NH_4OH will cause more damage than MAP, $(NH_4)_2SO_4$, and NH_4NO_3. Broadcast application or placement to the side and below the seed are effective methods of avoiding salt injury.

The salt index of a fertilizer is determined by placing the material in the soil and measuring the osmotic pressure of the soil solution in atmospheres. The salt index is the ratio of the increase in osmotic pressure produced by the fertilizer to that produced by the same weight of $NaNO_3$, based on a relative value of 100 (Table 10.3).

N and K salts have much higher salt indices and are much more detrimental to germination than P salts when placed close to or in contact with the seed (Fig. 10.22). These elevated conductivities are inversely related to the clay content of soil. Consequently, potential problems related to fertilizer salts are likely to be greatest in coarse-textured soils. High initial NH_4^+ concentrations resulting from ammoniacal sources will also increase osmotic suction and favor the temporary accumulation of NO_2^--N, which is toxic to plants. The effect of seed placed and 1.5×1.5-in. placement of N-P-K fertilizer is shown in Figure 10.23.

Mixed fertilizers of the same grade may also vary widely in salt index, depending on the carriers from which they are formulated. Higher-analysis fertilizers generally have a lower salt index per unit of plant nutrient than lower-analysis fertilizers because they are usually made up of higher-analysis materials. For example, to furnish 50 lb of N, 250 lb of $(NH_4)_2SO_4$ would be

TABLE 10.3 Salt Index per Unit of Plant Nutrients Supplied for Representative Materials

Material	Analysis*	Salt Index per Unit of Plant Nutrients
Nitrogen carriers		
Anhydrous ammonia	82.2	0.572
Ammonium nitrate	35.0	2.990
Ammonium sulfate	21.2	3.253
Monammonium phosphate	12.2	2.453
Diammonium phosphate	21.2	1.614
Potassium nitrate	13.8	5.336
Sodium nitrate	16.5	6.060
Urea	46.6	1.618
Phosphorus carriers		
Superphosphate (single)	20.0	0.390
Superphosphate (triple)	48.0	0.210
Monoammonium phosphate	51.7	0.485
Diammonium phosphate	53.8	0.637
Potassium carriers		
Manure salts	20.0	5.636
Potassium chloride	60.0	1.936
Potassium nitrate	46.6	1.580
Potassium sulfate	54.0	0.853
Potassium magnesium sulfate	21.9	1.971

*By analysis is meant the percentage of N in N carriers, of P_2O_5 in P carriers, and of K_2O in K carriers.

required, whereas with urea, 110 lb would be required. Hence, the higher-analysis fertilizers are less likely to produce salt injury than equal amounts of lower-analysis fertilizers. In addition, increasing the row width increases the quantity of fertilizer applied in a row, assuming that equal rates are applied. For example, with the same fertilizer rate, fertilizer placed per unit length of row is twice as great in 30-in. rows than in 15-in. rows.

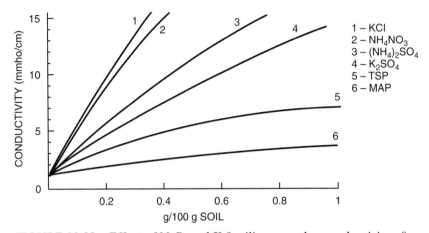

FIGURE 10.22   Effect of N, P, and K fertilizers on the conductivity of saturation extracts of a silt loam soil.   *Chapin et al.,* Soil Sci. Soc. Am. J., *28:90, 1964.*

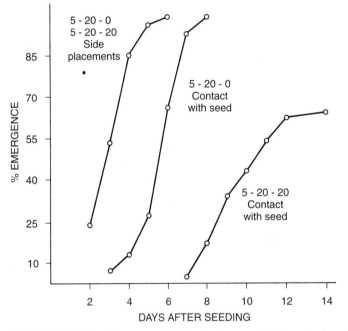

FIGURE 10.23   Placement 1.5 in. below and 1.5 in. to the side resulted in faster emergence of wheat seedlings under greenhouse conditions in contrast to placement in contact with the seed.   *Lawton et al.,* Agron. J., *52:326, 1964.*

Broadcast Applications

Broadcast applications usually involve large amounts of lime and/or nutrients in buildup or maintenance programs. With the trend to reduced tillage, more nutrients remain near the surface (Fig. 10.24). The advantages of broadcast application of nutrients include the following:

1. Application of large amounts of fertilizer is accomplished without danger of plant injury.
2. If tilled into the soil, distribution of nutrients throughout the plow layer encourages deeper rooting and improved exploration of the soil for water and nutrients.
3. Labor is saved during planting. The fertilizer marketing season is spread out through fall, winter, or early-spring applications.
4. This method can be a practical means of applying maintenance fertilizer, especially in forage crops and in no-till cropping systems.

Uniform and accurate spreading of fertilizers and lime is essential for effective utilization by the crop. The effects of uneven application of recommended rates of N-P-K fertilizer on yield of crops are shown in Table 10.4. As might be expected, fertilizer rates well below those needed for low-fertility soils will result in significant yield losses. There may also be yield reductions from overfertilization on soils medium or higher in fertility.

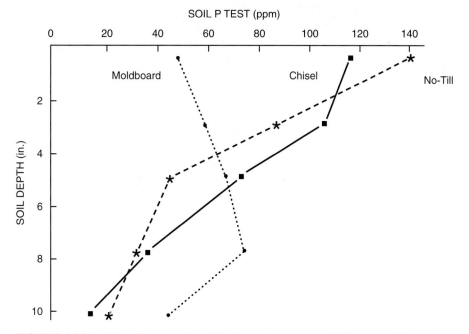

FIGURE 10.24   Distribution of soil P after three years of tillage and fertilization. *Randall, Univ. of Minnesota.*

TABLE 10.4   Effects of Uneven Application of
Recommended Rates of N-P-K Fertilizer on Crop
Yields

| | *Yield (kg/ha)* | | |
Spread Pattern	*Soybeans*	*Corn*	*Barley*
No fertilizer	1,278	2,059	592
Uniform	1,345	8,060	2,809
Nonuniform	1,264	7,271	2,540

SOURCE: Lutz et al., *Agron. J.*, 67:526 (1975).

Specific Nutrients

NITROGEN   Small amounts of N are important in early seedling vigor, but because of its mobility and potential salt effects, high rates of N fertilizers should be
applied before planting and at some distance from the seed or seedling, especially on sandy soils. The total amount of N could be reduced with surface band
applications rather than broadcast. Both downward and lateral movement of N
from the fertilized zone, combined with root extension into the areas of high N
concentration, compensate for lower N rates applied in a band. Similar results
from strip or dribble application of fertilizer on forages have occurred.

The addition of NH_4^+-N to the fertilizer at planting has beneficial effects on
absorption of P by the plant (see Chapter 5). Although dual application of N
and P may not increase the yield in all soils, positive responses have frequently
been observed, especially with winter wheat in the Great Plains. The data in Figure 10.25 show that dual-applied N, as either UAN or anhydrous NH_3, increased

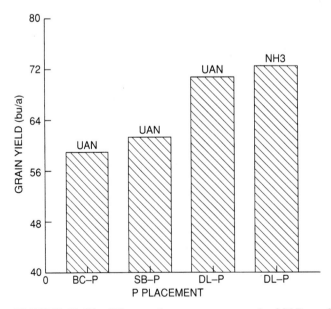

FIGURE 10.25   Winter wheat response to dual N-P application compared with
N and P applied separately. BC and SB represent broadcast and seed banded P,
respectively, while the UAN was knife applied separately. DL represents dual N-
P.   *Leikam, SSSAJ, 47:530–36, 1983.*

dryland winter wheat grain yield compared with the yield achieved with P placed with the seed (SB) or broadcast, but separately from the N as UAN.

It is usually undesirable to apply all N in the row at planting because of possible injury to the crop. Thus, most N is applied either before planting or as a top- or sidedressing after the crop is growing. Water movement carries the N down to the plant roots. If NH_4-N is used, it must be nitrified before it moves down in appreciable quantities, except on low-CEC soil. More efficient use of N for row crops can be obtained by sidedressing part of the N. This is particularly applicable on coarse-textured soils but can also be important on medium- and fine-textured soils. Prediction of the quantity of sidedress N rate can be improved by use of the presidedress NO_3^- soil test, as discussed in Chapter 9.

Starter N applications can be important to enhancing early growth and final yield (Table 10.5). Most starter materials contain multiple nutrients, because crop response to mobile and immobile nutrients can occur in high testing soils in cool, moist conditions. Under conditions conducive to nitrification, the addition of a nitrification inhibitor can improve the crop response to starter N (Table 10.6).

Production from permanent grassland and native range in semiarid regions is limited primarily by moisture and N availability. Low rates of N (i.e., less than 150 lb/a) are generally ineffective, because considerable N is immobilized by the high C/N grass residue (Fig. 10.26).

PHOSPHORUS   Since P is immobile in the soil, placement near roots is usually advantageous. Surface applications after the crop is planted will not place P near the zone of root activity and will be of little value to annual crops in the year of application.

TABLE 10.5   Effect of Starter Fertilizer on No-Till Corn Yield in Continuous Corn and Corn-Soybean Rotations in Illinois

Starter Fertilizer (lb/a)			Corn Yield (bu/a)	
N	P₂O₅	K₂0	Corn-Corn	Corn-Soybean
0	0	0	131	131
25	0	0	141	136
25	30	0	147	141
25	30	30	146	143

SOURCE: Hoeft, *Proc. Illinois Fert. Conf.* (1988).

TABLE 10.6   Corn Response to N and NP Starters with and without Nitrification Inhibitor

Starter				
Nitrogen	Phosphate	N-Serve	Yield (bu/a)	Grain Moisture (%)
	lb/a			
0	0	0	91	19.7
16	0	0	90	17.8
16	0	0.4	111	22.3
0	54	0	85	20.8
16	54	0	90	17.7
16	54	0.4	109	21.4

SOURCE: Anderson, *Better Crops* V. 72. (1988).

An exception to the inefficiency of surface application is with forage-crop fertilization. Topdressed P for maintenance purposes is an efficient method of placement. Some of the P is absorbed by the crowns of the plant, as well as by very shallow roots. Surface banding can be effective on low-P or low-K soils.

In establishing forage crops, surface or subsurface band-applied P and/or K is generally superior to broadcasting (Figs. 10.27 and 10.28). This is especially true in low-P and low-K soils.

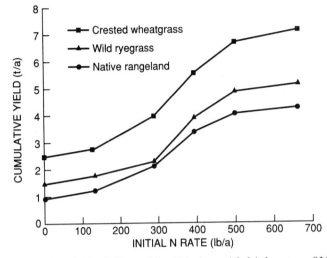

FIGURE 10.26   Effect of fertilization with high rates of N on production from permanent tame grassland and native range.   *Leyshon and Kilcher,* Proc. 1976 Soil Fert. Workshop, Soil Manag. 510, *Publ. 244. Univ. of Saskatchewan, Saskatoon, 1976.*

(a)

FIGURE 10.27   Alfalfa response to P and K placement in the year of establishment; 80 lb/a of P_2O_5 and K_2O were applied by (a) surface broadcast, (b) surface band or dribble, and (c) subsurface band or knife before planting. Early alfalfa growth is greater with band-applied P and K compared with broadcast.   *Courtesy of D. Sweeney, 1989.*

(b)

(c)

FIGURE 10.27 b, c

The question of band versus broadcast application is very important. When all of the P is either banded or broadcast, the relative efficiency is related to both the P status of the soil and the rate of application. In general, differences between seed-placed and broadcast P decline with increasing levels of available soil P (Fig. 10.29).

P placement for small grains is often more critical than for row crops and perennials. Limited root systems, shorter growing seasons, and cooler temperatures enhance the response to band over broadcast P, especially in low-P soils (Fig. 10.30). When high P rates are used in dry and/or coarse-textured soils, banding away from the seed at planting may be superior to banding with the seed (Fig. 10.30). The

FIGURE 10.28   Forage plants on the left resulted from broadcasting both seed
and fertilizer. Plants on the right resulted from drilling seed and banding fertilizer 1 in. below the seed. Both plots were planted September 17 and pho-
tographed March 31.   *Courtesy of Wagner et al., Natl. Fert. Rev., 29:13, 1954.*

highest yields were obtained when P was banded below the seed rather than being
placed with the seed or banded to the side and below the seed row at rates greater
than 60 kg/ha of P_2O_5. The loss in efficiency of high rates of seed-placed P supplied as MAP or DAP is probably due to NH_4^+ toxicity.

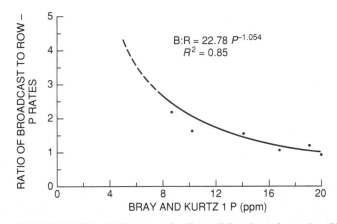

B:R = 22.78 $P^{-1.054}$
$R^2 = 0.85$

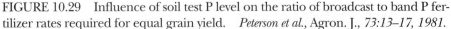

FIGURE 10.29   Influence of soil test P level on the ratio of broadcast to band P fertilizer rates required for equal grain yield.   *Peterson et al.,* Agron. J., *73:13–17, 1981.*

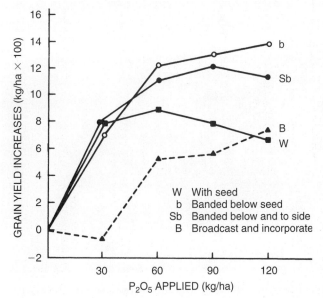

W   With seed
b   Banded below seed
Sb Banded below and to side
B   Broadcast and incorporate

FIGURE 10.30   Effect of P placement on yields of barley (avg. check yld. 1,695 kg/ha) in Manitoba.   *Bailey et al.,* Proc. Western Canada Phosphate Symp., *pp. 200–209, Alberta Soil and Feed Testing Laboratory, Edmonton, Alberta, 1980.*

When properly placed, band-applied P can enhance plant growth and yield potential compared with broadcast P (Table 10.7). In these data band-applied P greatly increased tiller number in wheat, which is directly related to number of heads and final grain yield. Increasing broadcast P to 3 × band P did not produce the same growth at late tiller stage. The influence of P management on root growth is also evident in Figure 10.31. Similar responses to P placement in corn and soybeans occur (Table 10.8). Notice the decrease in grain moisture content with starter P, which illustrates more advanced maturity with improved P nutrition.

Even with band placement, crops during any one season are generally able to recover only a small fraction of fertilizer P, usually less than 25%. This is in

TABLE 10.7   No-Till Winter Wheat Response to P Rate and Placement

P Treatment (lb/a of P_2O_5)	Number of Tillers (no./ft²)	Dry Matter (lb/a)		Number of Heads (no./ft²)	Grain Yield (bu/a)
		Full Tiller	Harvest		
0	11	280	4,978	43	34
BC 15	14	370	5,906	51	40
BC 45	17	475	8,131	65	53
KN 15	29	1,053	7,754	66	49
KN 45	32	1,096	8,641	70	56

*BC, broadcast unincorporated; KN, knifed 2 in. below seed; full tiller, Feekes growth stage 5 (see Fig. 10.2).

SOURCE: Havlin, *Proc. Great Plains Soil Fert. Conf.* (1992).

FIGURE 10.31   Influence of P placement on root growth of dryland wheat grown on a low-P soil. Plant on the left is unfertilized. Plant in the middle received 15 lb/a of P_2O_5 broadcast, while the plant on the right received 15 lb/a of P_2O_5 banded 2 in. below the seed.   *Havlin, 1988.*

marked contrast to the recovery of N and K, which may be 50 to 75%. Band placement of P reduces contact with the soil and should result in less fixation than broadcast application. With broadcasting and thorough mixing, fertilizer P comes into intimate contact with a large amount of soil. Therefore, band-applied P should increase crop recovery compared with broadcast P.

Although it is generally true that the most efficient use of limited quantities of P is at planting and that the highest return will be obtained by band applications, there may be some advantage in building up soil fertility in a long-term fertilizer program. The beneficial effect of building up soil test P on the value of four successive crops is shown in Table 10.9. When P is applied only once, higher net re-

TABLE 10.8   Effect of Rate and Placement on Plant Height, Grain Yield, and Grain Moisture

Placement	P rate (lb/a)	Plant height (in.)		Grain Yield (bu/a)		Corn grain moisture (%)
		Corn	Soybeans	Corn	Soybeans	
Broadcast	0	14.6	7.1	115	37	27.0
Broadcast	20	16.5	8.3	124	40	26.1
Broadcast	40	15.4	7.5	119	40	26.6
Broadcast	80	17.3	9.8	123	40	25.6
Starter	0	14.2	7.1	117	35	27.0
Starter	20	27.2	10.2	135	39	24.8
Starter	40	26.8	10.2	132	39	22.5
Starter	80	30.7	10.2	146	45	24.2

SOURCE: Fixen, Gerwing, and Faber, *Better Crops* V.72 (1984).

TABLE 10.9   Crop Values from P Fertilization over a 4-Year Cropping Period under Irrigation in Idaho

P Applied (lb/a), Fall, 1972*	Gross Crop Value ($/a)					P Fertilizer Costs ($/a)	4-Year Net Returns ($/a)
	1973 Sugarbeets	1974 Spring Wheat	1975 Potatoes	1976 Silage Corn	Total		
0	577	240	1,214	232	2,263	0	—
60	661	267	1,204	250	2,382	24	95
150	655	291	1,277	239	2,462	60	139
500	660	306	1,512	252	2,730	200	267

*Initial soil test P level of 5.6 ppm P (0–12 in.).

SOURCE: Westermann, *Proc. 28th Annu. Northwest Fert. Conf.*, pp. 141–46 (1977).

turns were achieved when high-value crops such as potatoes were grown after P fertilization. These results demonstrate that high P rates may be profitable over several crops. When high-value crops are grown on low-P soils, it is advisable to increase the soil test P with a buildup program.

POTASSIUM   K salts are much less mobile than NO_3^- but more mobile than $H_2PO_4^-$. Although some leaching on sandy soils may occur, losses from most soils are negligible. Because fertilizer K salts cannot be placed in contact with the seed in great quantity, they should be placed in a band to the side and below the seed (Fig. 10.16). In contrast, salt-tolerant crops such as barley and other small grains respond to rates of 15 to 30 lb/a of K_2O placed with the seed.

Broadcast K is usually less efficient than banded K; however, as soil test K increases, there is generally less difference between placement methods. The importance of placement also decreases as higher rates of fertilizer are used.

Starter responses from K, similar to those from N and P, occur with many crops planted under cool, wet conditions even on high-K soils (Table 10.10). This response is unlikely to occur every year, and it should be noted that the results here are for only 1 year. Nevertheless, this example shows the value of using K on a high-K soil.

TABLE 10.10   Starter K Overcomes Cold, Wet Soil in a
Normal Spring to Produce More Corn on a Kansas Soil
Testing Very High in Available K (>700 ppm)

Nitrogen (lb/a)	Yield (bu/a) with		
	No K_2O	20 lb/a Banded K_2O	Increase from Starter
0	72	80	8
75	128	137	9
150	167	182	15
225	166	182	16
300	167	185	18

SOURCE: H. Sunderman, *Report of Progress 382.* Colby Branch
Exp. Sta. Kansas State Univ. (1980).

MICRONUTRIENTS   Micronutrient deficiencies are increasing and can be expected to continue. Higher yields are being obtained and are putting a greater
demand on all nutrients. Interaction among macro- and micronutrients will assume greater importance. Specific micronutrients are applied in areas known to
be severely deficient or to crops known to have especially high micronutrient requirements. The micronutrient may be added to a mixed fertilizer or applied separately as a broadcast application or foliar spray.

Micronutrients added to N-P-K fertilizer should be placed in bands 2 in. away
from the seed to prevent fertilizer injury. B should not be band applied to crops
such as beans or small grains.

Time of Application

Application timing depends on the soil, climate, nutrients, and crop. However,
fertilizers are applied at times during the year that may not be the most efficient
agronomically but that are better suited to workload or distribution constraints
of both the grower and the dealer. Despite these considerations, growers should
apply nutrients at a time that will maximize recovery by the crop and reduce the
potential for environmental problems.

NITROGEN   N loss mechanisms (see Chapter 4) must be considered in selecting the time of application. Theoretically, it would be desirable to apply N as close
as possible to the time of peak N demand of the crop; however, this is seldom feasible except with sidedress N.

Because of N mobility in soils, the amount and distribution of rainfall are important considerations. The greater the rainfall, the greater the possibility of N loss
through leaching if the crop is not growing vigorously or if the land is not protected by a plant cover. In addition, conditions conducive to denitrification are
likely to occur when soils become waterlogged due to excessive amounts of water.

In warmer climates the temperatures are more optimum for nitrification during a greater portion of the year. N applied before planting would thus be more
subject to nitrification and leaching (see Chapter 4). In cooler climates, NH_4^+-N
application is generally recommended in the fall, when soil temperature drops
below 50°F, except on sandy or organic soil. However, compared with fall-applied

NH_4^+-N, spring applications are 5 to 10% more efficient on fine- and medium-textured soils and 10 to 30% more efficient on coarse-textured soils. Nitrification inhibitors are recommended for N applied to warm, sandy soils. Many growers apply N in the fall; however, sidedressing after crop emergence can often be superior to the fall application.

With fall-planted small grains, all or most of the N is applied in late summer or fall. In warmer, more humid regions, yields will be somewhat below those obtained by top dressing N in late winter because of leaching or gaseous losses. However, there are several important advantages to fall applications on small grains. In late winter the ground may be too wet for machinery to be operated, and spring N application after jointing is usually too late for small grains to respond in yield to the applied N.

PHOSPHORUS   In general, P should be applied just before or at planting because of the conversion of soluble P to less available forms. The magnitude varies greatly with the fixing capacity of the soil (see Chapter 5). On soils of low to moderate fixing capacity, broadcasting in the fall for a spring-planted crop is one of the most effective methods. On low-P and/or high-P fixing soils, band-applied P as close to planting as possible is the most efficient and should maximize recovery of P by the crop. On medium- to high-P soils, the time and method are less important and applications every 2 to 4 years may be recommended.

POTASSIUM   K is commonly applied and incorporated before or at planting, which is usually more efficient than sidedressing. Because K is relatively immobile, sidedressed K is less likely to move to the root zone to benefit the current crop. Fall-applied K is even more dependable than either P or N applied in the fall because fewer loss mechanisms exist with K.

Under some cropping practices, K fertilizers may be broadcast once or twice in the rotation. Fall incorporation of K is generally made before planting K-responsive crops, such as corn and legumes.

Maintenance application on forage crops can be made at almost any time. Fall applications are generally desirable, because the K will have had time to move down into the root zone. On hay crops, application is recommended after the first cutting and/or before the last cutting.

Foliar Applications

Certain fertilizer nutrients that are soluble in water may be applied directly to the aerial portion of plants. The nutrients must penetrate the cuticle of the leaf or the stomata and then enter the cells. This method provides for more rapid utilization of nutrients and permits the correction of observed deficiencies in less time than would be required by soil treatments. However, the response is often only temporary. When problems of soil fixation of nutrients exist, foliar fertilization constitutes an effective means of fertilizer application.

So far, the most important use of foliar sprays has been in the application of micronutrients. The greatest difficulty in supplying N, P, and K in foliar sprays is in the application of adequate amounts without severely burning the leaves and without an unduly large volume of solution or number of spraying operations. Nutrient concentrations of generally less than 1 to 2% are employed to avoid injury to foliage. Nevertheless, foliar sprays may be excellent supplements to soil applications.

Micronutrients readily lend themselves to spray applications because of the small amounts required. Foliar applications have been found to be many times more efficient than soil applications for fruit trees and other crops. Soil-applied Fe is often not effective on high-pH soil because of precipitation of $Fe(OH)_3$ (see Chapter 8). Efforts to correct Fe chlorosis have not always been successful, and more than one application may be needed on some crops. Chlorosis is a common problem on soybeans grown on high-pH soils under low-rainfall conditions.

Foliar application of urea has been successful in apples, citrus, pineapple, and other similar crops, because N is absorbed more rapidly than with soil applications.

Foliar applications of P are used less than foliar applications of N, largely because most P compounds are damaging to leaves when sprayed in quantities large enough to make the application beneficial. The maximum concentration of P is 0.5% for corn and 0.4% for soybeans.

Various environmental factors, including temperature, humidity, and light intensity, also affect the rate of absorption and translocation of nutrients applied to the foliage. To be most effective, two or three spray applications repeated at short intervals may be needed, particularly if the deficiency has caused severe stunting. Care must be taken to identify the nutrient needed.

Fertigation

Fertigation is the application of fertilizer in irrigation water. N and S are the principal nutrients applied by fertigation. P fertigation has been less common because of concerns about the precipitation of P in high-Ca and high-Mg waters. Application of soluble S through irrigation systems is effective.

Application of anhydrous NH_3 or other fertilizer materials such as UAN solutions containing free NH_3 to irrigation waters high in Ca^{2+}, Mg^{2+}, and HCO_3^- may result in precipitation of $CaCO_3$ and/or $MgCO_3$, causing scaling and plugging problems in equipment. Their formation can be prevented or corrected by the addition of H_2SO_4 or other acidic solutions.

The advantages of fertigation are that (1) nutrients, especially N, can be applied close to the time of greatest plant need and (2) one or more field operations are eliminated. Corn, for example, has two periods of high N uptake, vegetative growth stages V12 to V18 and reproductive growth or grain fill. Providing adequate N at these stages is important for maximizing yield. Midseason deficiencies in crops can also be corrected by fertigation.

The question of uniformity of application of nutrients in irrigation water is sometimes raised. Uniform application should not be a problem with skilled irrigation management and properly designed irrigation systems, since the dissolved nutrients accompany the water wherever it goes. However, unsatisfactory distribution of nutrients can occur under some conditions and with low rates of fertilization. Under row irrigation a large proportion of the nutrients may be deposited near the inlet.

To prevent nutrients from being leached beyond the root zone or from accumulating near the surface, inaccessible to the crop, they should not be introduced at the initiation of irrigation. Best results are obtained when the fertilizer materials are supplied toward the middle of the irrigation period and their application terminated shortly before completion of the irrigation.

Variable Nutrient Management

As discussed in Chapter 9, site-specific nutrient management is an emerging technology that improves nutrient use efficiency by distributing nutrients based on the variation in yield potential, soil test levels, and other appropriate spatial data. Figure 10.32 shows the spatial distribution in Bray-1 P and the associated P sufficiency levels (see Chapter 9). From these spatial data, the variable P recommendations are developed. Thus, fertilizer P is applied through a computer-controlled variable rate applicator according to the P recommendation map.

Infrared aerial crop photography is a reliable means of monitoring crop production systems. Near-infrared film is used to take photographs at variable altitudes. The photographs often can cover individual fields or as much as 1 sq mi of land.

Healthy green plants reflect a large amount of infrared light, whereas plants under stress caused by such factors as drought, nutrient deficiency, disease, weeds, and chemical spray damage do not reflect infrared light. These differences can be clearly recorded on infrared film. Other crop management problems that can be detected by this technique include inadequate drainage, poor stand establishment, and uneven application of fertilizers and herbicides. A number of such problems are identified in Figure 10.33, which is a black-and-white reproduction of an aerial infrared photograph.

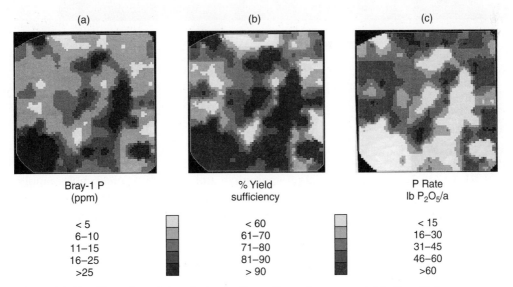

FIGURE 10.32   Use of spatial variation soil test P produces a variable P sufficiency map used to develop variable P recommendations. Note that the low P rates are applied to areas of high Bray-1 P and percentage of sufficiency.

DESCRIPTION	ACRES
1 LOW VIGOR AREA	15.2
2 HIGH VIGOR AREA	26.8
3 MEDIUM GROWTH AREA	89.8

FIGURE 10.33   Near-infrared aerial photograph of an irrigated cornfield shows a range in plant growth or vigor. The light areas (low vigor) are caused by water deficiency and stress. Notice the center pivot at the left side of the circle (8 o'clock position moving north toward 9 o'clock position). The darker region below the pivot shows reduced water stress from the irrigation.

Nutrient Management in Turf

Turfgrass production is a rapidly growing component of agriculture. Regardless of its use in residential or recreational environments, effective nutrient management is essential to turfgrass quality, durability, and aesthetic appeal. Like other crops discussed throughout the text, the relative nutrient content of turfgrasses is very similar (Table 9.1). The primary difference between most agricultural crops and turfgrass is that most of the nutrients applied to turf systems are not removed from the field. Turfgrass clippings are usually left on the field, and the nutrients cycle through soil OM components as the residues are degraded (Fig. 4.1). Most of these nutrients ultimately become plant available in subsequent years. In addition to soil tests, plant tissue can be sampled to assess nutrient status and adequacy of the fertilization program. Samples are collected by clipping leaves slightly above the soil surface two days after regrowth. Nutrient sufficiency ranges were provided in Table 9.3.

NITROGEN   Nitrogen requirements of turfgrass are greater than any other nutrient, which is similar to other agronomic crops (Table 9.1). Adequate N maintains a desirable dark-green leaf color, prolific tillering or shoot density, and some tolerance to other nutrient and pest stresses. Excessive N accumulation in leaf tissue results in increased growth and demand for water, which may cause moisture stress. Too much N enhances susceptibility to certain diseases and reduces tolerance to high temperature stress. Reduced root, stolon, and rhizome growth with increased heat and water stress results in thin, uneven growth patterns. If these symptoms are misdiagnosed as N stress, additional N applications can severely reduce turf growth and aesthetics and increase the opportunity for fertilizer N runoff into nearby streams. The goal of an efficient N management program is to provide adequate N to support vigorous growth without overfertilization.

Recommended annual N rates depend on the turfgrass species but usually range between 1 and 8 lb/1,000 ft^2 (40 to 350 lb/a of N) (Table 10.11). Because N is mobile in the soil, two to four applications throughout the season are recommended. More frequent applications result in higher quality and longer periods of dark-green color. Because of the midspring to midsummer and mid- to late-fall active growth pattern in cool-season grasses (e.g., bluegrass, ryegrass, and fescue), maintaining high forage quality requires three to four applications of 1 lb/1,000 ft^2 in late fall and early spring (Table 10.11). Warm-season grasses (e.g., Bermuda grass, and zoysia) exhibit active growth from midsummer through midfall. N is applied in

TABLE 10.11   Optimum N Application Rates and Timing for Turf*

Turf Species	Annual N Rate (lb/1,000 ft^2/yr)	Number of Applications			
		1	2	3	4
Fine-leaf fescue	1–2	EF	EF, ES	EF, ES, LF	EF, ES, MR, LF
Tall fescue	2–4	EF	EF, ES	EF, ES, LF	EF, ES, MR, LF
Perennial ryegrass	2–4	EF	EF, ES	EF, ES, LF	EF, ES, MR, LF
Kentucky bluegrass	2–4	EF	EF, ES	EF, ES, LF	EF, ES, MR, LF
Bermuda grass	4–8	ES	ES, MR	ES, ER, LR	ES, ER, MR, LR
Saint Augustine grass	2–4	ES	ES, MR	ES, ER, LR	ES, ER, MR, LR
Zoysia	2–4	ES	ES, MR	ES, ER, LR	ES, ER, MR, LR

*E, early; M, mid; L, late; S, spring; R, summer; F, fall.

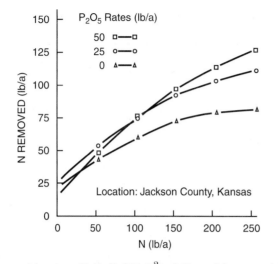

FIGURE 10.34   N recovery in bromegrass is much higher with applications of P to a low-P soil.

midspring (1 lb/1,000 ft^2), followed by monthly applications through early fall. Low N rates (<1.5 lb/1,000 ft^2) should be used with soluble N sources to maximize N recovery by the plant and to minimize N leaching. Higher rates can be used with slow-release N sources such as S-coated urea (Chapter 4).

PHOSPHORUS   As discussed in Chapter 5, P is important for early seedling vigor and stand establishment. In low-P soils, increasing P availability improves N utilization and recovery (Fig. 10.34). P deficiencies are uncommon in established turfgrass, because clippings are usually left on the surface and many mixed turf fertilizers contain some P. Soil testing is the best tool to identify low-P soils and the need for P fertilization. With high-P soil tests (>25 ppm Bray-1 P) fertilizers that contain P are not necessary. The common P sources were listed in Table 5.11.

POTASSIUM   Turfgrass can require as much K as N, although N is usually the first limiting nutrient. An N:K ratio of 2:1 in leaf tissue is considered normal. Using adequate N without K enhances plant susceptibility to diseases and drought stress. Balanced N and K nutrition provides the good root, stolon, and rhizome growth important for maintaining optimum turf density, water use efficiency, winter hardiness (in northern climates), and tolerance to heavy use.

K applications preceding heat or water stress periods, as well as an early fall application to improve winter hardiness, are critical. Fertilizers that contain a 1:1 ratio of N:K will supply adequate K in most cases. K fertilizers have higher salt indices than most N and P sources; thus, caution is recommended with applications at germinating and seedling growth stages (Table 10.3). Like N, K is mobile in the soil, and K leaching can occur in sandy soils. The salt index with K_2SO_4 is lower than other K sources, and there is some evidence that K leaching is also less with K_2SO_4 (Table 10.12). The common K sources were listed in Table 6.7.

SULFUR   Turfgrass usually requires more S than P. Adequate S nutrition is important for protein and chlorophyll synthesis; S greatly contributes to a healthy, dark-green color. S-deficiency symptoms are often mistaken for N stress. S is also essential for maximizing recovery of N and K (Table 10.12). Increased N use efficiency ultimately reduces N requirements and N leaching potential. Annual S rates

TABLE 10.12   Dry Matter (DM) Forage Yields and Recovery of N and S with Coastal
Bermuda Grass

N Rate	S Rate	Forage Yield (DM, lb/a)	N		S	
—lb/a—			Uptake (lb/a)	Recovery (%)	Uptake (lb/a)	Recovery (%)
0	0	4,743	81	0	6	0
	90	5,274	88	0	23	18
200	0	9,216	186	53	12	0
	90	10,426	223	72	33	39
400	0	10,259	236	39	26	0
	90	12,238	306	56	36	47

SOURCE: Phillips, *Better Crops* V.73 (1989).

are 0.5 to 2 lb/1,000 ft^2, either as a single application in early spring or split applied with N in the spring and fall. Split applications reduce the potential for S leaching, especially in sandy soils (Chapter 7). S fertilizer sources can be found in Table 7.5.

MICRONUTRIENTS   Dark-green turf color is also related to Fe and Mg nutrition, since these nutrients function in chlorophyll synthesis. Early-spring and midsummer applications are recommended. Soil testing provides the best guide to identifying soils low in plant available micronutrients. Micronutrient fertilizer sources can be found in Chapter 8. As discussed in Chapter 8, soil-applied Fe and other micronutrients may not be as efficient as foliar applications.

Conservation Tillage

With reduced tillage systems, nutrients concentrate in the upper 2 to 4 in. of soil (Fig. 10.24). Periodic tillage (i.e., every 5 or 6 years) will distribute these nutrients more uniformly throughout the root zone. Wherever feasible, soils low in fertility should be brought up to medium or higher fertility before initiating no-till operations.

Broadcast-applied P and K are effective under many conditions, particularly in the more humid areas. With surface residues there is more moisture near the surface and increased root growth; however, under low-fertility conditions and/or in cooler and drier areas, surface-applied P and K may not be sufficiently available (Table 10.13). Lower K in corn leaves at silking occurs with no-tillage systems, while K application increases K in the leaves under both tillage systems and reduces the yield loss from no-till operations. The same principle holds for P.

Yield increases from band-applied fertilizer are generally greater under no-till systems than under plowed systems (Fig. 10.35). Conservation tillage results in greater amounts of surface residues, which leads to cooler and wetter conditions at planting and lower nutrient availability in the soil.

As indicated earlier, a large portion of broadcast-applied N in reduced-tillage systems can be immobilized by the surface crop residues. Therefore, maximizing crop recovery of fertilizer N requires placement below the residue. The data in Figure 10.36 illustrate increased N response and recovery of applied N with subsurface N compared with broadcast and dribble N.

TABLE 10.13   K Fertilization Helps Compensate for Losses in Corn Yields Due to Reduced Tillage of a Wisconsin Soil Medium in Available Soil K

K_2O Applied Annually 1973–76 (lb/a)	Yield Loss (bu/a) from Not Plowing (Plowed-Unplowed)	Percentage of K in Ear Leaf Tissue	
		Plowed	No-Till
0	37	0.73	0.59
80	26	1.40	1.04
160	13	1.71	1.42

SOURCE: Schulte et al., in Soils, Fertilizer and Agricultural Pesticides Short Course, Minneapolis, Minn. (December 12–13, 1978).

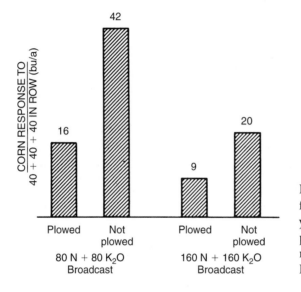

FIGURE 10.35   Use of row fertilizer greatly reduced the yield difference between plowed and unplowed treatments.   *Schulte,* Better Crops Plant Food, *63:25, 1979.*

Residual Fertilizer Availability

Applications of nutrients will result in a certain portion of these nutrients being left in the soil after harvest. The amounts remaining depend on the amounts added, the yield, the portion of the crop harvested, and the soil.

Long-term residual benefits of N do not usually receive as much recognition as do those of P and K. However, residual effects of fertilizer N applications are related to buildup of soil OM with intensively managed cropping systems (Chapter 13). The residual effects of P and K are well known. The availability of large initial applications of P can be observed for many years, depending on P rate and the P fixation potential of the soil (Fig. 10.37). Beneficial carryover effects can also occur with other nutrients. S applied as gypsum at seeding of winter wheat in the first year of a wheat-pea rotation lasted for 8 years (Fig. 10.38).

As the fertilizer application rate increases, the residual value also increases. In many cases, the cost of fertilization is charged to the crop treated. However, residual fertilizer availability should be included in evaluation of fertilizer economics.

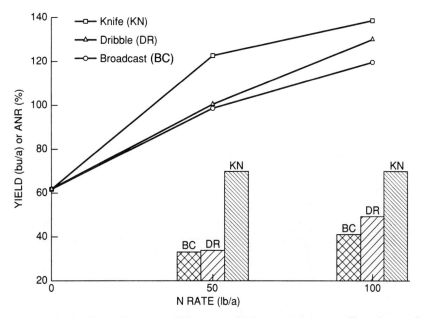

FIGURE 10.36   Influence of N rate and placement on no-till grain sorghum yield and apparent N recovered (ANR) by the grain. Placing the N below the crop residue increased crop yield and fertilizer N recovery compared with broadcast or dribble N. Lines represent yield, and bars represent ANR. *Lamond et al.*, Rep. of Prog., *Kansas State Univ., 1989.*

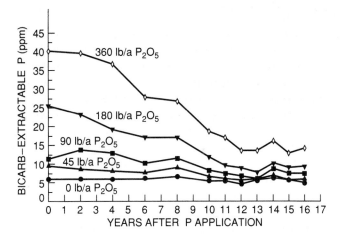

FIGURE 10.37   Effect of high initial broadcast P applications on soil test P levels.   *Halvorson and Black*, SSSAJ, *49:928–33, 1985.*

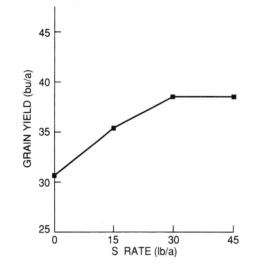

FIGURE 10.38 Yield response (bu/a) of the fourth crop of soft white winter wheat to S applied at the seeding of the first wheat crop in a wheat-pea rotation. The fourth wheat crop was the seventh crop grown in a wheat-pea rotation following S application. Yields are derived from 13 field experiments conducted over an 8-year period. *Ramig and Rasmussen,* Proc. 23rd Annu. Northwest Fert. Conf., *pp. 125–37, Boise, Idaho, July 17–20, 1972.*

Utilization of Nutrients from the Subsoil

The utilization of nutrients from the subsoil depends on numerous factors, including soil structure, aeration, pH, drainage, and root distribution.

Most humid-region subsoils are acidic and low in fertility. Low-fertility subsoil contributes very little to total nutrient supplies (Fig. 10.39). Deep-rooted crops such as alfalfa or sweet clover increase available P in the surface by upward transfer from the subsoil as the organic residues are returned to the surface soil and decomposed. The surface horizons of forest soils are commonly higher in nutrients than the subsoil horizons because of upward transfer and accumulation.

Loess or alluvial soils can be high in K and P throughout the profile and can be utilized by deep-rooted plants. In some areas, difficulty is experienced in correlating soil test results for P or K. When the content of P or K in the subsoil is considered, the relation between extractable P or K and crop response can be improved. Some states have made a systematic analysis of the subsoils of major series (Fig. 10.40). These data could be helpful in making more accurate fertilizer recommendations.

In calcareous soil, soil test K is usually high in both surface soil and subsoil, but most subsoils are low in plant available P. In addition, subsoil pH can be very high, where micronutrients (Zn and Fe) can be low. P and micronutrient fertilization of the surface soil is generally adequate to increase P and micronutrient availability.

Deep application of nutrients to a deficient soil zone will greatly enhance root development in the treated zone. Lime added to an acidic subsoil will not only supply Ca and Mg but will also reduce the quantities of Al, Fe, and Mn in solution.

In some cases, subsoiling alone can increase crop yields, although subsoil incorporation of fertilizers can further increase yields (Table 10.14). Subsoiling to a depth of 11 to 22 in. increased the 4-year mean yield of barley by 24%, and subsoil incorporation of P and K increased barley yield an additional 20%.

In some soils, deep tillage (24 to 36 in.) can improve root growth and crop yield without subsoil fertilization. In this situation, the benefit is related to more effi-

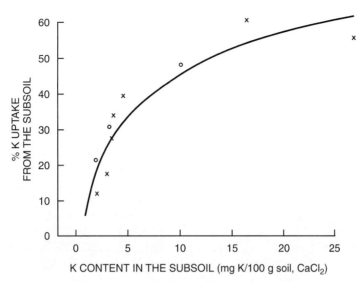

FIGURE 10.39   The effect of K content of the subsoil on the uptake of K by spring wheat from the subsoil, expressed as a percentage of total uptake (topsoil = 9 mg/100 g of soil, 10 field experiments in 1984 and 1986 on loess soils). *Kuhlman,* Plant & Soil *127:129,1984.*

cient use of subsoil water. If the plant is utilizing water from 24 to 36 in. in contrast to only 12 in. deep, the probability of drought stress will be reduced. Under some conditions, turning up heavy clay subsoil material may cause the surface soil to seal off more rapidly and decrease water intake. Deep plowing or chiseling to break up plowpans and improved management practices to encourage deeper rooting are important for improved productivity.

Fertilization with Manure

Greater attention is being given to effective disposal of animal manures because of (1) increased use of confinement production systems and associated manure-handling problems and (2) increased concern over contamination of groundwater and surface water by NO_3^- and other compounds originating from the manure.

Some of the beneficial effects of manure use are

1. An additional supply of NH_4-N.
2. Greater movement and availability of P and micronutrients due to complexation.
3. Increased moisture retention.
4. Improved soil structure, with a corresponding increase in infiltration rate and a decrease in soil bulk density.
5. Higher levels of CO_2 in the plant canopy, particularly in dense stands with restricted air circulation.
6. Increased pH and buffer capacity (BC).
7. Complexation of Al^{3+} in acidic soils.
8. Increased soil OM.

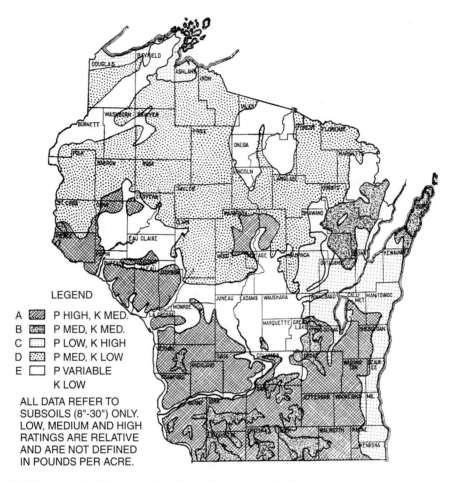

LEGEND

A ▨ P HIGH, K MED.
B ▨ P MED, K MED.
C ☐ P LOW, K HIGH
D ▨ P MED, K LOW
E ☐ P VARIABLE
      K LOW

ALL DATA REFER TO
SUBSOILS (8"-30") ONLY.
LOW, MEDIUM AND HIGH
RATINGS ARE RELATIVE
AND ARE NOT DEFINED
IN POUNDS PER ACRE.

FIGURE 10.40   General subsoil fertility groups in Wisconsin have been established.   *Univ. of Wisconsin.*

TABLE 10.14   Effect of Subsoiling and Deep Incorporation of P and K on the Yield of Barley Grain

Treatment	*Grain Yield (t/ha at 85% dry matter)*				
	1974	*1975*	*1976*	*1977*	*Mean*
None	4.89	2.30	3.43	2.90	3.38
Subsoiled alone	5.23	3.79	4.46	3.53	4.25
Subsoiled + P and K	6.21	4.69	4.51	4.34	4.94
P and K to topsoil	4.53	1.90	3.77	3.20	3.35

SOURCE: McEwan and Johnston, *J. Agr. Sci. (Camb.)*, 92:695 (1979).

The favorable effect of manure on increasing the available moisture content of soils is shown in Table 10.15.

Methods for handling and storing manure will affect its nutrient content. Previously, the common method of disposal was to collect the manure or manure plus bedding and spread it on fields. Liquid waste systems have been developed in which the manure is diluted with water and stored in pits or lagoons.

N losses from various systems are given in Table 10.16. P and K losses are only 5 to 15% under all but the open-lot and lagoon waste systems. In an open lot, about 50% of these nutrients is lost. In a lagoon, much of the P settles out and is lost from the liquid applied on the land.

The composition of animal manure varies according to the type and age of animal, feed consumed, bedding used, and handling system (Tables 10.17 and 10.18). As might be expected, dry matter is highest in the solid waste system, whereas N, P, and K are highest when manure is handled as a liquid. Four principal methods used for field application of manure are

1. Spreading of solid material when weather, soil, and crop permit.
2. Injecting the slurry of water and manure into the soil or spraying it on the surface.
3. Injecting the slurry into a sprinkler irrigation system.
4. Surface band applied under crop canopy.

N loss is greatly affected by the method of application (Table 10.19). Immediate incorporation will minimize N volatilization. The effectiveness of injected liquid

TABLE 10.15   Laboratory Measurements of Field Capacity, Wilting Point, and Available Water of the Topsoil from the Nil and Barnyard Manure Treatments (Average of 6 Years, 1966–71)

	No Manure	Manure
	—— % moisture ——	
Field capacity	21.6	24.8
Wilting point	7.1	8.2
Available water	14.4	16.6

TABLE 10.16   Effect of the Method of Handling and Storing on N Losses from Animal Manure

Handling and Storing Method	N Loss* (%)	Handling and Storing Method	N Loss* (%)
Solid systems		Liquid systems	
Daily scrape and haul	15–35	Anaerobic pit	15–30
Manure pack	20–40	Oxidation ditch	15–40
Open lot	40–60	Lagoon	70–80
Deep pit (poultry)	15–35		

*Based on composition of waste applied to the land versus composition of freshly excreted waste, adjusted for dilution effects of the various systems.

SOURCE: Sutton et al., *Univ. of Minn. Ext. Bull. AG-FO-2613* (1985).

TABLE 10.17   Approximate Dry Matter and Fertilizer Nutrient Composition and Value of Various Types of Animal Manure at the Time Applied to the Land

| | | | Nutrient (lb/t Raw Waste) | | | |
| | | | N | | | |
Type of Livestock	Waste-Handling System	Dry Matter (%)	Available*	Total[†]	P_2O_5	K_2O
Solid Handling Systems						
Swine	Without bedding	18	6	10	9	8
	With bedding	18	5	8	7	7
Beef cattle	Without bedding	15	4	11	7	10
	With bedding	50	8	21	18	26
Dairy cattle	Without bedding	18	4	9	4	10
	With bedding	21	5	9	4	10
Poultry	Without litter	45	26	33	48	34
	With litter	75	36	56	45	34
	Deep pit (compost)	76	44	68	64	45
*Liquid Handling Systems**						
Swine	Liquid pit	4	20	36	27	19
	Oxidation ditch	2.5	12	24	27	19
	Lagoon	1	3	4	2	0.4
Beef cattle	Liquid pit	11	24	40	27	34
	Oxidation ditch	3	16	28	18	29
	Lagoon	1	2	4	9	5
Dairy cattle	Liquid pit	8	12	24	18	29
	Lagoon	1	2.5	4	4	5
Poultry	Liquid pit	13	64	80	36	96

*Primarily NH_4-N, which is available to the plant during the growing season.

[†]NH_4-N plus organic N, which is slow releasing.

Application conversion factors: 1,000 gal = about 4 tons; 27,154 gal = 1 a–in.

SOURCE: Sutton et al., *Univ. of Minn. Ext. Bull. AG-FO-2613* (1985).

TABLE 10.18   Nutrient Value of Manure per Animal Unit (lb/1,000 lb Live Weight) per Year

Handling and Disposal Method	Swine			Beef			Dairy			Broilers		
	N	P_2O_5	K_2O	N	P_2O_5	K_2O	N	P_2O_5	K_2O	N	P_2O_5	K_2O
Manure pack												
Broadcast	84	107	124	63	77	99	77	50	112	215	200	149
Broadcast/ incorporation	102	107	124	77	77	99	91	50	112	263	200	149
Open lot												
Broadcast	58	61	80	44	45	64	51	30	59	—	—	—
Broadcast/ incorporation	70	61	80	53	45	64	61	30	59	—	—	—
Manure pit												
Broadcast	95	111	119	69	82	95	87	54	107	—	—	—
Knifing	124	111	119	94	82	95	114	54	107	—	—	—
Irrigation	92	111	119	65	82	95	84	54	107	—	—	—
Lagoon												
Irrigation	24	25	89	18	18	71	23	14	80	—	—	—

SOURCE: Sutton et al., *Univ. of Minn. Ext. Bull. AG-FO-2613* (1985).

TABLE 10.19   Effect of the Method of Application of
Manure on Volatilization Losses of N

Method of Application	Type of Waste	N Loss* (%)
Broadcast without cultivation	Solid	15–30
	Liquid	10–25
Broadcast with cultivation†	Solid	1–5
	Liquid	1–5
Knifing	Liquid	0–2
Irrigation	Liquid	30

*Percentage of total N in waste applied that was lost within 4 days
after application.

†Cultivation immediately after application.

SOURCE: Sutton et al., *Univ. of Minn. Ext. Bull. AG-FO-2613* (1985).

manure has been improved by maintaining the NH_4^+-N form by adding nitrification inhibitors. In large operations in which animals are confined to feedlots, manure dried and bagged for the speciality turf and garden trade is a valuable by-
product.

The availability of manure N depends on the length of time the material remains on the soil surface before incorporation (Table 10.20). In most cases, little
or no N is available if incorporation occurs later than 5 to 8 days after application.
Subsurface application maximizes N availability from manure.

The fertility program on many livestock farms includes manure; however, unless an unusually large quantity of legumes is produced on the farm and considerable commercial feed is consumed, thereby capitalizing on fertility from else-
where, a livestock program in itself will tend to gradually deplete soil fertility.

Many comparisons have been made between the effects of manure on crop production and those obtained from the application of equivalent amounts of N, P,
and K in commercial fertilizers. Long-term studies comparing manure and fertilizer use in Missouri showed variability in wheat yield response, but the total over
the 100-year period was similar (Fig. 10.41).

Distribution of manure by grazing animals presents a problem in the maintenance fertilization of pastures. For N, which does not remain in effective concentrations for more than a year, about 10% of a grazed area is effectively covered
in 1 year. On the other hand, with P, which is not leached or removed in large
quantities, some effect might be obtained from a given application as much as 10
years later. In general, nearly all of a pasture area will receive deposits of manure
in a 10-year period. K is intermediate between N and P in retention in the soil,
and manure-deposited K is effective to some degree for at least 5 years. During
this period, about 60% of a pasture will have been covered.

With low stocking rates, animal excreta will essentially have no effect on soil fertility. On highly productive pastures with a high carrying capacity, excreta may
have a beneficial effect on soil fertility over a period of time. Grain feeding on pastures has a considerable effect on soil fertility, and each increase of 4.5 tons of
grain fed per acre results in an increase of 53 steer-days of grazing.

TABLE 10.20   Availability of NH$_4$-N within the First Year of Manure Application

Season of Spreading	Days from Spreading to Incorporation	Percentage Available
Spring	Same day (0)	80
	1 day	65
	2 days	50
	3–4 days	30
	5–6 days	15
	7 or more/unincorporated	0
Fall	Within 2 days	20
	Unincorporated	0
Crop growing season	Injected or same-day incorporation	100

Producers interested in using manure as the nutrient source in cropping systems should consider the following:

1. The high cost of transportation potentially results in continued application of manure close to the source, where overapplication is common.
2. Nutrient content is highly variable, which makes it difficult to accurately supply required quantities for crop yield potential.
3. Variability of N mineralization of the manure N can reduce available N at periods of high N demand.
4. Increased soil compaction can occur with manure application equipment.
5. Possible nutrient imbalances (e.g. S supplementation can be beneficial with lagoon-stored hog waste.

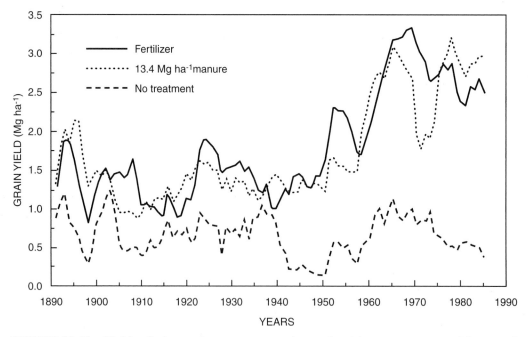

FIGURE 10.41   Yields of winter wheat grown continuously with no treatment, with annual treatment of 13.4 mg ha^{-1} manure, and full fertilizer as 5-year moving averages.

Application of Sludge

Disposal of processed sewage materials from treatment plants is of increasing concern because of population pressures, more stringent laws, and increases in energy costs. Use of sludge on agricultural land has both benefits and problems. Sludge is a source of OM containing macro- and micronutrients, and is a cost-effective alternative to more costly methods of disposal, such as burning or burying.

Sludges contain both inorganic and organic N (Table 10.21). Because of possible volatilization losses of up to 20 to 50% of the NH_4^+ from surface application of sludge and only partial mineralization (20 to 30%) of the organic N, exact application rates are difficult to determine. However, if the sludge is incorporated into the soil, very little NH_4^+ will be lost. After the first year, when 20% of the organic N is mineralized, about 3% of the remaining N will be released annually during the next few years.

N and P additions to soil from a single application of sludge can be as high as 800 lb/a (900 kg/ha) of total N in some areas, with about 50% as NH_4^+, and 1,025 lb/a (1,150 kg/ha) of P_2O_5. Use of fertilizer N and P on sludge-applied lands is generally not required for at least 2 years following sludge application. However, imbalances of other nutrients may have to be corrected.

It is essential that appropriate application and soil management techniques be used to protect the environment and the health of human beings and animals. Because of the possibility of applying excessive N and subsequent movement of NO_3-N into surface water and groundwater, careful monitoring is necessary.

Because of the potential for buildup of toxic levels of selected heavy metals in soils from frequent application of sludge, limits have been established (Table 10.22).

Care must be taken to avoid cadmium (Cd) contamination of crops. Although there is no immediate human health problem, the U.S. Environmental Protection Agency has established recommendations concerning the maximum amount of Cd that can be applied.

When Cd is the main concern, sludge should be applied only to soils of pH 6.5 or above. The maximum allowable rate for crops must not exceed 4.4 lb/a of Cd on coarse-textured soils. The life of a sludge application site is determined by the cumulative addition of metals. Guidelines for the buildup of Cd and other heavy metals in soils are summarized in Table 10.23. Use of CEC as a controlling soil fac-

TABLE 10.21   Typical Analysis of Sewage Sludge

Component	Concentration on Dry Weight Basis (%)	Component	Concentration on Dry Weight Basis (ppm)
Organic carbon	50	Fe	40,000
N		Zn	5,000
Ammonium	2	Cu	1,000
Organic	3	Mn	500
Total	5	B	100
P_2O_5	6.8		
K_2O	0.5		
Ca	3	Cd	150
Mg	1	Pb	1,000
S	0.9	Ni	400

SOURCE: *Univ. of Illinois Soil Manag. Conserv. Ser. Bull. SM-29* (1975).

TABLE 10.22   Maximum Allowable Concentrations of
Metals in Sewage Sludges in Different Countries

Metal	EEC	USA	Norway
	—mg kg^{-1} dry matter—		
Cd	20–40	20	4
Cu	1,000–1,750	1,200	1,500
Cr	—	—	125
Hg	16–25	—	5
Pb	750–1,200	300	100
Ni	300–400	500	80
Zn	2,500–4,000	2,750	700

TABLE 10.23   Maximum Amount of Metal
Suggested for Agricultural Soils Treated with
Sewage Sludge

Metal	Maximum Amount of Metal (lb/a) When CEC meq/100 g) is		
	<5	5–15	>15
Pb	440	880	1,760
Zn	220	440	880
Cu	110	220	440
Ni	110	220	440
Cd	4.4	8.8	17.6

SOURCE: Sommers et al., *Purdue Univ. Bull. AY-240* (1980).

tor does not necessarily mean that all of these metals are retained on the exchange complex. Rather, CEC was chosen as a single soil property that can be easily measured and one that is positively related to soil components that may minimize plant availability of metals in sludges added to soils. On soils of little agricultural value, higher levels of Cd may be permitted.

Soil testing and fertilizer recommendations are used in conjunction with sewage sludge characteristics to determine application rates. The annual rate of sludge is based on either the lowest tonnage that will satisfy the N requirements of the crop or the maximum quantity that can be used without exceeding permissible limits for Cd.

Some regulatory agencies confine the application of sludges to forages, oilseed crops, small grains, commercial sod, and trees. Unacceptable crops may include root crops, vegetables and fruit, tobacco, and dairy pastures. Direct grazing of sludge-treated forage lands is not usually recommended for 3 years immediately following application. Wheat is preferable to barley. Oats are not recommended in the first two growing seasons following sludge treatment.

The response of crops to sewage sludge is at least equal to the response to commercial fertilizer in the first year after application and may be somewhat greater in subsequent years because of residual effects of the added plant nutrients. Also,

TABLE 10.24   Four-Year Annual Dry Matter Yield of Five Forage Species as a Function of Wastewater and Fertilizer Levels

Wastewater Irrigation (cm/yr)	Fertilizer N-P (kg/ha/yr)	Forage (metric t/ha)					
		Alfalfa	Reed Canary	Brome	Altai Wildrye	Tall Wheat	Mean
	0–0	8.7	5.8	6.0	5.8	5.3	6.3
62.5	56–48	9.4	8.8	8.9	8.4	7.7	8.6
	0–0	10.2	9.4	10.0	9.6	7.9	9.4
125	56–48	10.1	11.5	11.1	11.7	9.8	10.8
Mean		9.6	8.9	9.0	9.0	7.7	8.8

SOURCE: Bole and Bell, *J. Environ. Qual.,* 7(2):222 (1978).

many of the favorable effects of the OM in sludges, probably very similar to the ones associated with manure, may be long lasting. Additionally, it may take one or more years for the applied nutrients to become effectively distributed in the root zone.

Sewage Effluent

Sewage effluent can be either a valuable water and nutrient resource for crops or a pollutant to land and waters. Large quantities of water are generally involved; thus, it is essential that the soil be (1) internally well drained and medium textured, having a pH of between 6.5 and 8.2, and (2) be supporting a dense stand of trees, shrubs, or grasses. The groundwater should be monitored periodically for NO_3^-.

Forage crops are commonly used for effluent application because of their long growing season, with the resultant high seasonal evapotranspiration, their high nutrient uptake, and their capacity to stabilize the soil and prevent erosion. Because forages are not eaten directly by human beings, the transfer of human diseases is unlikely.

Yields of five forage crops irrigated with sewage effluent are shown in Table 10.24. Alfalfa was the most suitable forage crop when the system was operated for optimum utilization of the wastewater. The water did not supply sufficient N for high production of the grasses.

The previous discussion indicates that there is much to be learned about the use of sewage sludge or effluent on agricultural crops. Society will benefit from wise application of this material and thus from the recycling of a valuable resource.

Selected References

FOLLETT, R. H., L. S. MURPHY, and R. L. DONAHUE. 1981. *Fertilizers and Soil Amendments.* Prentice-Hall, Inc., Englewood Cliffs, N.J.

RANDALL, G. W., K. L. WELLS, and J. J. HANWAY. 1985. Modern techniques in fertilizer application. *In* O. P. ENGELSTAD (Ed.), *Fertilizer Technology and Use.* Soil Science Society of America, Madison, Wisc.

RENDIG, V. V., and H. M. TAYLOR. 1989. *Principles of Soil—Plant Interrelationships.* McGraw-Hill Publishing Co., New York.

PIERCE, F. J. and E. J. SADLER (ed). 1997. The state of site specific management for agriculture. Amer. Soc. Agronomy. Madison, WI.

CHAPTER 11

Nutrients, Water Use, and Other Interactions

Two or more growth factors are said to interact when their influence individually is modified by the presence of one or more of the others. An interaction takes place when the response of two or more inputs used in combination is unequal to the sum of their individual responses. There can be both positive and negative interactions in soil fertility studies (Fig. 11.1). In addition, there can be circumstances in which there is no interaction, with the action of factors being only additive.

In negative interactions, the two factors combined increase yields less than when they are applied separately. This kind of interaction can be the result of substitution for and/or interference of one treatment with the other.

Positive interactions follow Liebig's law of the minimum. If two factors are limiting, or nearly so, the addition of one will have little effect on growth, whereas provision of both together will have a much greater influence. In severe deficiencies of two or more nutrients, all nutrient responses will result in strong positive interactions.

Yield increases from an application of one nutrient can reduce the concentration of a second nutrient, but the higher yields result in greater uptake of the second nutrient. This is a dilution effect, which should be distinguished from an antagonistic effect.

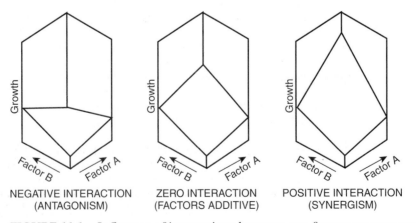

NEGATIVE INTERACTION (ANTAGONISM)   ZERO INTERACTION (FACTORS ADDITIVE)   POSITIVE INTERACTION (SYNERGISM)

FIGURE 11.1   Influence of interactions between two factors on crop growth.

Interactions can be between

1. Two or more nutrients.
2. Nutrients and a cultural practice such as planting date, placement, tillage, plant population, or pest control.
3. Nutrient rate and hybrid or variety.
4. Hybrid or variety and row width or plant population.
5. Nutrients and the environment (e.g., water and temperature).

Many interactions are not significant with average yields; however, with high yields, increasing stress is being placed on the plant, and interactions between the various factors contributing to those yields are often observed.

Future increases in agricultural productivity will likely be related to manipulation of the interactions between the numerous management inputs and factors. It is essential that growers and consultants recognize and take advantage of these interactions.

Nutrient-Water Interactions

Even in regions where annual precipitation exceeds growing season evapotranspiration, water stress frequently limits crop production. Stresses caused by nutrient deficiencies, pests, and other factors reduce the plants' ability to use water efficiently, which reduces productivity and profit. As pressures grow for increased industrial, recreational, and urban use of water, agriculture will have less access to water for irrigation.

Increasing water use efficiency is a major challenge to agriculture. It is estimated that overall efficiency of water in irrigated and dryland farming is 50%. In general, any growth factor that increases yield will improve the efficiency of water use.

Water Use Efficiency

Water use efficiency (WUE) is the yield of crop per unit of water—from the soil, rainfall, and irrigation. When management practices increase yields, WUE is increased. Yields of crops have increased greatly in the past 30 years on essentially the same amount of water, which is directly related to improved soil and crop management practices. For example, tillage systems that leave large amounts of surface residues conserve water by

1. Increased water infiltration.
2. Decreased evaporation from the surface.
3. Increased snow collection.
4. Reduced runoff.

In many parts of the world irrigation has stabilized production, but yields per unit of land have not increased greatly. After the lack of moisture is eliminated by irrigation, many factors may limit yields (Fig. 1.7). Because of these other factors, there can be many disappointments. If yields of 300 bu/a rather than 150 bu/a of corn or 14 t/a rather than 7 t/a of alfalfa are to be obtained, the nutrient removal is at least doubled. Consequently, the crop must obtain more nutrients from some source, whether from native soil supply, manures, or fertilizers.

How Water Is Lost from the Soil

Water in a soil is lost by (1) evaporation from the soil surface, (2) transpiration through the plant, and (3) percolation below the rooting zone.

The sum of transpiration and evaporation from soil plus intercepted precipitation is called *evapotranspiration*. With more complete cover, less water evaporates from the soil and more goes through the plant. Adequate fertility and satisfactory stands are among the factors that help to provide more plant cover rapidly and thus realize more benefit from the water.

With a sparse stand or growth, more sunlight will reach the soil, and a considerable amount of water may be evaporated from a moist soil. With a heavy crop canopy, the surface is shaded and less evaporative energy reaches the soil. The soil temperature is reduced, and the crop provides insulation to maintain higher humidity just above the soil because of less air movement. These three effects reduce evaporation from the soil. Even with a heavy canopy, a considerable amount of energy still reaches the soil.

Nutrient availability affects plant size, total leaf area, and often the color of the foliage. Close rows and adequate stands, along with adequate nutrition, provide a heavy crop canopy. For example, water use would be less in 21-in. rows than in 42-in. rows. Differences in evapotranspiration among crops may be small once a complete cover is developed. Daily crop water use varies greatly from one day to another, depending on soil and environmental conditions (temperature, moisture, and wind); however, daily losses of 0.1 to 0.3 in. of water/a are common.

Evaporation from the soil may account for 30 to 60% of the total water loss in a crop year in humid areas where the soil is wet. With local droughts or in arid regions the soil surface is dry, and very little water is lost from the soil. The moisture films between the particles are thin, and little water is transported to the soil surface by capillarity or diffusion of water vapor. Hence, in dry soils most of the water use is by transpiration, although most of the water received in a light shower would be evaporated quickly.

Heat advection, in which there is horizontal and vertical movement of air in a turbulent fashion, brings in more heat. In a hot, dry area with a strong wind, the heat from the air may contribute up to 25 to 50% of the total evapotranspiration. In arid and semiarid areas advection is great, and thus quite variable evapotranspiration may occur.

Fertilization and Water Extraction by Roots

Most crops use water more slowly from the lower root zone than from the upper soil. The surface soil is the first to be exhausted of available water; the plant must then draw water from the lower three-fourths of the root depth (Fig. 11.2).

The favorable effects of fertilization on the mass and distribution of roots when soils are nutrient deficient were illustrated in Figures 10.13 and 10.14. Under nutrient stress the plant may extract water from a depth of only 3 to 4 ft. With fertilization the plant roots may be effective to a depth of 5 to 7 ft or more. If the plant can utilize an extra 4 to 6 in. of water from the subsoil, the crop can endure droughts for a longer period of time without reducing yield. It should be emphasized that in areas in which the subsoil is dry, increased fertilization will not help crops penetrate the soil farther to get more water.

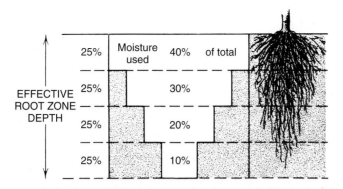

FIGURE 11.2 The top one-quarter of the root zone is the first to be exhausted of available moisture. Certain management practices, including adequate fertilization, help to develop a deeper root system to use the moisture from the lower root zone. *USDA,* SCS. Bull., *1972.*

The importance of adequate fertility for efficient crop water use and improvement of crop tolerance to low-rainfall conditions can be explained as follows:

1. *Root exploration of the soil is increased.* Adequate fertility favors expanded root growth and proliferation. When roots explore the soil 1 ft deeper, another 1 to 2 in. of water will be obtained.
2. *The major portion of P and K moves to roots by diffusion through the water films around the soil particles.* Under moisture stress the films are thin and path length increases, reducing P and K diffusion to the roots. Increasing the concentration of P and K in the soil solution increases their diffusion to the roots.
3. *Increased soil moisture tension (lower moisture) exerts a physiological effect on the roots.* Elongation, turgidity, and the number of root hairs decrease with increasing tension. Mitochondria development slows, and *carrier concentration* and *phosphorylation* decrease, which reduces nutrient uptake.
4. *Adequate fertility decreases the water requirement.* K has been shown to aid in closing the stomata, thus reducing water loss by transpiration.
5. *The foliage canopy is increased and the soil is covered more quickly.* Rapid canopy development reduces water evaporation from the soil, which increases water availability to the plant.
6. *Adequate fertility advances maturity.* Advanced maturity in sorghum and corn helps ensure pollination before summer drought periods. Similarly, small grains are adversely affected when growth is delayed, so that summer drought occurs during and following heading.
7. *The amounts of plant and root residues are increased.* With any given tillage practice, a higher amount of residues will break the impact of raindrops, reduce runoff, increase water infiltration, and reduce the erosive effect of wind and water on soil.

Soil Moisture Level and Nutrient Absorption

Water is a key factor in nutrient uptake by root interception, mass flow, and diffusion. Roots intercept more nutrients, especially Ca^{2+} and Mg^{2+}, when growing in a moist soil than in a drier one because growth is more extensive. Mass flow of soil

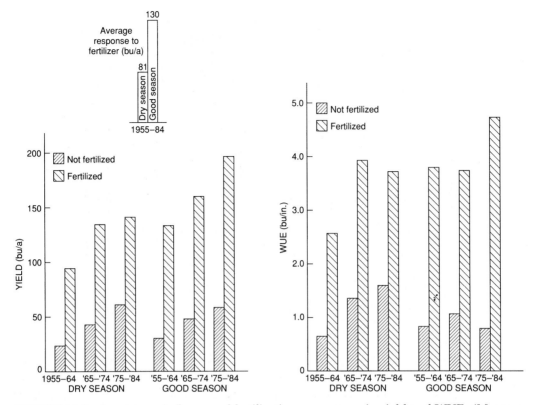

FIGURE 11.3   Long-term influence of fertilization on corn grain yield and WUE. (Morrow plots, Univ. of Illinois, 10-year averages).   *Adapted from Potash & Phosphate Inst. Fert. Improves Water Use Eff., 1990.*

water to supply the transpiration stream transports most of the NO_3^-, SO_4^{2-}, Ca^{2+}, and Mg^{2+} to roots. Nutrients slowly diffuse from areas of higher concentration to areas of lower concentration but at distances no greater than 1/8 to 1/4 in. The rate of diffusion depends partly on the soil water content; therefore, with thicker water films or with a higher nutrient content, nutrients diffuse more readily.

Nutrient absorption is affected directly by the level of soil moisture, as well as indirectly by the effect of water on the metabolic activity of the plant, the degree of soil aeration, and the salt concentration of the soil solution.

Of course, crop yield potential is greater with normal or higher moisture availability (Fig. 11.3). Substantial responses in grain yield and WUE to fertilization occurred in the dry years, as well as in normal rainfall years. Although the response to fertilization was less in dry years, adequate nutrient availability greatly reduced drought-related yield losses.

Dryland Soils

Moisture is the most yield limiting factor in semiarid and arid regions. In crop-fallow systems, conserving soil water may not always increase grain yield in some crops, but increased soil water conservation will reduce the dependence on fallowing through more intensive cropping (Fig. 11.4). These data illustrate that

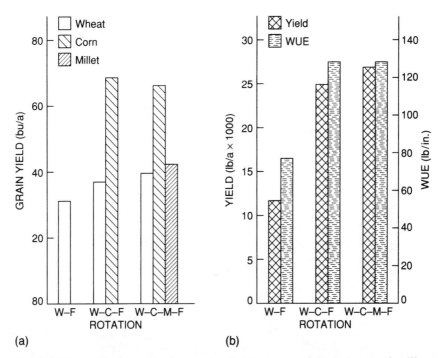

FIGURE 11.4   Influence of cropping intensity on wheat, corn, and millet grain yield (a) and on total grain production and water-use efficiency in a 12-year cycle (b). W, wheat; C, corn; M, millet; F, fallow.   *Peterson et al.,* Proc. Great Plains Soil Fert. Conf., *pp. 47–53, 1992.*

wheat yields in a wheat-fallow rotation are not greatly increased due to the extra water conserved in a no-till system. Although wheat yields were not increased, the additional water enabled production of wheat-corn-fallow and wheat-corn-millet-fallow rotations (two crops in 3 years and three crops in 4 years, respectively, versus one crop in 2 years). Thus, total WUE increased more than 50% in the 3-year rotation compared with the 2-year rotation.

NITROGEN   Although absorption of N is definitely reduced on dry soils, it is usually not reduced as much as P and K absorption. Under drought conditions, N mineralization is reduced. Also, when water is limiting, uptake of soluble nutrients in the water is reduced. NH_4-N does not move readily, but NO_3-N moves with the soil water. In heavy rains NO_3^- moves downward in the soil profile and is available for later use, unless it moves below the root zone.

The data in Figure 11.5 show that (1) fertilizer N will not increase yield without sufficient plant available water and (2) increasing stored soil water by conservation practices will not increase production without adequate fertility.

PHOSPHORUS   Crop yield response to P and other nutrients varies from year to year and can be related to the amount of rainfall (Fig. 11.6). The lower the rainfall, the greater the percentage response to P. The same relationship was found for K.

In low-P soils the majority of wheat response to N-P fertilization in dry years is due to P. In wet years, wheat yields dramatically increase, with both N and P contributing to the wheat response. Figure 11.7 indicates the inverse relation between the response of cereals to P and rainfall. The percentage yield response to P is greater with low rainfall.

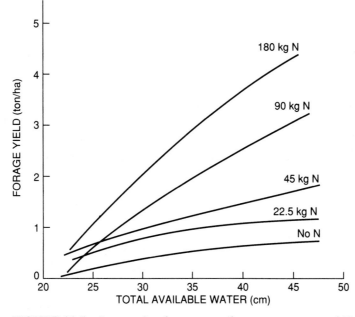

FIGURE 11.5   Interaction between soil water content and N fertilization on native grass forage production.   *Smika*, Agron. J., *56:483–86, 1965.*

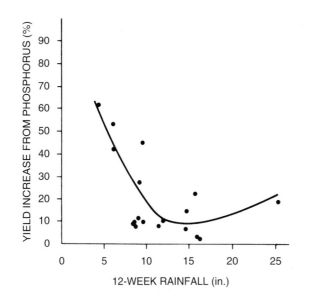

FIGURE 11.6   Soybean response to P is greater with low rainfall.   *Barber*, Better Crops Plant Food, *1971.*

POTASSIUM   The effect of decreasing soil water on corn growth is shown in Figure 11.8. However, as % K saturation of CEC was increased, growth increased at all three moisture levels; thus, adequate K availability reduced some of the water stress. Generally, the lower the rainfall, the greater the K response, which is related to the following:

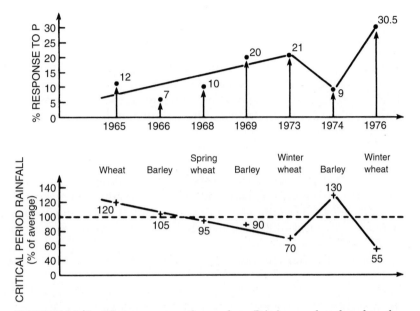

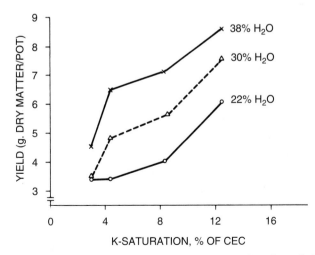

FIGURE 11.7   The response of cereals to P is inversely related to the amount of rainfall.   *Ignaze,* Phosphorus Agr., *70:85, 1977.*

FIGURE 11.8   Dry matter yield of corn after 3 weeks' growth in relation to K saturation and water supply.   *Grimme et al.,* Int. Symp. Soil Fertil. Eval. Proc., *Indian Society of Soil Science, New Delhi, 1:33, 1971.*

1. Most of the K absorbed moves to the roots by diffusion through the water films, and with low water content, K diffusion is reduced. Therefore, K fertilization will increase the K content in the water films and increase diffusion. The same is true for P.
2. In some soils the subsoil contains less K than the surface. When the surface soil is exhausted of water in dry periods, the plant roots must feed in the subsoil, where they cannot absorb as much K; this is also true for P.

In wet periods, the K response can also be large and is related to restricted aeration. Plant roots respire to obtain energy to absorb nutrients. Respiration requires O_2. Adequate K helps to meet the needs of the plant even when root respiration is restricted.

An example of water-K interaction is shown in Table 11.1. On a medium-K soil there was little or no yield response or profit in years of good rainfall. In stress years, K gave a 48-bu/a response on corn and an 18-bu/a response on soybeans, with excellent profits. In dry years, the K content in the corn or soybean leaves is below the sufficiency K level even with high K rates. The inability to take up adequate K probably contributes to lower yields in dry years.

MICRONUTRIENTS   Since transport of micronutrients to plant roots is by diffusion, low soil moisture content will reduce micronutrient uptake in dry weather, as with P uptake. The only difference is that plants require a much smaller quantity of micronutrients than P; thus, drought stress effects are not as great for micronutrients as for P.

Temporary B deficiency during dry weather is common and is related to the following:

1. Much of the B is in the OM, and under dry conditions mineralization is reduced.
2. In some areas the subsoil B is lower than the surface soil B. Under dry conditions, water uptake is predominantly from the subsoil; thus, plants take up less B. In contrast, in sandy soils, excessive rainfall may leach some of the available soil B.

Low soil moisture can also induce deficiencies of Mn and Mo, although Fe and Zn deficiencies are often associated with high soil moisture levels. Increased soil

TABLE 11.1   Effect of K on Corn and Soybean Yields and Profits in Years of Good Rainfall and in Dry Years

K_2O (lb/a)	Corn (bu/a)		Soybeans (bu/a)		K Soil Test, Initial 162 (lb/a)
	Good Year	Stress Year	Good Year	Stress Year	
0	163	81	56	30	129
50	163	113	59	42	152
100	167	121	60	48	196
200	163	129	58	48	236
Response	0	48	4	18	—
Profit ($/a)	0	87	18	104	—

SOURCE: Johnson and Wallingford, *Crops and Soils*, 36(6):15 (1983).

moisture results in greater amounts of Mo uptake. Mn becomes more available under moist conditions because of conversion to reduced, more soluble forms.

Irrigated Soils

Fertility is one of the important controllable factors influencing water use in irrigated soils. Figure 11.9 shows that adequate N increased forage yield, while water use decreased from 18 in./t with no N to 3 in./t with 1,000 lb/a of N.

When N is deficient, increasing N fertilization will increase yield and water use. Figure 11.10 shows that more irrigation water was required with increasing N rate; however, the WUE also increased.

The effects of irrigated and dryland moisture regimes and of increasing N rates on the yield of rapeseed are shown in Figure 11.11. Increasing N rates up to 168 kg/ha under both moisture conditions increased yields; however, optimizing available soil water with irrigation greatly increased the yield response to N compared with dryland conditions. This result illustrates the limiting effect of moisture stress on plants' response to applied nutrients. It should also be noted that it is necessary to provide sufficient nutrients to make the greatest use of available water.

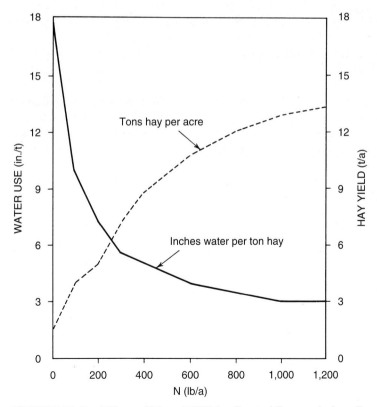

FIGURE 11.9   Effect of N on WUE by Coastal Bermuda hay. P and K were applied in liberal amounts.   *Texas A & M College Prog. Rep. 2193, 1961.*

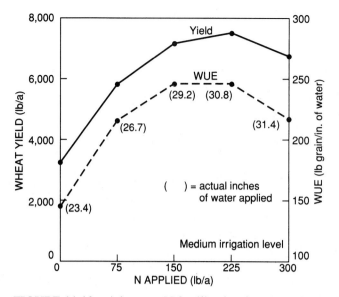

FIGURE 11.10   Adequate N fertilization increases irrigated wheat yield and WUE (Mesa, Ariz.).   *Taken from PPI, 1990.*

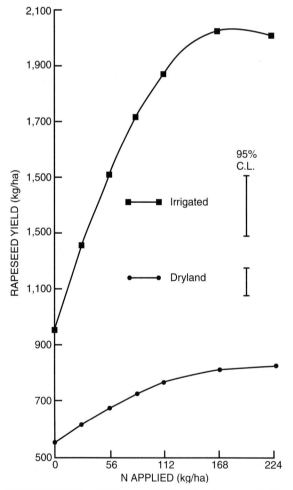

FIGURE 11.11   Effects of N fertilization on yield of rapeseed under irrigated and dryland conditions.   *Henry and MacDonald,* Can. J. Soil Sci., *58:305, 1978.*

Other Interactions

Interactions between Nutrients

N-K and N-P interactions are commonly observed. For example, under low-yield conditions when other nutrients are limiting or management practices are inadequate, plant growth is slow, and unless K is seriously limiting, some soils will release K at a rate adequate to meet the needs of the crop. With adequate N and P and improved management practices, there is more rapid growth, and the potential response to K, S, and other nutrients is greater (Fig. 11.12). With 30 kg/ha of N there was little response to K; however, with 90 kg/ha of N the response to K was linear up to the highest rate applied.

Interactions with micronutrients can be dramatic. On a low-P, low-Zn soil leveled for irrigation, adding P or Zn separately decreased corn yields (Fig. 11.13). When both were applied a substantial positive interaction occurred, increasing yield by 44 bu/a.

The effect of soil pH on the response of corn to P_2O_5 banded beside the row is shown in Figure 11.14. With a pH of 6.1 there was little response to P, but at pH 5.1 there was about a 20-bu/a response to 70 lb/a of P_2O_5. Also, liming alone increased yields substantially.

Crop response to N is greatly reduced when P is limiting. The data in Figure 11.15 illustrate that the N rate required for an optimum yield is considerably higher with 40 lb/a of P_2O_5 (160 lb/a of N) compared with no P (80 lb/a of N). When both N and P were adequate, crop recovery of fertilizer N was approximately 75% compared with about 40% without adequate P fertilization.

Maximizing crop recovery of fertilizer N reduces the quantity of profile NO_3^- after harvest (Fig. 11.15). The rooting depth is about 6 ft, and a significant quantity of fertilizer N moved below the root zone and could potentially reach the groundwater. Thus, adequate N and P fertilization will optimize yield and

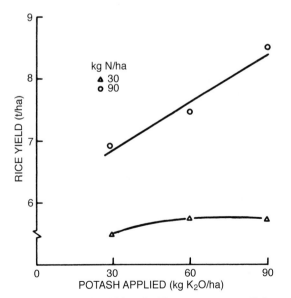

FIGURE 11.12   N level affects response of rice to K.   *Malavolta,* Nutrição mineral e adubação de arroz irrigado. *Ultrafertil S.A., São Paulo, Brazil, 1978.*

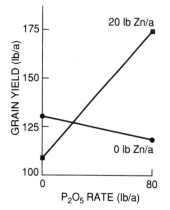

FIGURE 11.13   Interaction of P and Zn fertilization on corn yield.   *Ellis,* Kansas Fert. Handbook, *Kansas State Univ., Manhattan, Kans., 1967.*

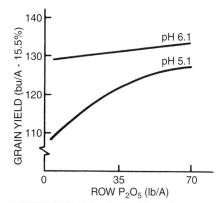

FIGURE 11.14   Soil pH and row P_2O_5 interact on corn.   *Schulte,* Better Crops Plant Food, *66:10, 1982.*

profitability (see Fig. 12.6 for an economic analysis of these same data) and maximize the fertilizer N recovered while minimizing the environmental impact of fertilizer N use.

As previously discussed, the positive interaction of N and P has also been shown in wheat with N-P placement in the same band compared with separate placement (Fig. 10.25).

Many nutrient interactions occur in soils; only a few examples have been provided. The most probable nutrient interactions in a given cropping system involve nutrients that are deficient or marginally deficient. For example, N-P or P-Zn interactions frequently occur on soils marginally deficient in P or Zn, respectively. Therefore, a good soil-testing program will enable the grower or consultant to anticipate potential nutrient interactions.

Interactions between Nutrients and Plant Population

Increasing the plant population may not optimize yield unless there is an adequate quantity of available plant nutrients. Similarly, increasing plant nutrients without a sufficient number of plants will not maximize the return. For example, increasing the plant population with 80 lb/a of N increased the corn yield 46

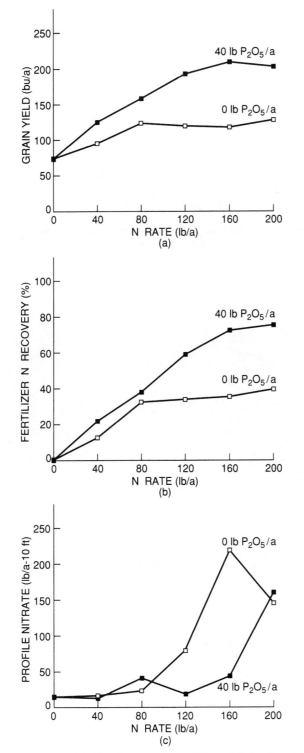

FIGURE 11.15   Interaction of N and P fertilization on irrigated corn grain yield (a), fertilizer N recovered in the grain (b), and profile NO₃ after harvest (c).   *Schlegel et al.,* Proc. Great Plains Soil Fert. Conf., *pp. 177–87, 1992.*

bu/a; however, with 240 lb/a of N, increasing the plant population increased yield 76 bu/a (Fig. 11.16). At 12,000 plants/a, increasing N to 240 lb/a resulted in a 37-bu/a increase, but with 36,000 plants the increase was 67 bu/a.

Interactions between Plant Population and Planting Date

Plant population interacts with planting date. Generally, plants are shorter with earlier planting, and a higher population can be utilized (Fig. 11.17). With a later planting date, there was a decrease in corn yield. Plants are taller with the May 30 planting date, and competition for light at this date would be higher with higher populations.

Interactions between Nutrients and Planting Date

Planting date has a marked effect on the response to nutrients (Table 11.2). Earlier planting dates for spring-planted crops result in higher yields. Note the greater response of soybeans to increased K soil test level with earlier planting.

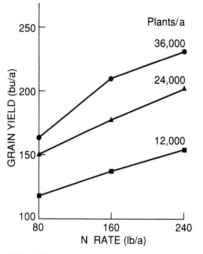

FIGURE 11.16   Interaction of N and plant population on corn yield.   *Rhoades, Quincy Res. Rep., 1978.*

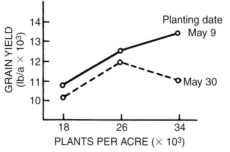

FIGURE 11.17   Planting date and population interact on corn.   Arjal et al., California Agriculture, *Univ. of California, March 1978.*

TABLE 11.2 Effect of the Planting Date on the Response of Soybean to K

| Planting Date | Yield (bu/a) at Soil K Level | | | Increase (bu/a) |
	Low	Medium	High	
May 27	40	47	53	13
June 16	40	44	46	6
July 8	31	36	37	6

SOURCE: R. Peaslee, Univ. of Kentucky, personal communication.

TABLE 11.3 Effect of Row Width and Soybean Varieties on Yield

| Variety | Yield (bu/a) at Row Width of | | Increase (bu/a) |
	30 in.	7 in.	
A	58	68	10
B	66	79	13
C	63	83	20

SOURCE: R. L. Cooper, Ohio State Univ., personal communication.

Also, the K level had a greater effect on increasing seed size and on decreasing the incidence of seed disease.

Similar planting date interactions with both N and P fertilization have been observed. The increased N and P response is related to the increased yield potential associated with timely planting and the longer growth period.

Interactions between Variety and Row Width

Varieties or hybrids may vary in their response to plant spacing. Note in Table 11.3 that soybean variety A gave a 10-bu/a response to 7-in. rows over 30-in. rows, while variety C gave a 20-bu/a response, raising the yield to 83 bu/a.

Interactions between Nutrients and Placement

Crop response to fertilization can be greatly increased if nutrients are applied properly (see Chapter 10). Examples were provided in Figures 10.27, 10.28, and 10.31.

Interactions between Nutrient Placement and Tillage

Surface accumulations of residues and nutrients, cooler temperatures, and higher moisture in the spring can influence nutrient use. In some situations, higher levels of nutrients applied below the soil surface may be needed. For example, a 42-bu/a corn response to 40-40-40 banded beside the row occurred where the soil was not plowed compared with 16 bu/a where the soil was plowed (Fig. 11.18). Even with higher broadcast rates, the responses were 20 and 9 bu/a, respectively.

In general, higher rates of N and perhaps S are required under no-till systems than under conventional tillage (Fig. 11.19). Under no-till operations, the broadcast NH_4NO_3 is partially immobilized and/or denitrified. To avoid fertil-

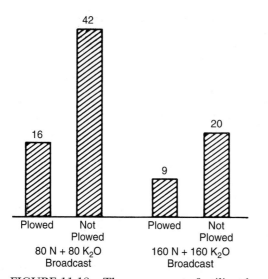

FIGURE 11.18   The response to fertilizer banded beside the row at planting was greater in the unplowed areas.   *Schulte,* Better Crops Plant Food, *63:25, 197.*

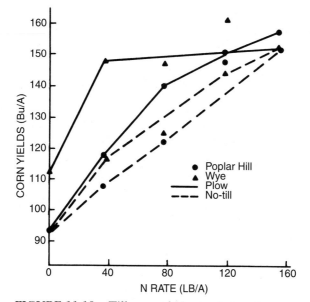

FIGURE 11.19   Tillage and N rates interact on corn yields.   Bandel et al., Agron. J., *75:782, 1975.*

izer N interactions with surface residues, N must be placed below the residue. The data in Table 11.4 show increased grain yield with N placed below the surface (knife) compared with broadcast N. Surface band-applied N (dribble) was only partially effective in minimizing N losses. These data also show that reducing immobilization or denitrification losses by subsurface N placement increased the percentage of fertilizer N recovered by the crop. Increasing fertilizer N efficiency will greatly reduce the residual profile N content after harvest and the potential for NO_3^- movement to groundwater.

TABLE 11.4   N Placement Effect on Average Grain Sorghum Yield and Apparent Fertilizer N Recovery at Two Locations in Kansas

	Riley Co. (1986–88)		*Greenwood Co. (1987–89)*	
N Placement	*Yield (bu/a)*	*AFNR* (%)*	*Yield (bu/a)*	*AFNR* (%)*
Broadcast	110	64	78	51
Dribble	117	70	81	56
Knife	130	87	89	65

*AFNR, apparent fertilizer N recovery.

SOURCE: Lamond et al., *J. Prod. Ag.,* 4:531–35 (1991).

The following points summarize the available information on N management in reduced-tillage systems:

1. Subsurface placement of N can prevent N losses and increase N removal by the crop.
2. After several years in no-tillage systems, the differences in N needs between no-tillage and conventional-tillage operations diminish.
3. Under some conditions, the yield potential is greater under no-till systems, which thus require more N.
4. Soil sampling for profile NO_3^- before planting can help predict the fertilizer N need.

Interactions between Nutrients and Hybrid or Variety

Within a given environment, one hybrid or variety may produce a greater response to applied nutrients than another. In Figure 11.20 the Dare soybean variety produced a higher yield and responded more to K than did the Bragg variety on this very-low-K soil.

Some corn hybrids are genetically able to produce greater yields than other hybrids from higher rates of applied nutrients (Table 11.5). At the lower fertility level corn hybrids differed by 19 bu/a; at the higher fertility level the difference was 85 bu/a. Selection of hybrids or varieties that respond to a high-yield environment is essential for maximum productivity.

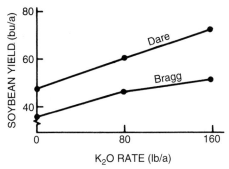

FIGURE 11.20   Interaction between soybean varieties and the response to K. *Terman,* Agron. J., *69:234, 1977.*

TABLE 11.5   Interaction of Fertility and Corn Hybrids to Increase Yields

N (lb/a)	P_2O_5 (lb/a)	K_2O (lb/a)	Yield (bu/a) of		Increase (bu/a)
			Hybrid A	Hybrid B	
250	125	125	199	218	19
500	300	300	227	312	85

SOURCE: R. L. Flannery, New Jersey Agricultural Experiment Station, personal communication.

The importance of exploiting interactions in maximizing productivity and profitability cannot be overemphasized. When one practice or group of practices increases the yield potential, the nutrient requirement will be increased. Also, as breakthroughs occur in genetic engineering, rhizosphere technology, plant growth regulators, and related areas, they will be successful only if the technology is integrated in a manner that allows the expression of positive interactions.

Selected References

JACKSON, T. L., A. D. HALVORSON, AND B. B. TUCKER. 1983. Soil fertility in dryland agriculture. In H. E. DREGNE AND W. O. WILLIS (Eds.), *Dryland Agriculture.* American Society of Agronomy, Madison, Wisc.

THORNE, D. W., AND M. D. THORNE. 1979. *Soil, Water, and Crop Production.* AVI Publishing Co., Westport, Conn.

CHAPTER 12

Economics of Plant-Nutrient Use

Despite the rise in use of plant nutrients, reliable estimates show that much land is still underfertilized and that increased nutrient use would be profitable. For developing countries, economic development must give high priority to agriculture. As higher rates of plant nutrients are required, it becomes more important that the nutrients be applied so that they will be utilized most efficiently.

Higher crop yields represent the greatest opportunity for reducing per-unit production costs. Successful growers have the attitude of all good business managers: they must spend money to make money. This adage certainly applies to soil amendments. The demand for fertilizer is evidence that growers recognize the returns realized from plant nutrient inputs.

To obtain a given level of production, farmers can vary the inputs of land, fertilizer, labor, machinery, and so on. The actual use of each depends on relative costs and returns. Production costs can vary from year to year, but costs gradually increase over time. The relative costs of many farm inputs have increased more than the costs of fertilizers and chemicals (Fig. 12.1). Although the price of fertilizers and lime will continue to rise, it may not rise as fast as other input prices.

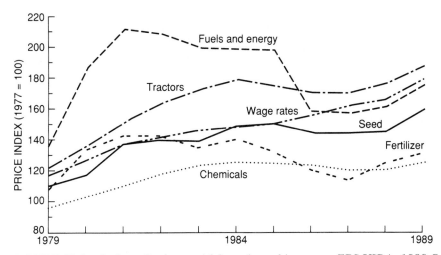

FIGURE 12.1   Index of prices paid for selected inputs.   *ERS-USDA, 1989 Costs of Production—Major Field Crops, EC1FS 9-5.*

425

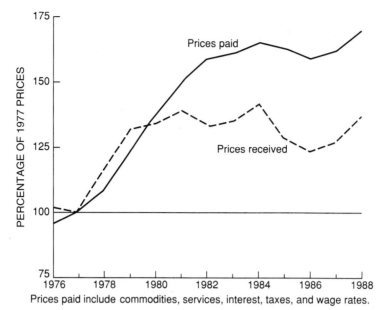

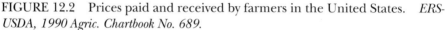

FIGURE 12.2   Prices paid and received by farmers in the United States.   *ERS-USDA, 1990 Agric. Chartbook No. 689.*

Unfortunately, the input prices paid by farmers have increased much more than the output prices received (Fig. 12.2). Therefore, it is imperative that growers achieve optimum productivity through efficient and cost-effective use of only those inputs that will ensure adequate returns on the investment.

Maximum Economic Yield

Maximum economic yield is somewhat lower than maximum yield and is the point at which the last increment of an input pays for itself (Fig. 12.3). Maximum economic yields vary among soils, although on most farms they are much higher than those generally achieved, regardless of the soil. Growers want to maximize profits, and higher yields are the key to this goal; however, the need for increased production and yield is based on other factors.

Yield Level and Unit Cost of Production

Practices that increase yield per unit of land lower the cost of producing a unit of crop, since it costs just as much to prepare the land, plant, and cultivate a low-yielding field as it does a high-yielding field. As Table 12.1 shows, increasing the yield raised the total cost of production per acre but decreased the cost per bushel and increased net profit. Land, buildings, machinery, labor, and seed will be essentially the same, whether production is high or low. These and other costs are called *fixed* and must be paid regardless of yield. *Variable costs* are those that vary with the total yield, such as the quantity of fertilizer applied, pesticides, harvesting, and handling.

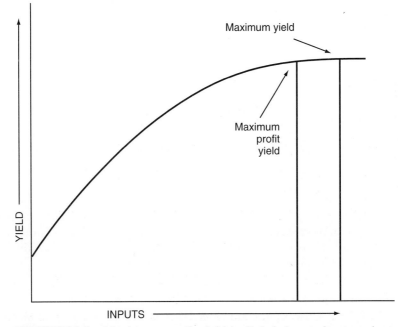

FIGURE 12.3   Maximum profit yield is slightly lower than maximum yield.

TABLE 12.1   Effect of Increasing Yields on the Cost per Bushel of Corn and the Net Profit per Acre ($2.75/bu)

Yield bu/a)	Production Costs		Net Profit ($/a)
	$/a	$/bu	
100	331	3.31	−56
125	343	2.74	1
150	359	2.39	54
175	383	2.18	100

SOURCE: Adapted from Hinton, *Farm Economics Facts and Opinion,* Univ. of Illinois (January 1982).

Key factors in obtaining the most efficient use of inputs are the weather and the ability of the farmer. Time of planting, proper variety selection, tillage, plant spacing, pest control, and timely harvesting are factors controlled by farmers. As a grower aims for increased yields, much of the initial increase will come from improved practices, not just additional nutrients. Many of these practices cost little or nothing. Some examples follow.

1. *Timeliness.* Timeliness is important in planting, tillage, equipment adjustment, pest control, observations, and harvesting.
2. *Date of planting.* Delaying soybean planting beyond the optimum date reduces yield 10 to 20 bu/a in some areas, while corn yields can be reduced 1 to 2 bu/a for each day of delay in planting.
3. *Pest control.* Identifying pest problems early will allow application of the most effective control practice.

4. *Variety selection.* Large differences in productivity, disease resistance, quality, and responsiveness to inputs exist among varieties and hybrids. Proper variety or hybrid selection can substantially affect yield and profitability.

5. *Plant spacing.* Planting the appropriate population for the productive capacity of the soil and environment is critical. Many growers plant below optimum populations. Row spacing influences yield; for example, narrowing soybean rows from 30 to 7 in. increases yields 10% or more in many areas. In humid regions, reducing wheat row spacing from 8 to 4 in. can increase yields 10 to 20%.

6. *Rotation.* Rotating crops is a valuable, no-cost input to increase crop yields and profit. Rotation may not only reduce weed, disease, and insect problems but also improve soil structure.

With superior management, higher rates of nutrients can and must be used. The general relationship in Figure 12.4 shows that *A* is the most profitable rate with average management and *B* is the most profitable rate with improved management.

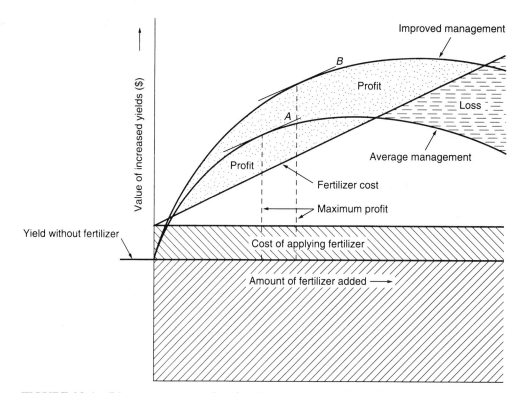

FIGURE 12.4   Diagram representing fertilizer economics associated with average management and improved management. The fertilizer rate for maximum yield occurs where the slope of the response curve is equal to 0 or is parallel with the *x*-axis. The fertilizer rate for maximum profit occurs where the slope of the response curve is parallel to the fertilizer cost line.   *After Miller,* Soils, *Prentice-Hall, Englewood Cliffs, N.J., 1990.*

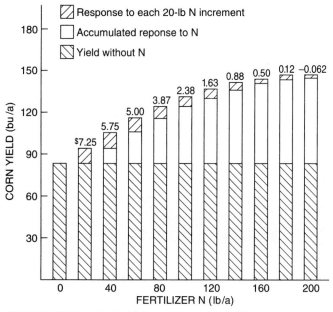

FIGURE 12.5 Diminishing returns in yield response of corn to fertilizer N. The dollar values on top of each bar represent the net rate per added dollar invested.

Returns per Acre

In general, as the nutrient rate increases, the return per dollar spent decreases. This decrease is the result of a reduced response for each successive incremental input. Eventually, the point is reached at which there is no further response to increasing amounts of nutrient, a principle called the *law of diminishing returns.*

When the soil is deficient in a nutrient needed for a desired crop yield, the first added increment of the nutrient will result in a large yield increase. The next added increment may also give an increase, but not as large proportionately as the first (Fig. 12.5). Consequently, responses to fertilizer increments continue diminishing to the point at which the last incremental yield value just equals the cost of input. This application rate gives the maximum profit.

Progressive growers recognize that although returns per dollar spent are important, the significant figure is the net return per acre. With adequate cash or credit, the farmer must select the input levels that will earn the greatest net return per acre. A low rate of fertilizer or another needed input may result in a high unit cost of production.

Profitable Nutrient Rate

The most profitable nutrient rate can be determined by calculating the maximum net profit or minimum cost per bushel of production. Figure 12.6a shows that the N rate for maximum net revenue was about 10 lb/a less than the N rate for maximum yield and 10 lb/a more than the N rate for the least cost per bushel.

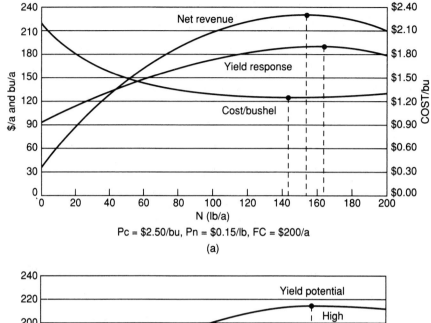

Pc = $2.50/bu, Pn = $0.15/lb, FC = $200/a

(a)

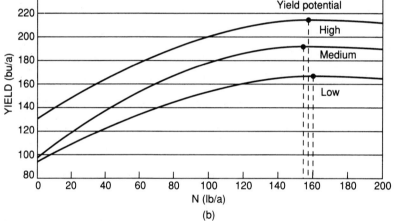

(b)

FIGURE 12.6   Influence of N rate on irrigated corn yield, net revenue, and cost per bushel (a) and the economic optimum N rate for three corn yield potentials (b).   *Schlegel,* J. Prod. Agric., *9: 114–118, 1996.*

Although these values represent a range of 20 lb/a of N, the most profitable N rate is essentially the same regardless of which parameter is used.

Although yield potential is an important parameter in determining the recommended N rate (see Chapter 9), the economic N rate usually does not vary greatly, even over a fairly wide range in actual yield potential. In the 30-year study mentioned previously, the data were grouped according to yield potential (Fig. 12.6b). Using the same costs and prices as before, the economic optimum N rate was 155 to 160 lb/a for the three yield potentials.

When too much fertilizer is added, the economic loss is not as great as when a crop is underfertilized by the same proportion (Table 12.2). The residual effect must also be considered and helps to compensate for the extra fertilizer cost. Over a period of years, it appears more profitable to use the optimum amount,

TABLE 12.2   Effect of Underfertilizing versus Overfertilizing on the Net Return from Added Fertilizer

	Yield of Corn (bu/a)	Net Return from Fertilizer (bu/a)	Difference from Optimum (bu/a)
1/4 less	142	50.4	−5.8
Optimum	151	56.2	0
1/4 more	153	52.0	−4.2
None	79	—	—

SOURCE: S. A. Barber, Purdue Univ., personal communication.

even if the rate would be more than optimum in unfavorable years. This guideline, of course, applies to nutrients that do not leach from the soil.

Although the cost per pound of nutrients may fluctuate, these variations are much less than the fluctuations in crop prices. There are several methods that take into consideration both the cost of the nutrient and the price of the crop. In any case, the response of the crop to increasing nutrient rates must be known for the general soil condition.

To calculate the fertilizer rate required for maximum yield and maximum profit, the equation that describes the yield response is needed. The hypothetical yield response function for yield response Y (Fig. 12.7) is

$$Y = 70 + 1.0\,X - 0.0025X^2$$

where
Y = grain yield (bu/a)
X = N rate (lb/a)

N rate for maximum yield:

1. Set the first derivative of the response function equal to zero.
2. Solve for X.

$$\frac{dY}{dX} = 1.0 - 0.005X$$

$$0 = 1.0 - 0.005X$$

Thus,

$$X = \frac{1.0}{0.005} = 200 \text{ lb/a of N}$$

3. The N rate for maximum yield is shown in Figure 12.7 and represents that point on the curve at which the slope equals 0 $(dY/dX = 0)$.

N rate for maximum profit:

1. Set the first derivative of the response function equal to the ratio of fertilizer cost (i.e., \$0.20/lb N) to grain price (i.e., \$2.50/bu).

2. Solve for X.

$$1.0 - 0.005X = \frac{\$0.20}{\$2.50}$$

$$1.0 - 0.005X = -0.92$$

$$X = 184 \text{ lb/a}$$

3. The N rate for maximum profit is shown in Figure 12.7 and represents the point on the curve at which the slope is parallel to the fertilizer cost line.

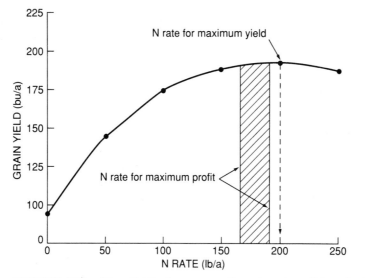

FIGURE 12.7 Hypothetical response function describing the influence of N rate on grain yield. N rate of maximum yield and maximum profit are shown (see the text for the calculation). The shaded bar represents the range in N rates for maximum profit, which varies with crop price received and fertilizer cost (see Table 12.3).

TABLE 12.3   Effect of Crop Price and Fertilizer Cost on the N Rate for Maximum Profit*

Crop Price ($/bu)	N Fertilizer Cost ($/lb N)			
	0.15	0.20	0.25	0.30
	lb/a of N			
4.00	192	190	188	185
3.50	191	189	186	183
3.00	190	187	183	180
2.50	188	184	180	176
2.00	185	180	175	170

*Based on the response function shown in Figure 12.7.

Of course, the fertilizer rate required for maximum profit depends on the fertilizer cost and crop price (Table 12.3). As the fertilizer cost increases at constant crop price, the fertilizer rate for maximum profit decreases. Alternatively, as the crop price increases at a constant fertilizer cost, the fertilizer rate for maximum profit increases. These data also illustrate that although differences exist, changes in crop price and fertilizer cost have relatively minor effects on the fertilizer rate for maximum profit. The largest differences in optimum N rate occur when the crop price is low and the fertilizer cost is high compared with a high crop price and a low fertilizer cost. Since it is difficult to predict the crop price at the time of fertilization, it is advisable to fertilize for near-maximum yields and not consider crop price and fertilizer cost factors.

Residual Effects of Fertilizers

Although discussed in Chapter 10, the residual effects of fertilizers are mentioned here because they should be included in fertilizer economics. Usually the entire cost of fertilization is charged to the current crop, whereas the lime cost is amortized over 5 to 10 years. With high rates of fertilization, however, residual effects can be substantial, especially with immobile nutrients.

At optimum fertilization rates, it has been shown that on some soils about 30% of the N may be residual for next year's crop, provided that it is not leached below the root zone. The residual value of P and K depends on the soil and how the crops are managed, but it can vary from 25 to 60%. The lower figure would apply when hay, straw, or stover is removed from the land.

High crop yields are impossible with low levels of fertility. The fertility level of soils is a plant growth factor that is easily controlled, and up to a certain point the increasing fertility level will be profitable. However, the initial cost of building soil fertility from low to high levels may discourage growers if viewed as an annual rather than long-term investment.

Immobile nutrients such as P and K can be built up in soils; thus, the buildup is a capital investment to be amortized over a period of years. The cost of building up soil P levels from 45 to 55 lb/a (P_1) is an example. Using a value of 9 lb of P_2O_5 to raise the P_1 test 1 lb, 90 lb/a of P_2O_5 would be required. The initial cost is $24.30/a with P_2O_5 at $0.27/lb. Using a payoff period of 15 years and an interest rate of 12%, the annual payment would be $3.57 (Table 12.4). A yield increase of 1.2 bu/a of $3/bu corn or 0.6 bu/a of $6/bu soybean would pay for this cost.

TABLE 12.4 Annual Payment Necessary to Amortize the $24.30/a Initial Cost of Buildup P with Various Interest Rates and Amortization Periods

Payoff Period (yr)	Annual Payment of Payoff at an Interest Rate of		
	8%	12%	16%
1	$26.24	$27.22	$28.19
5	6.09	6.74	7.42
10	3.62	4.30	5.03
15	2.84	3.57	4.36
20	2.47	3.25	4.10

SOURCE: Welch, *Better Crops Plant Food*, 66:3 (Fall 1982).

TABLE 12.5   Approximate Prices of Fertilizer Materials and Cost of Nutrients per
Pound (for Illustration Only)

Material	Analysis	Price per Ton ($)	Cost of N, P_2O_5, or K_2O (cents/lb)
Ammonium sulfate	20.5% N	95	23.2
Ammonium nitrate	33.5% N	170	25.3
Urea	46% N	215	23.3
Anhydrous ammonia	82% N	220	13.4
Nitrogen solution	28% N	135	24.1
Superphosphate (triple)	44% P_2O_5	205	23.2
Muriate of potash	60% K_2O	150	12.5
Sulfate of potash	52% K_2O	250	24.0

Price per Pound of Nutrients

Growers are interested in the most economical source but are accustomed to buying on the basis of cost per ton of fertilizer rather than cost per ton of plant nutrients (Table 12.5). Wide variations in the cost per unit of nutrient exist, and other factors, such as the cost of application and the content of secondary nutrients, must be considered.

Growers need to choose a fertilizer based on the cost per unit of nutrient in mixed fertilizers, mixed and straight materials, or all straight materials. A knowledge of the cost calculation is important. For example, a farmer has a choice of 12-24-24 or 6-24-24:

6-24-24 costs $194/ton

12-24-24 costs $206/ton

Assuming that the P and K cost is the same in both mixtures, the additional N in 12-24-24 amounts to $12, or 10 cents/lb of N.

Another calculation that farmers need to make relates to the economics of high-analysis fertilizer. In a comparison of 5-10-10 and 10-20-20 fertilizer, 2 tons of 5-10-10 are required to furnish the same amount of nutrients contained in 1 ton of 10-20-20. If 5-10-10 costs $100 and 10-20-20 costs $180 per ton, the 10-20-20 will be $20 cheaper than 2 tons of 5-10-10.

In addition to the actual cost of the material, farmers must consider the cost of transportation, storage, and labor used in applying the fertilizer. These costs may be difficult to evaluate, but if the actual price of the nutrients from one source is the same as that from another source, growers will take the one requiring less labor. The higher-analysis goods require less labor in handling. Time is also saved because fewer stops are made in applying the material.

Liming

The returns from liming are quite high when it is applied where needed (Table 12.6). These data illustrate the high net return to liming, which varies with lime

TABLE 12.6   Effect of Changing Soybean Prices, Limestone Rate, and Yield Response on Net Return to Liming, $/a*

Lime Needed (t/a)	Annual Yield Increase							
	3 bu		6 bu		9 bu		12 bu	
	$6	$8	$6	$8	$6	$8	$6	$8
1	14	20	32	44	50	68	68	92
2	10	16	28	40	46	64	64	88
3	6	12	24	36	42	60	60	84
4	2	8	20	32	38	56	56	80

*Limestone cost amortized over 5 years at 10% interest, assuming a total cost of $15/t applied, with net return being rounded to the nearest dollar.

SOURCE: Hoeft, *Nat. Conf. Agr. Limestone,* National Fertilizer Development Center, Muscle Shoals, Ala. (1980).

rate, lime cost, yield response to liming, and price received for the crop. In spite of a high return, however, lime is often neglected in the fertility program because (1) responses to lime are often not as visual as those obtained with N, P, or K unless the soil is particularly acidic and (2) liming effects last for several years and the returns are not all realized in the first year.

Lime is the first step in any sound soil management program, and broadcast applications in accordance with soil and plant requirements are essential for the greatest returns from fertilizer.

Animal Wastes

Benefits from manure, in addition to those from macro- and micronutrients, may be related to the organic components that might improve soil moisture relations and increase the downward movement of P, K, and micronutrients.

There is considerable variability in manure, depending on methods of storing and handling; however, with current fertilizer, labor, and equipment costs, it is usually profitable for the grower to use livestock manure. Because manure is largely an N-P-K fertilizer, the best returns are obtained by using it on nonleguminous crops. Hauling charges can be reduced by applying it on fields close to the source and using commercial fertilizer on more distant fields.

In some farm management programs it is imperative that the manure be disposed of, and its value may be balanced against equipment and hauling costs. The general feeling, however, is that it is worth a little more than the cost of handling. In areas or times when fertilizers are difficult or impossible to obtain, animal manures play an important role in supplying fertilizer needs.

Plant Nutrients as Part of Increasing Land Value

When buying land, the farmer may be faced with the possibility of choosing high-priced or low-priced property. Which is the better buy? The higher-priced land is generally more productive, fertile, and has better improvements. The lower-priced land may actually be a good buy, however, provided that it has not been severely eroded or has no other physical limitations. Low-priced land is usually infertile and may need considerable lime and/or nutrients. Adequate liming and fertilization, as indicated by soil tests and combined with other good practices, can rapidly increase productivity. Expenditures to improve fertility may be included as part of the cost of the land. If the problem is considered in this light, $100/a for liming and buildup fertility may be reasonable. Thus, with proper management, it is possible to increase land productivity and value, and the cost can be amortized over a period of years.

Additional Benefits from Maximum Economic Yields

Increase in Energy Efficiency

Higher yields are an effective means of improving energy efficiency in agriculture. Higher yields require more input energy per acre, but the energy cost per bushel or ton is less; some costs are the same regardless of yield level. For example, it takes just as much fuel to till a field yielding 40 bu/a of soybean as one yielding 60 bu/a.

Reduction in Soil Erosion

Raindrops strike the soil with surprising force, dislodging particles and increasing soil erosion. However, growing crops, crop residues, and roots absorb the impact and slow water movement, increase water intake, and decrease soil erosion. The damaging effects of wind erosion are also reduced by the presence of crops and their residues. Highly productive cropping systems match perfectly with soil conservation because

1. Crop canopy development is speeded.
2. Crop canopy density is increased.
3. More top and root residues are left.

Conservation tillage practices such as no-till systems and chisel plowing leave more residues on the surface than moldboard plowing (Chapter 13). However, with any given tillage practice, higher amounts of residues will generally decrease soil losses.

Increase in Soil Productivity

Increasing soil OM is a long-term process; however, the productivity benefits of raising OM can be substantial (see Chapter 13). In areas of higher temperatures and lower moisture, it is more difficult to increase OM; however, larger amounts of decomposing residues keep the soil in better physical condition. Infiltration of water is increased, and the supply for the plant is improved. Thus, plants are better able to withstand periods of drought. Also, less water runs off and erosion is reduced.

TABLE 12.7　Effect of Adequate K on Soybean Yield, Disease, and Dockage

K_2O (lb/a)	Yield (bu/a)	Diseased and Moldy Beans (%)	Dockage (cents/bu)	Value at $6 per Bushel ($/a)
0	38	31	54	207.48
120	47	12	22	271.66

SOURCE:　M. Kroetz, Ohio State Univ., personal communication.

Reduction in Grain Moisture

More adequate fertilizer, particularly N and P, decreases the amount of water in the grain at harvest and results in lower drying costs (Table 4.4; Table 10.6).

Improvement in Crop Quality

Adequate plant nutrition improves grain or forage quality. For example, increasing grain protein of wheat with N additions (Fig. 9.14) can increase the market value with protein premiums of 10 to 20 cents/bu/% protein.

As shown in Table 12.7, on a low-K soil, supplemental K not only increased soybean yields but also decreased the incidence of disease and mold in the seed.

Selected References

FOLLETT, R. H., L. S. MURPHY, AND R. L. DONAHUE, 1981. *Fertilizers and Soil Amendments.* Prentice-Hall, Englewood Cliffs, N.J.

TERMAN, G. L., AND O. P. ENGELSTAD. 1976. *Agronomic Evaluation of Fertilizers: Principles and Practices. Bull. Y-21.* Tennessee Valley Authority, National Fertilizer Development Center, Muscle Shoals, Ala.

CHAPTER 13

Agricultural Productivity and Environmental Quality

The objective of any soil and crop management program is *sustained* profitable production. The strength and longevity of any civilization depend on the ability to sustain and/or increase the productive capacity of its agriculture. Sustainable agriculture encompasses soil and crop productivity, economics, and environment and can be defined by

> The integration of agricultural management technologies to produce quality food and fiber while maintaining or increasing soil productivity, farm profitability, and environmental quality.

Achieving agricultural sustainability depends on many agronomic, environmental, and social factors. Of course, many criteria can be used to evaluate sustainable farming systems, such as

1. Maintain short-term profitability and sustained economic viability.
2. Maintain or enhance soil productivity.
3. Provide long-term environmental quality.
4. Maximize efficient use of resources.
5. Ensure food safety, quality of life, and community viability.

From a soil productivity–soil fertility standpoint, soil conservation is essential for long-term sustainability (Fig. 13.1). Soil management practices that contribute to or encourage soil degradation will reduce soil productivity and impair progress toward sustainability.

Soil erosion represents the greatest threat to sustained soil productivity. Physical removal of nutrient-rich, high-OM topsoil, and oxidation of OM with tillage, reduces the productive capacity of the soil. Exposed subsoil is often less productive because of (1) poor soil physical condition, (2) reduced water availability, (3) decreased nutrient supply, and (4) many other site-specific parameters.

Although growers, consultants, and others recognize erosion on the lands they manage, they may not be overly concerned because crop yields have substantially increased since the 1950s. One should not confuse increasing crop yields with in-

438

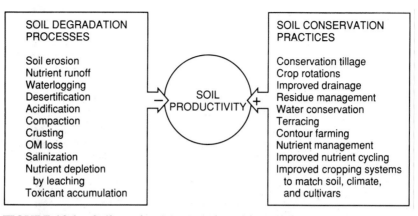

FIGURE 13.1   Soil productivity is reduced by soil degradation processes and improved by soil conservation practices.   *Parr et al.,* Advances in Sci., *13:1–7, 1990.*

creasing soil productivity, because yield increases are primarily due to technological advances in crop breeding and genetics, fertilizers and fertilizer management, pesticides and pest management, and other agronomic technologies. Soil conservation and good soil management embrace more than just the prevention of soil losses. Soil erosion is a *symptom* of poor soil management, whether it be inadequate plant nutrients or improper cropping systems.

Soil and Crop Productivity

The interest and concern for soil productivity are not new and have been recognized since the early 1900s. A historical explanation for the slow increase in yields before the 1950s is provided in Figure 13.2. Soil productivity decreased 40% in the 60-year period from 1870 to 1930, closely related to decreased fertility levels. Soil OM and the native nutrient supply, especially N, have decreased. The nutrients removed have generally been greater than the amounts returned to the soil in manure and fertilizers.

Degradation of native soil fertility by not returning nutrients removed in the crops is evident in the first 40 years of dryland production (Fig. 13.3). Technological developments in varieties, water conservation, and P fertilization increased productivity during the next 50 years; however, continued soil erosion and OM loss again limited productivity of the wheat-fallow-wheat cropping system. After 1950 growers adopted wheat-legume rotations to provide forage for livestock, and the increased N availability from the legume residue dramatically increased wheat yields. Soil erosion was reduced, OM increased, and these soils are much more productive now than in 1900.

Another example is provided in Figure 13.4, where, before 1940, wheat yields were less than 20 bu/a. Average annual precipitation remained relatively constant, but wheat yields tripled from 1940 to 1985. The major technological advances contributing to increased productivity were (1) improved varieties, (2) increased water conservation with stubble mulch tillage, (3) increased N and P fertilization, and (4) improved planting and harvesting methods.

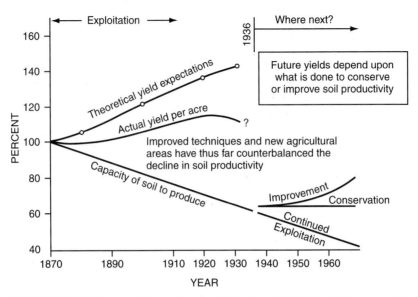

FIGURE 13.2 Improved practices in Ohio since 1870 should have resulted in yields 40 to 60% higher per acre in 1930, but the aggregate yield increased less than 15%. Improved practices only slightly more than counterbalanced the decline in the ability of the soil to produce. Yields can be increased only if proper soil management programs are adopted. *Ohio State Agr. Ext. Serv. Bull. 175, 1936.*

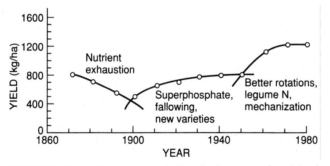

FIGURE 13.3 Changes in dryland wheat productivity in Australia. *Donald, Agriculture in Australian Economy, Sydney Univ. Press, 1981.*

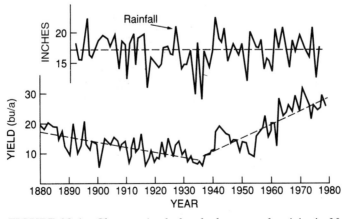

FIGURE 13.4 Changes in dryland wheat productivity in North Dakota. *Fanning and Reff,* NDSU Coop. Ext. Ser. Bull. *SC-710, 1981.*

Profitable crop production on eroded soils has been an important agricultural problem. The generally reduced crop yield of nonlegumes on subsoil is well known. On permeable soils this decrease is largely the result of less OM and the subsequent lower release of N.

Figure 13.5 shows that erosion must not be taken for granted and that it is imperative that crop production systems minimize the destructive effects of water and wind erosion. More than one-third of the cropland in the United States is subject to erosion severe enough to significantly reduce soil productivity. Despite the extent of soil erosion, losses have been significantly reduced over

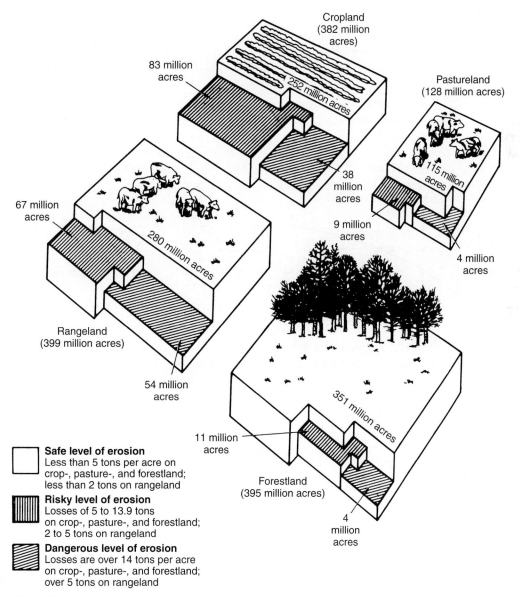

Safe level of erosion
Less than 5 tons per acre on crop-, pasture-, and forestland; less than 2 tons on rangeland

Risky level of erosion
Losses of 5 to 13.9 tons on crop-, pasture-, and forestland; 2 to 5 tons on rangeland

Dangerous level of erosion
Losses are over 14 tons per acre on crop-, pasture-, and forestland; over 5 tons on rangeland

FIGURE 13.5   How serious is erosion? Shown here are the results of the National Resource Inventory conducted by the Soil Conservation Service in 1992.

time as producers adopt technologies that conserve soil and enhance productivity (Table 13.1).

Water and wind erosion of topsoil can reduce productivity by exposing less productive subsoil (Fig. 13.6). The productive capacity of eroded Ulysses soil is less than that of eroded Harney soil because the latter is a deeper soil and has a greater OM content in the subsoil, which improves nutrient availability and water-holding capacity. As a result, the yield loss associated with increasing soil loss is also greater in the Ulysses than in the Harney soil. Specifically, in the Ulysses soil, 1.8-bu/a yield loss per inch of topsoil loss occurs compared with 0.8-bu/a yield loss per inch of topsoil in the Harney soil.

TABLE 13.1   Decrease in Soil Erosion in the United States, 1982–92

	Water Erosion		Wind Erosion	
	1982	1992	1982	1992
Total soil loss (billion tons)	1.7	1.2	1.4	0.9
Annual soil loss (t/a)	4.1	3.1	3.3	2.5

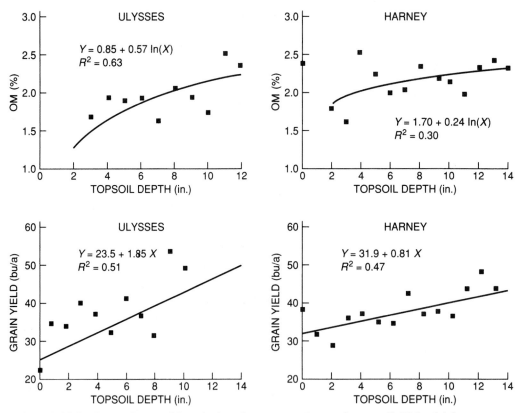

FIGURE 13.6   Loss of topsoil by wind and water erosion reduces soil OM, which contributes to wheat grain yield loss. Loss in productivity varies between soils, depending on initial topsoil depth and productivity of the subsoil. Compared with the Ulysses soil, the Harney soil has a deeper topsoil; thus, productivity is not reduced as much as the Ulysses soil under equivalent topsoil loss.   *Havlin et al.,* Proc. Great Plains Soil Fert. Conf., *1992.*

TABLE 13.2   Influence of Soil Erosion on Surface Soil Physical Properties of Several Sites and Their Relationship to Corn Grain Yield Loss

Site	Surface Texture	Erosion Phase	OM (%)	Dry Bulk Density (g/cm³)	Estimated Soil Profile Θva* (cm H₂O)
3	Fine	Noneroded	2.6	1.44	14.1
	Sandy	Severely eroded	0.7	1.46	12.8
	Loam	Depositional	5.5	1.23	18.1
6	Gravelly	Noneroded	3.7	1.37	7.4
	Sandy	Severely eroded	2.2	1.68	5.1
	Loam	Depositional	4.3	1.38	11.1

	1982		1983	
Site	Yield Reduction (%)	Major Cause(s)	Yield Reduction (%)	Major Cause
2	38	N, H₂O	52	H₂O
3	43	H₂O	38	H₂O
5	32	H₂O	35	H₂O
6	66	H₂O, N	59	H₂O

*Θva, estimated preseason plant available water in 39.4 in. of soil.

SOURCE: Battiston et al., *Erosion and Soil Products*, ASAE 8–85, pp. 28–320 (1985).

TABLE 13.3   Selected Indicators of Soil Quality and Some Processes They Impact

Measurement	Process Affected
OM	Nutrient cycling, pesticide and water retention, soil structure
Infiltration	Runoff and leaching potential, plant water use efficiency, erosion potential
Aggregation	Soil structure, erosion resistance, crop emergence, infiltration
pH	Nutrient availability, pesticide absorption and mobility
Microbial biomass	Biological activity, nutrient cycling, capacity to degrade pesticides
Forms of N	Availability to crops, leaching potential, mineralization and immobilization rates
Bulk density	Plant root penetration, water- and air-filled pore space, biological activity
Topsoil depth	Rooting volume for crop production, water, and nutrient availability
Conductivity or salinity	Water infiltration, crop growth, soil structure
Available nutrients	Capacity to support crop growth, environmental hazard

In some years, however, moisture may limit yield because of reduced water availability related to lower water infiltration and reduced water-holding capacity. The data in Table 13.2 show the lower OM and plant available water content on severely eroded soil compared with the noneroded or depositional areas in the field. Reduced N and water availability were primarily responsible for the 32 to 66% corn yield reduction on the eroded soil.

Soil Quality

Many interrelated factors in a soil influence its productive capacity or potential. These properties collectively represent *soil quality* (Table 13.3). Although all of these properties are important, the soil OM content is the most critical, because of its influence on many biological, chemical, and physical characteristics inherent in a productive soil (Table 13.4).

TABLE 13.4   General Characteristics of Soil OM and Associated Effects on Soil Properties

Property	Remarks	Effect on Soil
Color	The typical dark color of many soils is caused by OM	May facilitate warming
Water retention	Can hold up to 20 times its weight in water	Helps prevent drying and shrinking; improves the moisture-retaining properties of sandy soils
Combination with clay materials	Cements soil particles into aggregates	Permits exchange of gases, stabilizes structure, and increases permeability
Chelation	Forms stable complexes with Cu^{2+}, Mn^{2+}, Zn^{2+}, and other cations	May enhance the availability of micronutrients to higher plants
Solubility in water	Insolubility of OM is due to its association with clay	Little OM is lost by leaching
Buffer action	Exhibits pH buffering	Helps to maintain a uniform soil pH
Cation exchange	Total capacities of humus range from 300 to 1,400 meq/100 g	OM may increase the soil CEC from 20 to 70% of the CEC of many soils
Mineralization	Decomposition of OM yields CO_2, and nutrients	A source of nutrients for plant growth
Combines with organic molecules	Affects bioactivity, persistence, and biodegradability of pesticides	Modifies pesticide rates for effective control

SOURCE: Stevenson, *Humus Chemistry,* Copyright © 1982 John Wiley & Sons, Inc. (1982).

For example, increasing soil C increases the stability of soil aggregates, which can help soil to resist water and wind erosion (Fig. 13.7). Because of the increased aggregate stability, bulk density is lower with higher soil C, which improves root proliferation through the soil and ultimately productivity (Fig. 13.8).

The steady-state OM level depends on the soil and crop management practices utilized. If these practices are changed, a new OM level is attained that may be lower or higher than the previous level and depends entirely on management.

Maintenance of OM for the sake of maintenance alone is not a practical approach to farming. It is more realistic to use a management system that will give sustained profitable production without degradation of OM and productivity. The greatest source of soil OM is the residue contributed by current crops. Consequently, the cropping system and the method of handling the residues are equally important. Proper management and fertilization will produce high yields, which will increase the quantity of residue and organic C returned to the soil.

Tillage of the soil produces greater aeration, thus stimulating more microbial activity, and increases the rate of disappearance of soil organic C (Figs. 13.9 and 13.10). When a virgin soil is cultivated, the OM decline is rapid during the first 10 years and then continues at a gradually diminishing rate for several decades. Many studies have suggested that under continuous cultivation, soil OM declines approximately 50% in 40 to 70 years, depending on

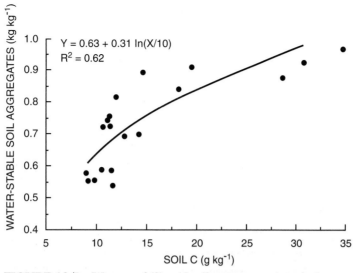

FIGURE 13.7　Water stability of soil aggregates in relation to soil C.

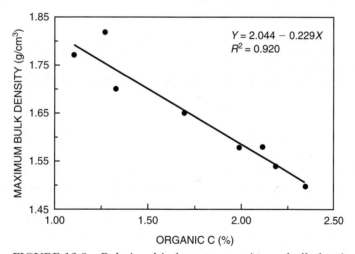

FIGURE 13.8　Relationship between maximum bulk density and organic C.

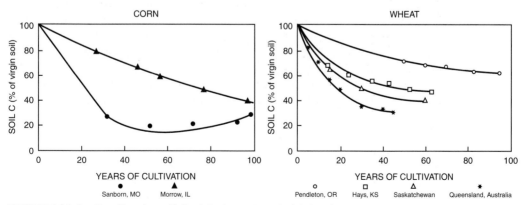

FIGURE 13.9　Decline in soil C with time since initial cultivation in corn and wheat cropping systems. *Paustian et al., 1997,* Mgmt. Controls on Soil C. *p. 25, CRC Press.*

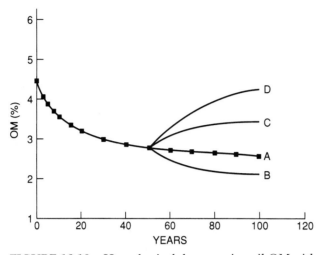

FIGURE 13.10   Hypothetical decrease in soil OM with time. At 50 years, changes in soil and crop management system can either continue (A), decrease (B), or increase (C, D) soil OM. *A* represents no change in cropping system, while *B* represents a change that would accelerate OM loss (i.e., more intensive tillage). *C* might represent adoption of reduced tillage or a crop rotation that produces more residue, whereas *D* might reflect the change in OM following adoption of a high–yield no-till system or rotations that return large quantities of residue.

the environment and the quantity of residue returned to the soil. Eventually, an apparent equilibrium is reached.

Several long-term studies demonstrate the exponential decline in soil OM after virgin soils are tilled (Fig. 13.11). These data show that OM decreased from 3.4 to 2.0% after 45 years of conventionally tilled wheat-fallow system. Soil OM increased with annual application of 10 t/a manure (C and N added), whereas 40 lb/a of N had little influence on OM. Reducing the C input by burning the crop residue further decreased soil OM. The influence of C and N balance on grain yield is also evident (Fig. 13.11). Similarly, the long-term Morrow plots in Illinois show the influence of increasing C and N on soil OM (a) and grain yield (b) through crop rotation compared with continuous corn (Fig. 13.12).

Generally, the quantity of residue returned to the soil will have a much greater effect on increasing soil OM than the residual N content. Figure 13.13 shows that even though the N content of alfalfa is much greater than that of corn, the original organic C content was maintained at 1.8% C by 2 t/a/yr of either corn or alfalfa residue. Increasing the residue produced and returned with either crop increased soil organic C. If all of the residue had been left on the soil surface instead of incorporated with tillage, the increase in organic C would have been greater or the original soil C content would have been maintained with lower residue mass.

Thus, added N will certainly encourage rapid residue degradation by microorganisms, with little or no immobilization of soil inorganic N (see Chapter 5); however, the quantity of residue added influences OM more than the N content. Ad-

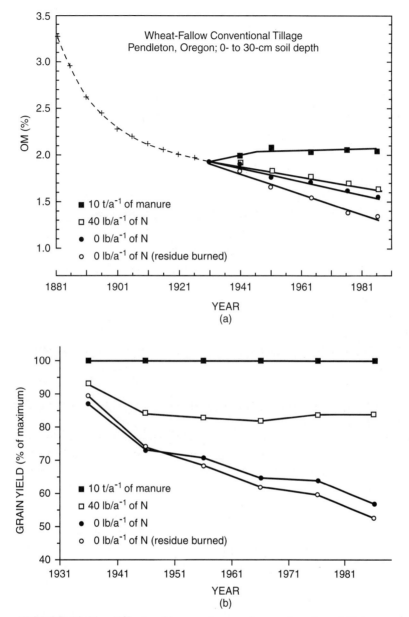

FIGURE 13.11 Effects of increasing or decreasing C and N (manure or fertilizer N) inputs on soil OM (a) and grain yield (b) in a wheat-fallow cropping system. *Rasmussen, USDA-ARS, Bull. 675, 1989.*

equate use of N, coupled with the return of crop residues, can not only maintain the level of soil OM but may actually increase it (Fig. 13.14).

The dryland cropping systems referred to in Figure 11.4 also showed increased soil OM content with increasing cropping intensity or reduced dependence on fallowing (Fig. 13.15). Soil OM increased as more residue was produced in the

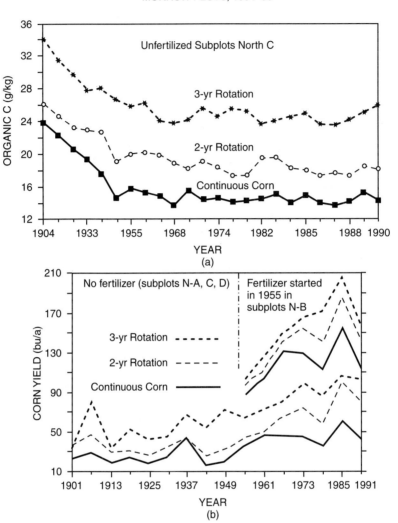

FIGURE 13.12 Effects of adding C and N on soil OM (a) and grain yield (b) in corn rotations compared with continuous corn cropping systems. *Darmody and Peck, 1997,* Soil organic C changes in Morrow plots. *p. 165–166. CRC Press.*

wheat-corn-fallow and wheat-corn-millet-fallow systems compared with the wheat-fallow-wheat system.

Soil OM transformations are very dynamic. Intensive tillage systems, fallowing, and low crop productivity, combined with physical soil loss by erosion, decrease the OM content over time (Fig. 13.10). Increasing soil OM requires reducing the tillage intensity and increasing the quantity of CO_2 fixed by plants and returned to the soil. Increasing the C input depends on the interaction between more productive rotations and reduced tillage.

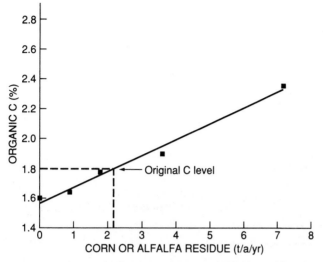

FIGURE 13.13   Influence of corn or alfalfa residue incorporated into the soil for 11 years.   *Larson et al.,* Agron. J., *64:204–8, 1972.*

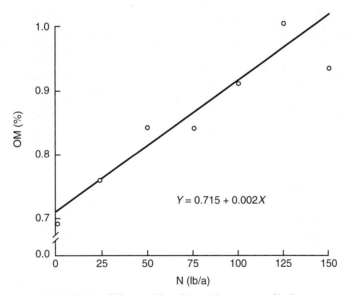

FIGURE 13.14   Effect of fertilizer N rates applied to cotton on soil OM content over an 11-year period in Arkansas.   Crops and Soils *37(9):34, 1985.*

Many factors determine whether the soil OM is increased or decreased by cropping systems. *The key is to keep large amounts of crop residues (stover and roots) passing through the soil. Continued good management, including adequate fertilization, helps to maintain the cycle.* Sustaining the productive capacity of soil for future generations ultimately depends on maintaining optimum soil OM levels.

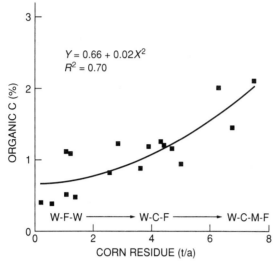

$$Y = 0.66 + 0.02X^2$$
$$R^2 = 0.70$$

FIGURE 13.15   Increasing no-till cropping intensity increased OM compared with wheat-fallow-wheat systems. W, wheat; C, corn; M, millet; F, fallow. *Peterson and Westfall,* Proc. Great Plains Soil Fert. Conf., *1990.*

TABLE 13.5   Conservation Tillage Methods

Row Crop Agriculture	Small-Grain Agriculture
Narrow strip tillage	Stubble mulch farming
No-till, zero-till, slot plant	Stirring or mixing machines
Strip rotary tillage	Disk-type implements
	• One-way disk
Ridge planting	• Offset disk
Till plant	• Tandem disk
Plant conventionally on ridge	Chisel plows
	Field cultivators
Full width—no plow tillage	Mulch treaders
Fall and/or spring disk	
Fall or spring chisel, field cultivate	Subsurface tillage
	Sweep plows
Full width—plow tillage	Rotary rodweeder
Plow plant	Rodweeder with semichisels
Spring plow–wheel–track plant	
	Ecofallow, no-till

SOURCE: Mannering and Fenster, *J. Soil Water Cons.,* 38:141–43 (1983).

Conservation Tillage

Over the years, growers have become interested in reducing tillage to give greater protection to the soil against soil and water losses. The amount of surface residue and surface roughness both have an effect. Crop residue management has been developed to leave more of the harvest residue, leaves, and roots on or near the surface.

Conservation tillage is a term used to describe any tillage system that reduces soil and/or water loss compared with clean tillage, where all residues are incorporated into the soil (Table 13.5).

Advantages

1. Higher crop yields, except in level, fine-textured, poorly drained soils.
2. Less soil erosion by water and wind.
3. Improved infiltration and more efficient use of water.
4. Increased acreage of sloping land that can safely be used for row crops.
5. Improved timing of planting and harvesting.
6. Lower labor, machinery, and fuel costs.

Disadvantages

1. More potential for rodents, insects, and diseases, in some systems.
2. Cooler soil temperatures in spring, resulting in slower germination and more stand problems.
3. Greater management ability is required.

Types of Conservation Tillage

CHISELING   A chisel implement may till 8 to 15 in. deep, with points 12 to 15 in. apart. A considerable amount of surface residue is left on the surface, and the surface is rough.

TILL PLANT AND RIDGE TILL   Till plant and ridge till is a once-over tillage-planting operation. Planter units work on ridges made the previous year during cultivation or after harvesting (Fig. 13.16). The planter pushes the old stalks, root clumps, and clods into the area between the rows. This practice is perhaps most useful on fine-textured, poorly drained soils. Ridges are retained in the same position year after year; hence, wheel tracks are in the same place.

STUBBLE MULCHING   Two types of tillage equipment are used—those such as disks, chisel plows, and field cultivators, which mix crop residues with the soil, and those such as sweeps or blades and rodweeders, which cut beneath the surfaces without inverting the soil. Increased moisture efficiency and reduced wind erosion are primary goals of stubble mulching.

NO-TILL   In the no-till system, all of the residue is left on the surface; therefore, the system has been most successful in the better-drained soils. A seed zone 2 in. wide or less is prepared in previously untilled ground. This zone is made by a fluted coulter running ahead of a planter unit with disk or hoe openers. Seeding by this method is successful in residues of many crops. Narrow hoe openers or narrow-angle, double-disk openers are also used for minimum- and zero-tillage seed placement.

Effects of Conservation Tillage

SURFACE RESIDUES   The approximate quantity of surface residue remaining after one tillage operation varies with the implement (Table 13.6). Subsurface implements that leave most of the residue on the soil surface help protect the surface against erosion. The quantity of residue incorporated or left on the surface

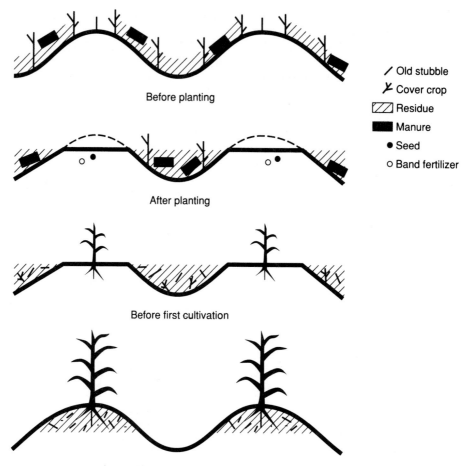

	Old stubble
	Cover crop
	Residue
	Manure
	Seed
	Band fertilizer

Before planting

After planting

Before first cultivation

Last cultivation builds new ridges

FIGURE 13.16   Ridge tillage advantages in production systems. The planter tills 2 to 4 in. of soil in a 6-in. band on top of the ridges. Seeds are planted on top of the ridges, and soil from the ridges is mixed with crop residue between the ridges. Soil on ridges is generally warmer than soil in flat fields or between ridges. Warm soil facilitates crop germination, which slows weed emergence. Crop residue between the ridges also reduces soil erosion and increases moisture retention. Mechanical cultivation during the growing season helps to control weeds, reduces the need for herbicides, and rebuilds the ridges for the next season.

after tillage also varies with the crop; thus, the percentage of surface residue cover also varies. Figure 13.17 illustrates the relationship between residue cover and residue mass. Crops such as soybean, sunflower, and cotton provide very little surface cover compared with small- and coarse-grain crops.

SOIL LOSS   The quantity of residue required to prevent soil erosion depends on

1. Soil characteristics (e.g., texture, OM, surface roughness, structure, depth, slope percentages, and slope length).

TABLE 13.6   Effect of Tillage Equipment on Surface Residue Remaining after Each
Operation

Tillage Machine	Approximate Residue Maintained (%)
Subsurface cultivators	
Wide-blade cultivator, rodweeder	90–95
Mixing-type cultivators	
Heavy-duty cultivator, chisel, other types of machines	50–75
Mixing and inverting disk machines	
One-way, flexible disk harrow; one-way disk; tandem disk; offset disk	25–50
Inverting machines	
Moldboard, disk plow	0–10

SOURCE: Anderson, *Great Plains Ag. Council Publ. No. 32* (1968).

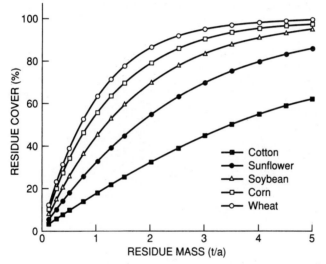

FIGURE 13.17   Relationship between residue mass and percentage of surface
cover for selected crops.

2. Residue characteristics (e.g., type, quantity, orientation).
3. Rainfall characteristics (e.g., quantity, duration, and intensity).
4. Wind characteristics (e.g., velocity, direction, gusts, and duration).

In general, as the percentage of surface residue cover increases, the potential for
soil loss decreases (Fig. 13.18). In addition, increasing the surface residue level
by reducing the tillage intensity drastically reduces soil loss under a rainfall sim-
ulator (Fig. 13.19). These data also show the value of farming on the contour
compared with up and down the slope. The same relationship can be seen in Tables 13.7 and 13.8. Soil loss was greater with all tillage systems when the crop was
planted up and down the slope (Table 13.7) compared with across the slope
(Table 13.8). In addition, surface cover was much less after planting soybean than
corn, resulting in considerably greater soil loss.

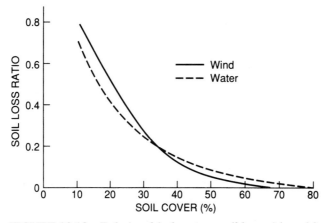

FIGURE 13.18   Relationship between soil loss with residue cover divided by soil loss from bare soil (soil loss ratio) and percentage of surface residue cover.
Adapted from Laflen et al., ASAE Publ. 7-81, 1981, pp. 121–33, and Fryrear, Sci. Reviews, Arizona Res., Scientific Publ., pp. 31–48, 1985.

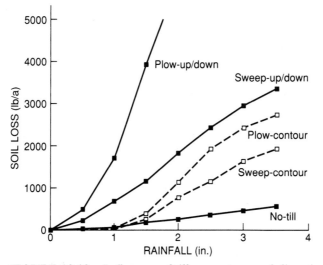

FIGURE 13.19   Influence of tillage system and direction (up or down hill or on the contour) on soil loss for a 4% slope. Initial residue level was 13,500 lb.
Nebraska, Dickey et al., 1981. Neb Guide 181–554 Univ. of Neb.

SOIL TEMPERATURE, MOISTURE, AND MICROBIAL ACTIVITY Soil temperatures early in the growing season are generally lower under conservation tillage than under conventional tillage due to the insulating effect of the unincorporated crop residues on the surface. Soil temperatures in the top 6 in. of the root zone during May and June can be several degrees lower under no-tillage than under conventional tillage systems.

The influence of surface residue cover on soil temperature also depends on the crop. As Figure 13.20 shows, surface soil temperature was lower following corn

TABLE 13.7   Surface Cover and Soil Loss from Various Tillage Systems of 4% Slope
Land Tilled up and down the Slope Following Corn and Soybeans (Harlan, Indiana)*

	Surface Cover		Soil Loss	
Tillage System	After Corn (%)	After Beans (%)	After Corn (mt/ha)	After Beans (mt/ha)
Fall moldboard plow	7	1	22.0	41.0
Fall chisel tillage	25	12	15.0	30.3
No-till	69	26	2.5	13.5

*Morley clay loam with a slope length of 10.7 m. Tests were made after overwinter weathering but
before spring tillage. Two storms were applied at 6.25 cm of rainfall each.
SOURCE: Mannering, *Crop. Ext. Serv. Publ. AY-222*, Purdue Univ. (1979).

TABLE 13.8   Surface Cover and Soil Loss from Various Tillage Systems of 5% Slope
Land Tilled across the Slope Following Corn and Soybeans (Urbana, Illinois)*

	Surface Cover		Soil Loss	
Tillage System	After Corn (%)	After Beans (%)	After Corn (mt/ha)	After Beans (mt/ha)
Fall moldboard plow	4	2	12.8	25.6
Fall disk-chisel tillage	50	11	1.3	7.4
No-till	85	59	1.1	3.8

*Catlin silt loam with slope length of 10.7 m. Tests were made after overwinter weathering but
before any spring tillage; 12.5 cm of simulated rainfall were applied in two storms.
SOURCE: Siemens and Oschwald, *Am. Soc. Agr. Eng. Paper No. 76-2552* (1976).

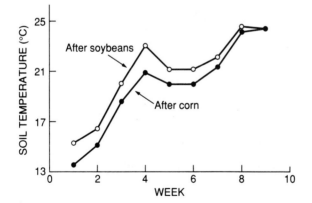

FIGURE 13.20   Weekly means for daily maximum soil temperatures in no-till corn.
Griffith et al., No Tillage and Surface Tillage Agriculture, *p. 34, John Wiley & Sons,*

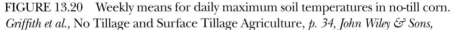

than soybean planting, which was related to the higher surface cover with corn
compared with soybeans.

Decomposition of crop residue and soil OM, with subsequent release of plant
nutrients, including N, P, and S, is restricted by low soil temperatures. Thus, recycling of essential nutrients may be delayed. Further, low soil temperatures retard root development and activity (Fig. 13.21). In tropical and semitropical regions the cooling effect of crop residue on soil temperatures may be beneficial.

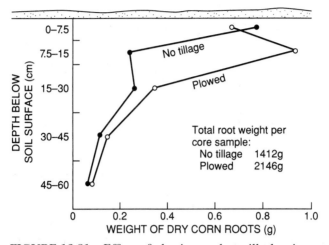

FIGURE 13.21   Effect of plowing and no-till planting on amount and distribution of corn roots.   *Griffith et al.*, No Tillage and Surface Tillage Agriculture, *p. 39, John Wiley & Sons, 1986.*

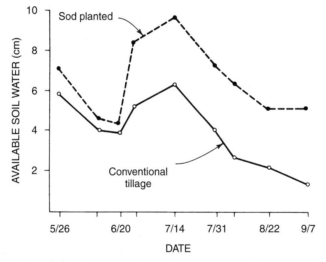

FIGURE 13.22   Available soil water in 0- to 60-cm profile as affected by tillage practice for corn with orchard grass.   *Bennett et al.,* Agron. J., *65:488, 1973.*

Increasing residue on the surface by reducing tillage reduces runoff and soil erosion while increasing infiltration. Soils with tillage pans may require chiseling to gradually create a deeper root zone with greater water intake and water-holding capacity. This deeper root zone may provide an extra inch or two of water at critical stages of growth (Fig. 13.22).

In semiarid regions, water conservation increases with maintenance of surface residue cover. Increasing residues with no-till systems increased the total water stored, consequently improving sorghum yield and water use efficiency (WUE) compared with residue incorporation with the disk (Table 13.9).

The interaction of soil temperature and moisture with tillage dramatically influences microbial activity (Fig. 13.23). When the soil is tilled, increased aeration

TABLE 13.9   Tillage Effects on Water Storage, Sorghum Grain Yields, and WUE in an Irrigated Winter Wheat–Fallow–Dryland Grain Sorghum Cropping System in Bushland, Texas, 1973–77*

| Tillage Method | Water Storage | | Grain Sorghum Yield | | | |
	Amount (mm)	Efficiency (% of Precipitation)	(Mg/ha)	(bu/a)	Total water use (mm)	WUE (kg/m³)
No-till	217	35.2	3.14	47	350	0.89
Sweep	170	22.7	2.50	37	324	0.77
Disk	152	15.2	1.93	29	320	0.66

*Precipitation averaged 347 mm during the fallow period.
SOURCE: Unger and Weise, *SSSA*, 43:582–88 (1979).

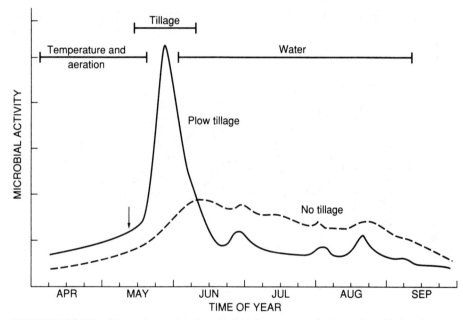

FIGURE 13.23   Hypothetical relationship between relative microbial activity and time of year in a plowed and a no-till soil; factors controlling activity are shown on top of the graph; the arrow indicates time of plowing.   *J. W. Doran, personal communication.*

(along with residue C mixed in the soil) encourages microbial activity and mineralization of OM, which eventually releases N and other nutrients. Microbial activity is lower early in the season because of lower temperature; however, it is slightly higher later in the season because of greater soil moisture.

The net effect of tillage is increased mineralization of OM, which causes gradual OM loss over time (Fig. 13.24). Reducing tillage intensity reduces OM mineralization; thus, soil OM levels can be sustained. The data in Figure 13.24 show the percentage of total N in the surface soil relative to the total N content in an undisturbed, noncropped soil (native prairie grass), as influenced by tillage. In-

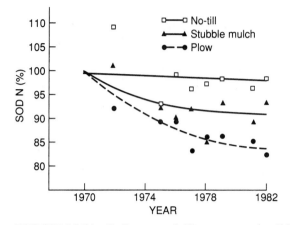

FIGURE 13.24   Influence of tillage on total soil N measured as a percentage of total N in undisturbed prairie soil.   *Lamb et al., SSSAJ, 49:352–56, 1985.*

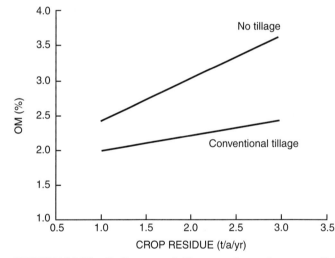

FIGURE 13.25   Influence of tillage and rotation on soil OM content. Crop residue returned per year for the three rotations is about 1, 2, and 3 t/a/yr for continuous soybean, soybean-sorghum, and continuous sorghum, respectively. No tillage and conventional tillage represent 90% and 0% surface residue cover, respectively.   *Havlin et al., SSSAJ, 54:448–52, 1990.*

creasing tillage intensity from no-till to plowing systems increased N mineralization, which reduced soil OM over time.

The relationships between tillage and residue on accumulation of OM have been well documented (Fig. 13.25). In this study three crop rotations—continuous soybean, continuous sorghum, and sorghum-soybean—were managed for 12 years under conventional and no-tillage systems (0 and 100% surface residue cover, respectively). The total residue returned to the soil increased with increasing frequency of sorghum in the rotation (continuous sorghum > sorghum-soybean > continuous soybean). Under conventional tillage, soil OM increased only slightly compared with no tillage, where all the residue was left on the soil

TABLE 13.10   Soil Organic C to 30-cm Depth after 5 (1975) and 20 Years (1989) of
Continuous Conventional Tillage and No-Tillage Corn

N Rate (kg ha^{-1})	1975 (Mg ha^{-1})			1989 (Mg ha^{-1})		
	CT*	NT*	Sod	CT	NT	Sod
0	39.7	46.8	53.4	48.9	55.3	55.5
84	47.8	48.4		56.2	58.3	
168	47.7	46.3		56.4	58.6	
336	45.9	52.8		61.3	66.2	

*CT, conventional tillage; NT, no tillage.

SOURCE: Frye and Blevins, *Soil OM under long-term tillage in Kentucky.* CRC Press, p. 233 (1997).

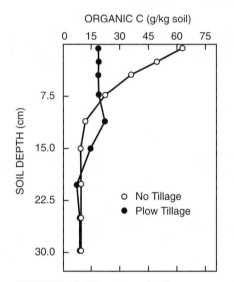

FIGURE 13.26   Organic C concentrations in soil profiles after 18 years of continuous plow or no tillage.   *Dick et al., 1997,* Continuous application of no-tillage to Ohio soils. *CRC Press, p. 178.*

surface. Under no tillage, soil OM increased 45% as the level of residue increased from 1 to 3 t/a/yr. These data illustrate that the quantity of residue returned is important to maintaining or increasing OM; however, reducing oxidation of the organic residue with no tillage had a greater effect on increasing OM. Although soybean residue has a lower C/N ratio (higher N content) than sorghum residue, the quantity of residue added is more important to increasing OM.

As discussed earlier, N applications combined with high-residue-producing cropping systems can increase soil OM, although 5 to 8 years of continuous no-till systems may be needed before effects are measurable (Table 13.10). In the first several years, the increase in soil OM occurs predominantly in the surface 2 in., although increased OM can be measured deeper in the profile after decades of continuous no-till cropping (Fig. 13.26).

Generally, slightly higher N rates are required for crops grown under no-tillage than under conventional tillage systems because of reduced N availability in no-till fields due to increased N immobilization, leaching, and/or denitrification. A number of studies have demonstrated more efficient utilization of fertilizer with no-till production compared with conventionally grown corn. Crop yields under no tillage systems are often equal to or better than those of conventionally tilled crops, particularly with optimum management of fertilizer N and other plant nutrients. In situations in which crop yields are initially reduced with no tillage, yields may improve with time because of reduced erosion; improved weed control; increased levels and/or quality of soil OM; improved availability of N, P, and other nutrients; higher cation exchange capacity (CEC); greater soil water-holding capacity; and better soil structure.

Adoption of No-Till Systems

Currently, about 50% of the cropland in North America (1990 estimate) is under some form of conservation tillage to reduce wind and water erosion, to increase precipitation efficiency, and to reduce fuel, labor, and equipment costs (Fig. 13.27). Compared with conventional tillage, there are fewer operations and less thorough incorporation of crop residue under conservation tillage systems. With zero tillage, all crop residue remains on the soil surface rather than being worked into the soil. The insulating and shading effect of crop residue lying on the surface reduces N and S mineralization because of lower soil temperatures. The physical nature of loose, coarse accumulations of crop debris is also far from ideal for rapid turnover of OM and release of N and S. This slow cycling or turnover of N and S under no-tillage systems favors maintenance of soil OM.

No-till cropping is destined to increase, and it is estimated that within the next 20 years, more than 50% of the cropland in the United States will be farmed by no-till systems. The attractive features of no-till systems are as follows:

1. Row crop production on sloping lands is more feasible, with less loss of nutrients, soil, and water.
2. WUE is increased.
3. No-till corn production can readily follow soybeans.
4. Double cropping of row crops such as soybeans, sorghum, or corn planted immediately after a wheat harvest is possible in many areas.

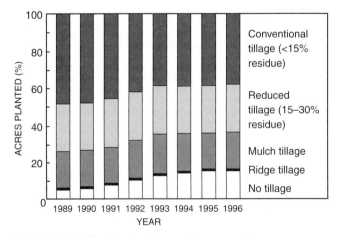

FIGURE 13.27   National use of crop residue management, 1989–96.   *USDA-ERS. 1996.*

5. Seeding legumes and/or nonlegumes in rundown pastures can be accomplished using the same principle. Fertilizer may be placed beneath the seed or broadcast. Herbicides are used to kill or retard existing grasses and weeds. Again, this technique is useful on sloping lands that should not otherwise be tilled, as well as on areas that are more level.
6. Energy, time, labor, and machinery costs are reduced.

Rotations versus Continuous Cropping

Although monoculture was once considered a sign of poor farming, the increased supply of fertilizer, especially that of N in the 1950s, encouraged continuous cropping on soils on which erosion was not a serious problem.

Most continuous cropping systems are currently used because of economics. In some regions, the crop options may be limited because of the environment or the market; thus, growers continuously produce a most adapted crop that maximizes the profit potential. Conventionally tilled, continuous dryland wheat, cotton, and irrigated corn are common examples.

Numerous long-term experiments have demonstrated that, in general, rotations increase long-term crop productivity compared with continuous cropping (Fig. 13.12). The reasons for the production advantage with rotation cropping compared with continuous cropping are not fully understood; however, the following discussion provides some insights.

A crop may have a harmful effect on the following crop, be it a different crop or the same crop. There is some evidence that substances released from roots or formed during the decomposition of residues are responsible for the toxicity. The comparison of continuous corn versus a corn-soybean rotation is an example. Seeding alfalfa following alfalfa is often unsatisfactory for unknown reasons. *Allelopathy* is the term used to describe the antagonistic action of one plant on another.

Advantages of Both Systems

ROTATIONS

1. Deep-rooted legumes may be grown periodically over all fields.
2. There is more continuous vegetative cover, with less erosion and water loss.
3. Tilth of the soil may be superior.
4. Crops vary in extracting range of roots and nutrient requirements: deep-rooted versus shallow-rooted, strong feeder versus weak feeder, and N fixer versus nonlegume fixer.
5. Weed and insect control are favored.
6. Disease control is favored. Changing the crop residue fosters competition among soil organisms and may help reduce pathogens.
7. Labor is more broadly distributed and income is diversified.
8. Erosion is reduced.

CONTINUOUS CROPPING OR MONOCULTURE

1. Profits may be greater, but they depend on the crops involved.
2. A soil may be especially adapted to one crop.

3. The climate may favor one crop.
4. Machinery costs may be lower.
5. The grower may prefer a single crop and become a specialist; however, mono-culture demands greater skills, including pest, erosion, and fertilization control.
6. The grower may not wish to be fully occupied with farming year-round.

Control of Disease, Weeds, and Insects

In most cases, crop rotation or initiation of other cropping practices will help control certain diseases, weeds, or insects. For example, reducing root-rot diseases in wheat and other cereals requires crop rotation, together with resistant varieties, clean seed, and field sanitation practices. Legumes, other dicotyledons, and even cereals such as oats, barley, or corn are often suitable alternate crops in place of wheat when take-all occurs. However, in some instances, this disease can be severe, even in wheat following alfalfa, soybeans, and grass crops. Corn root rots have been reduced by rotations, and the severity of several seedling diseases has been reduced by rotations combined with field sanitation. Susceptible crops should be grown on the same field only once in every 3 to 4 years.

Crop rotation is an important approach for the control of nematodes feeding on the roots of annual crops. Grass crops are commonly used in rotation to control root-knot nematodes. Acceptable yields of irrigated cotton can be obtained following 2 or more years of root-knot–resistant alfalfa. Two years of clean fallow also effectively controlled root-knot nematodes. Few important bacterial or viral diseases are controlled by crop rotation.

The role of crop rotation for weed control depends on the particular weed and the ability to control it with available methods. If all of the weeds can be conveniently and economically controlled with herbicides, then crop rotation is not a vital part of a weed control program. However, there are situations in which rotations are necessary for control of a troublesome weed.

For example, downy brome and jointed goat grass can severely reduce yields in a wheat-fallow-wheat system. Use of atrazine in a reduced-tillage wheat-sorghum/corn-fallow rotation will eliminate these weed problems.

Rotation was once a common practice for insect management, but its use declined with the development of economically effective insecticides. Interest in rotations has increased because of insect resistance to the chemicals and increased costs. Rotation can be helpful where the insects have few generations a year or where more than one season is needed for the development of a generation. For example, northern corn rootworm can be a serious problem in continuous corn cropping systems. Rotation of soybean and corn replaces the need for insecticide control of this insect. Rotation is only partially successful in reducing damage by cotton bollworm. Sorghum, when planted at the proper time, will protect cotton from worms.

Effect on Soil Tilth

Most recommended crop production practices provide good plant cover and return large amounts of crop residue to the soil. In addition, there is less tillage, which reduces the detrimental effects of compaction and deterioration of soil structure. The important issue is not one of monoculture versus rotation but rather involves two factors: the amount of residue returned to the soil and the nature of the soil tillage necessary in the rotation.

Rotation can greatly improve the soil structure and tilth of many medium- and fine-textured soils. Pasture grasses and legumes in rotation exert significant beneficial effects on physical properties of soil. When soils previously in sod are plowed, they crumble easily and readily shear into a desirably mellow seedbed. Plow draft is often reduced in fine-textured soils when less intensively tilled crops are grown in the rotation. Internal drainage can be improved so that ponding and the time needed for the soil to drain excess water are reduced. More information is required concerning the soil conditions under which deep-rooted legumes are needed in the rotation.

Corn in monoculture is unique, since on many soils it maintains reasonably acceptable soil physical conditions. The compensating factors are (1) the return of several tons of crop residue to the soil when corn is harvested for grain and (2) the corn crop is well adapted to reduced tillage and decreased damage from traffic on the soil.

Double cropping, such as small-grain–soybeans or small-grain–corn; triple cropping; and even quadruple cropping of rice in areas with long growing seasons and the possibilities of irrigation are becoming more common. With four crops a year, 27 t/ha of rice are possible. This necessitates utilizing soil, solar, and water resources to the maximum. If adequate fertility and pest control are provided and varieties are improved, soil productivity should gradually increase. Thus, more attention will be directed to measuring yield per unit area per year.

Crop Removal of Nutrients in the Root Zone

Crop plants vary considerably in their content of primary, secondary, and micronutrients. In addition, crops may absorb nutrients from different soil zones, thus making the choice of cropping sequences important to plant nutrition. Deep-rooted crops absorb certain nutrients from the subsoil. As their residue decomposes in the surface soil, shallow-rooted crops may benefit from the remaining nutrients. On a soil marginal in a particular micronutrient, it is possible that the preceding crop will have a considerable effect on the supply of this element to the current crop.

The net effect of cropping practices on P and K levels depends on the removal of nutrients by the harvested portion of the crop, nutrients supplied by the soil, and supplemental fertilization.

Effect of Rotation on Soil and Water Losses

Generally, increasing the crop yield and the residue produced and left on the soil surface will increase water infiltration and reduce runoff and soil loss. Some of the characteristics of cropping systems and/or fertility management practices related to soil losses are the following:

1. The denseness of the cover or canopy. The denseness affects the amount of protection from the impact of rain and evaporation. Residue and stems reduce the velocity of water and evaporation. Residue, when turned back to the soil, makes the soil more permeable to water.

TABLE 13.11   Effect of Management Level on Crop Yields, Runoff, and Erosion, 1945–68

	Prevailing Practices	Improved Practices
Corn (t/ha)	5.1	7.3
Wheat (t/ha)	1.5	2.3
Hay (t/ha)	4.3	7.8
Runoff, growing season (cm)	1.9	1.0
Peak runoff rate (cm/hr)	2.3	1.5
Erosion from corn (t/ha/yr)	10.6	3.1

SOURCE: Edwards et al., *SSSA Proc.,* 37:927 (1973).

TABLE 13.12   Hydrologic Data from a 6- to 7-a Watershed Comparing Various Cropping and Tillage Systems

Time Period	Cropping System*	Tillage	Annual Rainfall (in.)	Annual Runoff (in.)	Annual Soil Loss (t/a)
1972–74	Fallow/soybean	Conventional	54.0	8.7	11.6
1974–76	Barley/grain sorghum	No-till	52.0	3.5	0.2
1976–79	Barley/soybean	No-till + in-row chisel	46.5	0.8	0.06
1979–83	Crimson clover/grain sorghum	No-till + in-row chisel	43.7	0.2	0.002

SOURCE: Hargrove and Frye, *Role of Legumes in Conservation Tillage Systems,* Soil Conservation Society of America, Madison, Wisc. (1987).

2. The proportion of time that the soil is in a cultivated crop versus the amount of time in a close-growing crop such as small grains or forage.
3. The time that the crop grows in relation to the distribution and intensity of rainfall. The period from May to September is most vulnerable.
4. The type and amount of root system.
5. The amount of residue returned.

   Adoption of improved soil and crop management practices can increase yields and reduce runoff and erosion. As Table 13.11 shows, the improved management effects on reducing runoff and erosion were due in part to better surface protection through a quicker cover in the spring, a denser cover throughout the season, and a more extensive root system of the growing crop. The influence of maintaining a surface residue cover and crop rotation on reducing runoff and soil loss is demonstrated in Table 13.12.

Winter Cover–Green Manure Crops

Winter cover crops are planted in the fall and plowed down in the spring. These crops may be a nonlegume, a legume, or a combination grown together. There are several advantages to the last practice. A greater amount of OM is produced, the nonlegume can benefit from the N fixation, and because the nonlegume is usually more easily established, a stand of at least one crop is ensured.

TABLE 13.13   Dry Matter and N Concentration of Various Cover Crops and Influence of Cover Crops on Corn Grain Yield

Cover Crop	Dry Matter (lb/a)	N Concentration (%)	N Content (lb/a)	Yield by N Rate (lb/a of N)	
				0	200
				— bu/a —	
Fallow	—	—	—	63	161
Wheat	1,178	2.01	35	32	121
Winter pea	1,423	4.56	61	132	165
Hairy vetch	2,526	4.62	113	156	168
Crimson clover	2,883	3.67	102	143	172

SOURCE: Neely et al., *Role of Legumes in Conservation Tillage Systems*, p. 49, Soil Conservation Society of America, Madison, Wisc. (1987).

TABLE 13.14   Biomass Yield and Nutrient Accruement by Select Cover Crops

Cover Crop	Biomass*	N	K	Ca	P	Mg
				lb/a		
Hairy vetch	3,260	141	133	52	18	11
Crimson clover	4,243	115	143	62	16	11
Austrian winter peas	4,114	144	159	45	19	13
Rye	5,608	89	108	22	17	8

*Dry weight of above-ground plant material.

SOURCE: Hoyt, *Role of Legumes in Conservation Tillage Systems*, p. 96, Soil Conservation Society of America, Madison, Wisc. (1987).

Decomposition of green manure crops is rapid, but the residual effects are well recognized. The smallest residual effects generally are expected in areas in which the mean annual temperature is high and the soil is sandy.

Small grains or other crops can be grazed in late fall and winter when the amount of growth and soil conditions permit. Adequate fertility, either residual or added, is necessary, and extra N may be needed. Grazing allows additional return from cover crop inputs.

N and OM Added

One important reason for using green manure legume crops is that they supply additional N, depending on the yield and N content (Table 13.13). These data show that increasing the quantity of N produced in the legume cover crop increased the corn yield (unfertilized). The grain yield after fallowing was greater than following the wheat cover crop because of N mineralization during the fallow period. When a nonlegume is turned under, only the N from the soil or that supplied in fertilizer is returned.

The nutrient content in several legume cover crops is shown in Table 13.14. Increasing the cover crop yield or biomass will subsequently increase the quantity of N_2 fixed and the N returned to the soil (Fig. 13.28). Legume cover crops can contribute large quantities of N to subsequent nonlegume crops (Table 13.15).

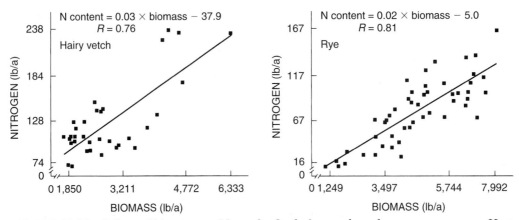

FIGURE 13.28   Effect of biomass on N uptake for hairy vetch and rye cover crops.   *Hoyt, Role of Legumes in Conservation Tillage Systems, Soil Conservation Society of America, Madison, Wisc. 1987.*

TABLE 13.15   Estimates of the N Contribution of Winter Legumes to the N Requirements of No-Till Corn, Grain Sorghum, and Cotton

Location	Crop	Cover Crop	Fertilizer N (lb/a)
Kentucky	Corn	Hairy vetch	85
		Big flower vetch	45
Georgia	Grain	Crimson clover	75
	sorghum	Hairy vetch	81
		Common vetch	53
		Subterranean clover	51
Alabama	Cotton	Hairy vetch	61
		Crimson clover	61

SOURCE: Hargrove and Frye, *Role of Legumes in Conservation Tillage Systems,* p. 2, Soil Conservation Society of America, Madison, Wisc. (1987).

One of the benefits attributed to winter cover crops is the OM supplied to the soil (Table 13.16). Green manures will help maintain the soil OM and will sometimes even increase it.

In rotations in which the crops return little residue, maintenance of soil productivity may be particularly difficult. The lengthening of the rotation to include green manure crops could be beneficial. The acreage of corn and sorghum silage is increasing in some areas, which leaves the soil with almost no surface residue. Oats or rye seeded immediately after harvesting or seeded by airplane before harvesting will help to protect the soil and increase the residue returned.

Protection of the Soil against Erosion

Protection against erosion is one of the most important reasons for winter cover crops. The benefits, however, should be related to the distribution of rain and the erosion potential during the year. The effect of cover crops on soil loss is generally small when winter cover crops are turned under in early spring.

TABLE 13.16   Influence of 5 Years of Various Cropping Sequences and Tillage on Soil Organic C and N Concentrations in the Surface 7.5 cm of Soil

Cropping Sequence	Tillage Treatment	Fertilizer N Rate (lb/a/yr)	Organic C	Organic N	C/N Ratio
			———%———		
Wheat/soybean	Conventional	70	1.4	0.12	11.7
Wheat/soybean	No-till	70	1.6	0.15	10.7
Clover/sorghum	No-till	0	2.2	0.17	13.0
Clover/sorghum	No-till	120	2.4	0.19	12.6

SOURCE: Hargrove and Frye, *Role of Legumes in Conservation Tillage Systems,* Soil Conservation Society of America, Madison, Wisc. (1987).

Surface residue from summer crops, if left undisturbed, may provide more protection than cover crops seeded in the fall. The greater the percentage of the soil surface covered by mulch, the less the soil loss. Freshly tilled land is quite susceptible to erosion, and considerable time is required before the cover crop can provide enough protection to have much effect on reducing soil loss. For example, rye cover has been grown after corn, but in comparison to heavy corn residue, it is generally not as effective for erosion control.

For perennial crops such as peaches and apples planted on steep slopes, continuous cover is helpful in reducing erosion. Since the trees and the cover crops occupy the land simultaneously, care must be taken, particularly in young orchards, to prevent competition for water and N. In some of the muck soils suitable for vegetables, a strip of small grain or a row of trees at intervals helps to reduce losses from erosion.

Environmental Quality

As plant nutrients cycle through the soil-plant-atmosphere continuum, some will be naturally removed from access by plants or recovery through plant uptake (refer to specific nutrient cycles in each chapter). Movement of nutrients dissolved in water away from the field occurs through runoff and subsurface lateral flow toward streams and rivers and leaching to the groundwater (Table 13.17; Fig. 13.29). Volatilization to the atmosphere also occurs with N and S, as discussed in Chapters 4 and 7. Nutrients that are most soluble and mobile in the soil exhibit the greatest

TABLE 13.17   Soil Processes Affecting Water Quality

Soil Processes	Impact on Water Quality
1. Soil erosion	Transport of dissolved and suspended sediments in surface runoff
2. Leaching	Movement of nutrients, agricultural chemicals, and dissolved organic carbon in percolating water
3. Macropore flow	Rapid transport of water and pollutants from surface to subsurface and into a drainage system
4. Mineralization of humus	Release of readily soluble compounds that are easily washed away or leached out

SOURCE: Lal and Stewart, *Soil Processes & Water Quality,* CRC Press, p. 4, (1994).

potential for movement to groundwater and surface water. This *nonpoint source* movement is the primary mechanism of nutrient loading into surface and groundwater in agricultural systems. Figure 13.30 shows that fine soil particles lost through erosion and runoff, and nutrients lost through runoff or subsurface flow, are the most common nonpoint source contaminants. Nutrients and pesticides are the primary contaminants of groundwater. The nutrient of primary concern in agriculture is N, but P is also a concern.

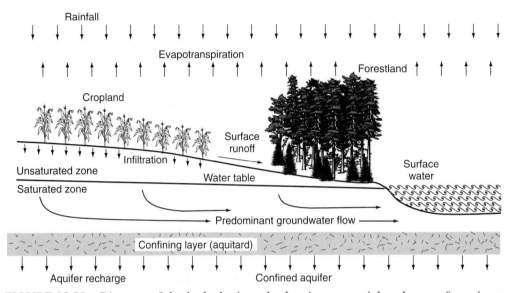

FIGURE 13.29   Diagram of the hydrologic cycle showing potential pathways of nutrients to surface and groundwater. The *aquitard,* or confining layer to downward movement of water (leaching), is common to the coastal regions of the southeast United States. Aquitards seldom occur in the Midwest and Great Plains regions of the United States. *Evans et.al., 1991. NC Agric. Ext. Service, AG-443.*

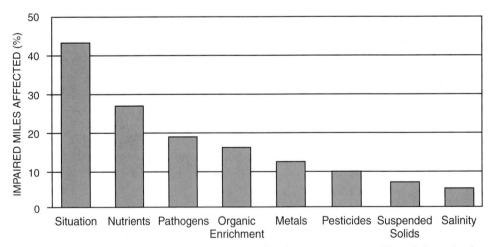

FIGURE 13.30   Most common constituents that impact water quality of rivers in the United States.   *US EPA, 1996.*

It is important to recognize that nutrient movement to surface and ground-water occurs in natural ecosystems. Nutrients are found in "background" levels in all waters. For example, typical background concentrations of NO_3 can range between 1 and 10 ppm N. Water quality standards for drinking water have been established for all elements that adversely affect health when present in high concentrations (Table 13.18). Approximately 2% of groundwater wells used for drinking water exceed the primary drinking water standard of 10 ppm N. Several adverse health effects occur when humans or animals access high-NO_3 water (Table 13.19). Although rare, the most notable health effect of high-NO_3 water is low blood oxygen levels in human infants, called *methemoglobinemia*. Concentrations of $NO_3 > 40$ ppm N are hazardous to adults and livestock (Table 13.20).

TABLE 13.18   Water Quality Standards for Human and Livestock Consumption

Chemical Element or Compound	Concentration (mg/l)	
	Human	*Livestock*
Pb	<0.1	0.05
Mo	—	0.01
As	<0.05	0.05
Se	<0.01	0.01
Zn	<15	<20
Cd	<0.01	0.01
Ba	<1.0	—
Ca	<200	<1,000
Hg	<0.01	0.002
NO_3	<45	<200
NH_4	<0.05	—
N	<10	<50
Cl	<400	<1,000

SOURCE: Lal and Edwards, *Soil Processes & Water Quality*, CRC Press, p. 5 (1994).

TABLE 13.19   Potential Adverse Environmental and Health Impacts of N

Impact	*Causative Agents*
Human health	
Methemoglobinemia in infants	Excess NO_3^- and NO_2^- in waters and food
Cancer	Nitrosamines from NO_2^-, secondary amines
Respiratory illness	Peroxyacyl nitrates, alkyl nitrates, NO_3^- aerosols, NO_2^-, HNO_3 vapor in urban atmospheres
Animal health	
Environment	Excess NO_3^- in feed and water
Eutrophication	Inorganic and organic N in surface waters
Materials and ecosystem damage	HNO_3 aerosols in rainfall
Plant toxicity	High levels of NO_2^- in soils
Excessive plant growth	Excess available N
Stratospheric ozone depletion	Nitrous oxide from nitrification, denitrification, stack emissions

SOURCE: Owens, *Soil Processes and Water Quality*, CRC Press, p. 138 (1994).

TABLE 13.20   Guidelines for Use of Water with Known NO_3 Content

NO_3 Level	NO_3-N Level	Interpretation*
(ppm)		
<45	<10	U.S. Public Health Service standard is 45 ppm NO_3 or 10 ppm NO_3-N; safe for humans and livestock
45–90	10–20	Generally safe for human adults and all livestock; should not be used by pregnant women or infants
90–180	20–40	Humans and some livestock at risk, especially young or those in high-risk category; monitor nitrates in livestock feed
>180	>40	Hazardous to humans and livestock; do not use for drinking or cooking without treatment

* Interpretations are primarily based on short-term effects. Chronic, long-term risks are not fully understood.

N Leaching

Although NO_3 naturally occurs in all water, NO_3 loading of surface and groundwater can be greatly elevated from N added to agricultural systems. As mentioned in Chapter 4, N sources utilized in increasing yield potential of crops include fertilizer, legumes, and manure. Because of its predominance as a N source, fertilizer N is often considered the primary cause of contamination of surface and groundwater. Little fertilizer N is lost directly through runoff, because most N is subsurface applied and its solubility and mobility result in immediate movement into the root zone in moist soils. In general, the crop recovers 40 to 60% of fertilizer N in the first year. The remaining N stays in the soil as NO_3, immobilized to organic N, leached as NO_3 below the root zone, or volatilized as N gases (Fig. 4.1). High N recovery by the crop (low residual N after harvest available for leaching) will occur with a readily available N source applied to a crop that can utilize it quickly. Slowly available N that is not used during the first crop year can be leached during noncrop periods. Thus, applied at appropriate rates, fertilizer N can exhibit a lower potential for leaching than manure, sludge, or legume N because a portion of the organic N in these materials will mineralize during periods of low plant growth and N uptake (Fig. 4.12). Thus, organic N can often contribute more to nonpoint source contamination of surface and groundwater than fertilizer N at equal application rates.

Leaching loss of NO_3 is normal; excessive NO_3 loss is unacceptable. For NO_3 leaching to occur, the soil water must contain NO_3 and move below the root zone. Water transport below the root zone generally occurs in regions in which rainfall or irrigation exceeds evapotranspiration (Fig. 13.31).

In addition, soil profile characteristics are important in determining the quantity of NO_3 transported below the root zone (Fig. 13.32). In these examples, the time required for NO_3 to enter the groundwater is very short (3 to 9 months) in a sandy soil with a shallow vadose zone, which represents the material below the rooting depth but above the aquifer. With similar vadose zone thickness, time required for transport to groundwater can be two to three times longer in fine-textured soils compared with sands (Fig. 13.32).

Irrigated cropping systems can contribute to nonpoint source NO_3 contamination, especially with excessive irrigation combined with high N application rates. With furrow irrigation systems, more water is applied at the beginning than

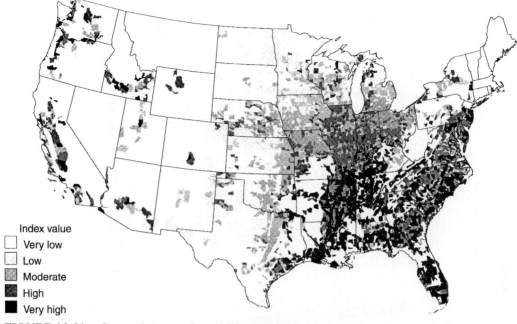

FIGURE 13.31 Groundwater vulnerability index for N. *USDA. 1996.*

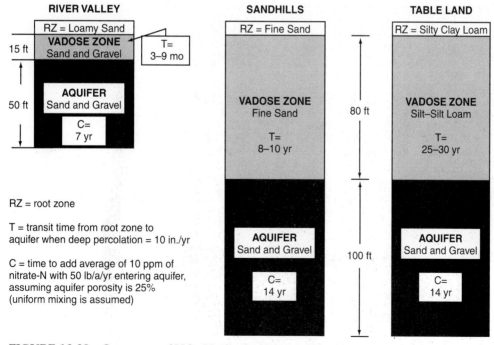

FIGURE 13.32 Summary of NO$_3$-N transit times and aquifer contamination times for three example situations. *Watts, Univ. of Nebraska, 1992.*

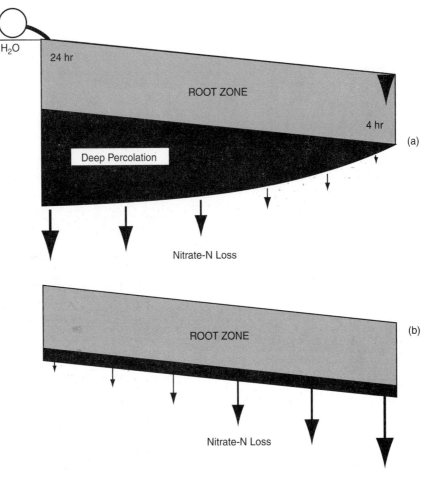

FIGURE 13.33   NO₃ leaching pattern during the irrigation season (*a*) and off-season (*b*) for long set times and/or long irrigation runs.   *Watts, 1992. Univ. of Nebraska, 1992.*

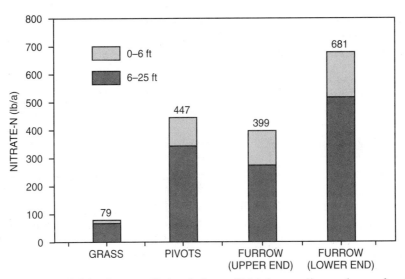

FIGURE 13.34   Average lb/a of nitrate-N in deep soil samples under 4 center-pivot and 10 furrow-irrigated corn fields.   *D. Watts, 1992, Univ. of Nebraska.*

at the end of the furrow, increasing the quantity of water transport below the root zone during irrigation (Fig. 13.33). After the irrigation season, NO_3 leaching is greater down field. NO_3 leaching can be as problematic under center-pivot systems with similar overirrigation and/or overfertilization (Fig. 13.34).

Best Management Practices (BMP)

Many site, environment, and management factors interact to reduce the potential for nonpoint source contamination of surface and groundwater. Understanding the principles involved in availability, detachment, transport, and deposition is essential for identifying the best management factors for reducing the impact of nutrient use on water quality. Figures 13.35 and 13.36 illustrate how

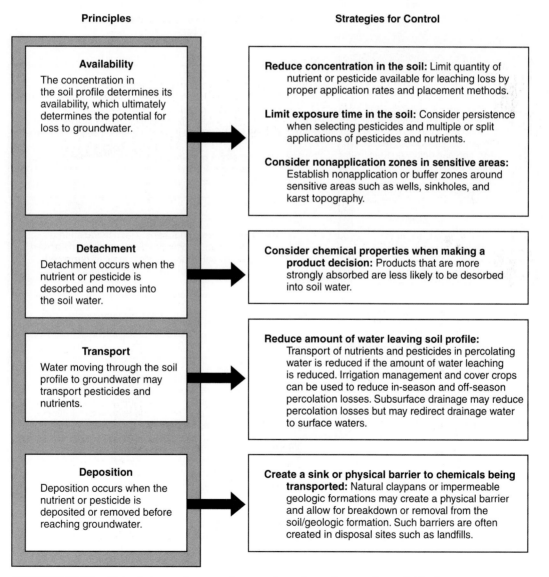

Principles

Strategies for Control

Availability
The concentration in the soil profile determines its availability, which ultimately determines the potential for loss to groundwater.

Reduce concentration in the soil: Limit quantity of nutrient or pesticide available for leaching loss by proper application rates and placement methods.

Limit exposure time in the soil: Consider persistence when selecting pesticides and multiple or split applications of pesticides and nutrients.

Consider nonapplication zones in sensitive areas: Establish nonapplication or buffer zones around sensitive areas such as wells, sinkholes, and karst topography.

Detachment
Detachment occurs when the nutrient or pesticide is desorbed and moves into the soil water.

Consider chemical properties when making a product decision: Products that are more strongly absorbed are less likely to be desorbed into soil water.

Transport
Water moving through the soil profile to groundwater may transport pesticides and nutrients.

Reduce amount of water leaving soil profile: Transport of nutrients and pesticides in percolating water is reduced if the amount of water leaching is reduced. Irrigation management and cover crops can be used to reduce in-season and off-season percolation losses. Subsurface drainage may reduce percolation losses but may redirect drainage water to surface waters.

Deposition
Deposition occurs when the nutrient or pesticide is deposited or removed before reaching groundwater.

Create a sink or physical barrier to chemicals being transported: Natural claypans or impermeable geologic formations may create a physical barrier and allow for breakdown or removal from the soil/geologic formation. Such barriers are often created in disposal sites such as landfills.

FIGURE 13.35   Principles and control strategies of nutrient and pesticide losses to groundwater. *Best Mgmt. Practices for Wheat, 1994 National Assoc. Wheat Growers, Wash., D.C.*

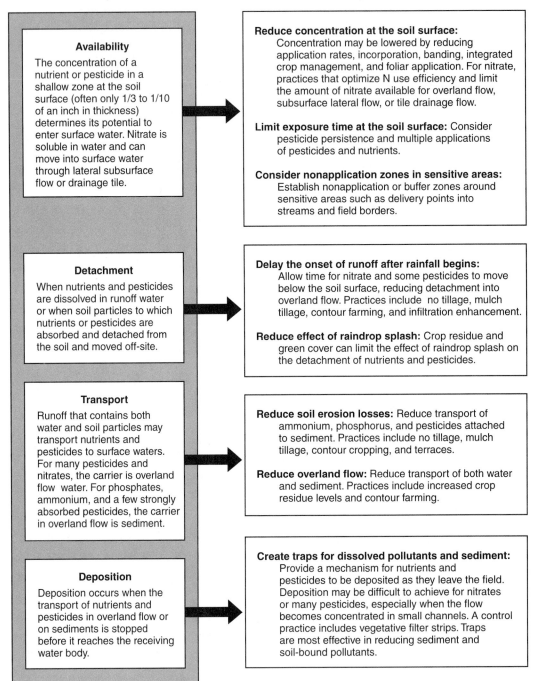

FIGURE 13.36   Principles and control strategies for nutrient and pesticide entry into surface water.   *Best Mgmt. Practices for Wheat, 1994 National Assoc. Wheat Growers, Wash., D.C.*

TABLE 13.21   Best Management Practices (BMPs) for Controlling the Entry of N Compounds into Surface and Groundwater

BMPs for Maintaining Surface Water Quality	BMPs for Maintaining Ground Water Quality
1. Apply appropriate N rate	1. Apply appropriate N rate (appropriate yield goal)
2. Timely fertilizer applications	2. Timely N applications
3. Incorporate fertilizers	3. Improved cropping/irrigation management
4. Proper cropping/residue management	4. Control N transformations
5. Control soil erosion–land structures	5. Foliar applications
	6. Cover crops to scavenge NO_3

SOURCE: Hergert, *Soil Sci. Soc. Amer. Spec. Publ. No. 21*, (1987).

TABLE 13.22   Typical Credits used for Estimating Crop N Requirements

N Source	N Credit
Soil OM	30 lb N/% OM
Residual soil NO_3	3.6 lb N/ppm NO_3-N
Manure	10.0 lb N/t manure
Irrigation water	2.7 lb N/acre-ft. $\times$ ppm NO_3-N
Previous alfalfa/sweet clover	50–100 lb/a of N
Other previous legume crop	30 lb/a of N

these principles direct the appropriate best management strategies. Although we focus on N, most of the BMPs are effective in ameliorating impacts of other nutrients on water quality. BMPs for N can be categorized into those essential for groundwater and surface water (Table 13.21).

Understanding how each N BMP works requires knowledge of N transformations in soils (Chapter 4), quantification of N availability and requirements for the soil-plant system involved (Chapter 9), and management of the nutrient source for optimum productivity and maximum recovery by the plant (Chapter 10). The material developed in these chapters provides the foundation for nutrient management plans that incorporate the appropriate BMPs.

In addition to N placement, timing, and other factors, an accurate N management plan involves evaluation of all N sources to accurately assess the N requirement for the crop. Thus, the ability to estimate N contribution from soil OM, irrigation water, legume N, and so forth is essential for accurate determination of N requirements (Table 13.22; see Chapter 9). An example of a N balance calculation is illustrated for several irrigation systems in Table 13.23. The estimated crop N requirement is the same in both systems; however, recognition that residual NO_3 is greater under one system than the other results in a 20-lb/a lower N recommendation.

N RATE   The most important N BMP is identifying the correct N rate required to maximize yield. When N rate exceeds yield potential, residual NO_3 may leach if water is sufficient to move below the root zone (Fig. 13.37). Using a linear-plateau model, Figure 13.38 shows that significant NO_3 accumulation occurred only when N rate exceeded optimum N rate by nearly 40%. These contrasting results illustrate

TABLE 13.23   An Example Data Set from Nebraska to Determine N Fertilizer
Recommendations for Corn with Two Irrigation Systems

Data	Conventional	Center Pivot
Residual soil N		
0–20 cm (mg kg^{-1} NO$_3$-N)	13.5	12.3
20–90 cm (mg kg^{-1} NO$_3$-N)	7.2	4.5
0–90 cm (wt. average, mg kg^{-1} NO$_3$-N)	8.8	6.2
OM		
0–20 cm (%)	1.77	1.9
Irrigation water (mg l^{-1} nitrate-N)	30	30
Expected yield (mg ha^{-1})	12.6	12.6
Estimated water application (cm)	25	25
Estimated crop N need (kg ha^{-1})	309	309
21.4 (expected yield) + *39*		
N credits		
Residual soil N (kg ha^{-1}) = *9* (mg NO$_3$-N kg^{-1})	79	56
OM (kg ha^{-1}) = *2.5* (expected yield) • (% OM)	56	60
Irrigation water (kg ha^{-1})	75	75
0.1 (cm depth) (mg l^{-1} NO$_3$-N)		
Others		
Soybean (50 kg ha^{-1})	0	0
Alfalfa (135–170 kg ha^{-1})	0	0
Manure (variable)	0	0
Fertilizer N required (kg ha^{-1})	99	118

SOURCE: Rice, Havlin, Schepers, *Fertilizer Research*, 42:89 (1995).

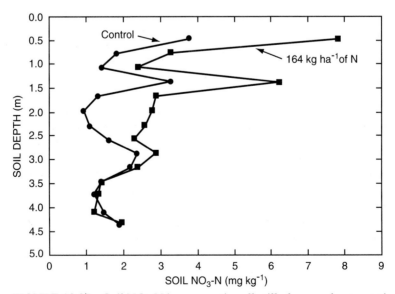

FIGURE 13.37   Soil NO$_3$-N in conventionally tilled corn-wheat rotations when
urea was applied at 0 or 164 kg ha^{-1} of N in April 1990 (sampled in March 1991).

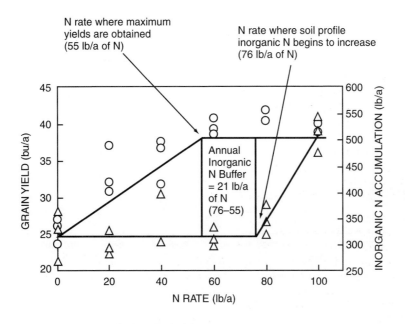

FIGURE 13.38 The relationship between inorganic N accumulation and annual N applied and the estimated soil-plant N buffering zone. *Raun and Johnson,* Agron. J. *87:827,* 1995.

that N recommendations should be evaluated for each individual situation. Remember that nutrient recommendations obtained from a laboratory represent an educated guess and thus should be adjusted for the specific field situation.

N TIMING The importance of application timing was discussed in Chapter 10. Figure 13.39 illustrates the relationship between single and split applications on the quantity of fertilizer N subject to N loss. These data demonstrate substantial reduction in total N susceptible to N loss by leaching with four split applications compared with a single application.

N PLACEMENT In most cases, subsurface application of N will reduce N volatilization losses (Chapter 4). In permanent pasture or turf systems, subsurface applications are less desirable than surface broadcast applied N; however, spoke-wheel and other innovative application technologies offer opportunities to maximize recovery of N by the crop (Chapter 10). In reduced-tillage systems, which are critical to ensuring long-term soil and crop productivity, subsurface application of N is essential to maximizing crop recovery of applied N and reducing N immobilization by high-C-containing crop residues. Table 13.24 illustrates the importance of subsurface application of N on grain yield. The N content in the leaf and grain is reduced with subsurface application because the higher growth and yield dilutes N concentration.

CROP ROTATION There are several issues relative to cropping systems that can influence nutrient management decisions. It is important to recall that NO_3

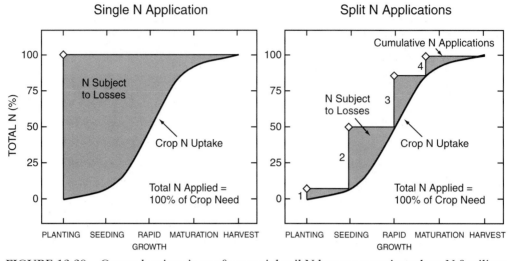

FIGURE 13.39   General estimations of potential soil N losses occurring when N fertilizer is applied in a single or in split applications.   *Waskom et al., 1994. BMP's for Irrigated Agric. Colorado Water Resources Institute Report No. 184.*

TABLE 13.24   Effect of N Source and Placement on Corn Yield, Leaf N Concentration, and Grain N Concentration in Both Plow and No-Till Production Systems, (Mengal, Indiana)

Nitrogen Source and Placement	Plow			No-Till		
	Yield (bu/a)	Ear Leaf (% N)	Grain (% N)	Yield (bu/a)	Ear Leaf (% N)	Grain (% N)
UAN broadcast, not incorporated	145	2.48	1.21	128	1.63	1.08
UAN broadcast, incorporated	153	2.34	1.23	—	—	—
UAN injected	149	2.44	1.29	156	2.13	0.94

leaching can occur when N is supplied with either fertilizer, manure, or legume N. Figure 13.40 illustrates that, except for fallow systems that produce the largest amount of leachable N (no N uptake during periods of N mineralization), rotations that include legumes may contribute more leachable N than nonlegume-based systems. The data in Table 13.25 show that soybean in the rotation increased the soil profile NO_3 concentration as well as the NO_3 concentration below the root zone.

Crops can be used to recover soil profile N or prevent NO_3 leaching in certain cropping systems. Whenever soil test NO_3 data suggest that significant residual NO_3 is present in the soil profile after harvest, cover crops may recover significant quantities of soil profile NO_3 to prevent N transport to groundwater. Figure 13.41 shows that as N applied to corn increased (with subsequent increase in residual NO_3), the N uptake to the successive oat crop increased. These data illustrate that cover crops can significantly reduce NO_3 concentration in the soil profile left after a high-N-requiring crop such as corn.

N FROM ORGANIC WASTES   As discussed previously, organic N sources can contribute to plant available N as efficiently as fertilizer N. Unfortunately, if applied at rates exceeding crop N requirement, N mineralization after the crop uptake pe-

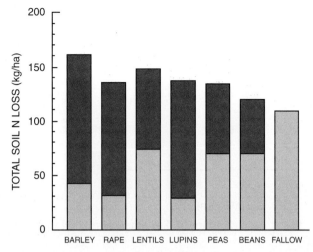

FIGURE 13.40   Contribution of net soil N uptake by grain crop sown in spring 1988 (■) and apparent leaching losses during the following autumn/winter (□) to the total loss of soil N from spring 1988 to spring 1989.

TABLE 13.25   Comparison of the NO_3-N in the Soil with that in the Soil Solution below the Root Zone

Treatment	NO_3-N	
	Soil 0–0.45 m May, June, July* (mg kg^{-1})	*Soil solution avg. of 1.2 and 1.5 m June, July, August (mg l^{-1})*
Corn (1990)		
No-till	14 (56)†	9
Tilled	19 (77)†	10
Wheat (1991)		
No-till	4 (16)	4
Tilled	6 (24)	10
Beans		
1990	26 (105)	12
1991	33 (133)	27

* Used different times to compensate for the time necessary for the soil solution to flow from the surface to the 1.2- to 1.5-m depth.

† Estimated NO_3-N in the soil solution (soil NO_3-N × 4.03 [1.33 Mg m^{-1} bulk density and 0.33 m^3 m^{-1} volumetric water content]).

SOURCE: Meek et al., *Soil Sci. Soc. Am. Jour.* 58:1464 (1994).

riod may contribute to leachable N. Thus, organic waste application can provide sufficient plant nutrients; however, because of the delayed N mineralization, organic N can contribute to N contamination of surface and groundwaters. Figure 13.42 illustrates that increasing poultry manure rate greatly increases NO_3 concentration in the soil profile. Application of sewage sludge materials can produce the same results (Fig. 13.43). Compared with fertilizer N, even split applied, application of manure N can result in greater profile N content after harvest (Table 13.26).

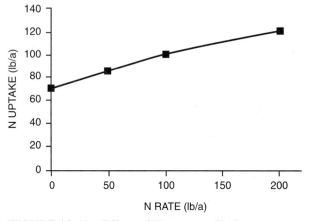

FIGURE 13.41   Effect of N rates applied to corn on total N uptake by the succeeding oat crop.

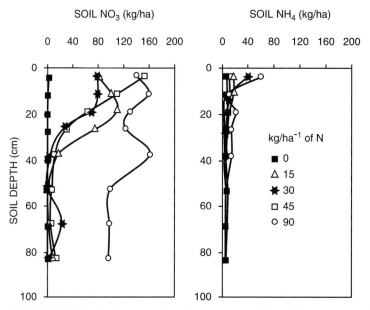

FIGURE 13.42   Influence of poultry manure N application on soil NO_3 and NH_4 in the soil profile. NH_4 concentrations are less than NO_3 concentrations because of the significant nitrification that occurs throughout the growing season, regardless of plant N uptake.   *Scott et. al., 1995. Arkansas Agric. Exp. Bull. 947.*

SOIL EROSION–RIPARIAN BUFFERS   Preventing soil loss through water and wind erosion is essential for maintaining surface water quality. Although retaining surface residue cover through conservation tillage systems can substantially reduce soil erosion, riparian buffer zones are effective in reducing NO_3 concentration in subsurface flow and in filtering sediments and nutrients in surface runoff water (Fig. 13.44). Depending on the width of the grass and/or forest buffer, 60 to 100% reduction in sediment can occur (Table 13.27). Reduction of

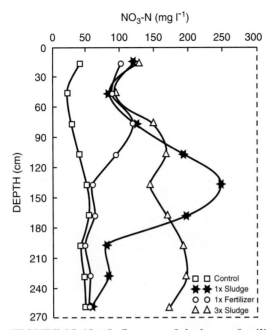

FIGURE 13.43 Influence of sludge or fertilizer treated soil on NO_3^- -N in soil profile (1990) of fertilizer or sludge. 1 ×, 3 ×, represents one and three annual applications, respectively. *Artiola and Pepper, 1992, Biology Fertility of Soils, 14:30.*

TABLE 13.26 Corn Yield and NO_3–N Concentration in Soil Water as Influenced by N Treatments

N Form	N Rate (lb/a)	Time Applied	2-yr. Average Yield (bu/a)	NO_3–N in Soil Water (ppm†)
Anhydrous NH_3	150	Spring	177	12
Anhydrous NH_3	75 + 75	Spring + sidedress	173	10
Hog manure	196*	Spring	184	41

* Estimated available N. Total N applied was 315 lb/a, half being inorganic and half organic. Researchers assumed 100% availability from inorganic N in year of application.

† Measured at 5-ft depth by suction lysimeters at end of second year.

SOURCE: W. Griffith, *Better Crops,* 73:23 (1989).

N and P in surface runoff ranges between 10 and 80%; the wider the buffer area, the greater the deposition in nutrients.

The reduction in NO_3 concentration in subsurface flow occurs through denitrification (Fig. 13.45). Anaerobic denitrifying microorganisms obtain their C from root mass in the buffer zone and convert NO_3 to N_2 gas. In many crop production fields, the soil profile NO_3 concentration can range between 15 and 30 ppm N after harvest. Figure 13.46 shows that denitrification reduces NO_3 concentration to <10 ppm N, depending on buffer width.

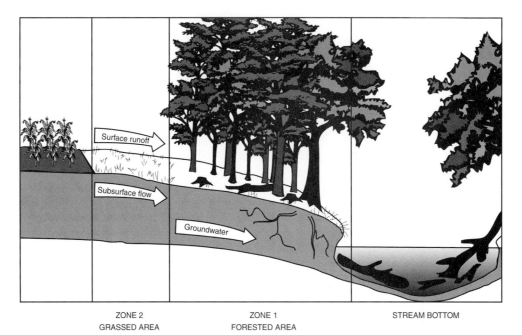

ZONE 2   ZONE 1   STREAM BOTTOM
GRASSED AREA   FORESTED AREA

FIGURE 13.44   Schematic of the two-zone riparian forest buffer system.   *Modified from Lowrance et al., U.S. EPA, Wash. D.C. 903-R-95-004, 1995.*

BMPs: Case Study   Although the range of options for BMPs should be evaluated for each site, the decisions on which management practices are utilized depend on the skill of the manager and the situation involved. Table 13.28 compares a conventional production system with one that uses BMPs appropriate for the local environment and cropping system. Although 30% higher N rates were applied, appropriate N management (soil and tissue testing, N timing, reduced tillage, crop rotation, etc.)increased N utilization and decreased the soil profile NO_3 after harvest. In fact, BMPs substantially increased yield and profit while reducing the quantity of leachable N.

This example illustrates why producer adoption of BMPs is essential to maximizing profit and minimizing the impact of nutrient use on the environment. It is also important to recognize that substantially less land was required to produce the same yield with the BMPs. Our continued ability to produce sufficient food for an expanding population depends on continued increases in crop yield per unit land area. Under increasing production pressure, conservation of our limited natural resources (quantity and quality) can only occur with full adoption of existing BMPs and continued development of new agricultural technologies.

TABLE 13.27   Effects of Different Size Riparian Buffers on Reductions of Sediment and Nutrients from Field Surface Runoff

Buffer Width (m)	Buffer Type	Sediment			Nitrogen			Phosphorus		
		Input Concentration	Output Concentration	Reduction[§] (%)	Input Concentration	Output Concentration	Reduction[§] (%)	Input Concentration	Output Concentration	Reduction[§] (%)
		—$mg\ l^{-1}$—			—$mg\ l^{-1}$—			—$mg\ l^{-1}$—		
4.6*	Grass	7,284	2,841	61.0	14.11	13.55	4.0	11.30	8.09	28.5
9.2*	Grass	7,284	1,852	74.6	14.11	10.91	22.7	11.30	8.56	24.2
19.90‡	Forest	6,480	661	89.8	27.59	7.08	74.3	5.03	1.51	70.0
23.6‖	Grass/ forest	7,284	290	96.0	14.11	3.48	75.3	11.30	2.43	78.5
28.2#	Grass/ forest	7,284	188	97.4	14.11	2.80	80.1	11.30	2.57	77.2

* Calculated from masses of total suspended solids, total N, total P, runoff depth, and plot size (22 × 5 m).

† Nitrogen = Nitrate-N + exch. part. ammonium + diss. ammonium + part. organic N + diss. organic N. Phosphorus = part. P + diss. P.

‡ Surface runoff concentrations at 19 m into forest reported by Peterjohn and Correll (1984). N and P constituents same as input.

§ Percentage of reduction = 100 * (input − output)/input.

‖ 4.6-m grass buffer plus 19 m of forest.

9.2-m grass buffer plus 19 m of forest.

SOURCE: Lawrence et al., *U.S. EPA*, Wash. DC. 903-R-95-004 (1995).

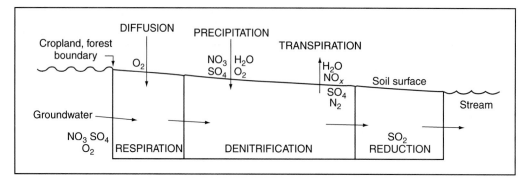

FIGURE 13.45   Conceptual model of below-ground processes affecting groundwater nutrients in riparian forest.   *From Correll and Weller, 1989. Freshwater, Wetlands and Wildlife. US Dept. Energy, p. 9–23.*

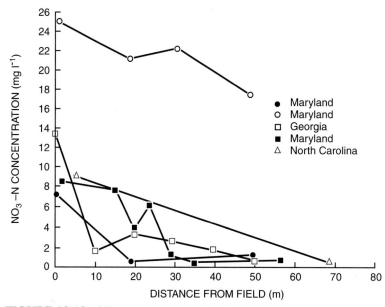

FIGURE 13.46   Nitrate concentrations in groundwater beneath riparian forests.   *Gilliam et al., Selected Agric. Best Mgmt. Practices to Control N. NC Agric. Res. Service Tech. Bull. 311 NC State Univ.*

TABLE 13.28   Best Management Practices for Soft Red Winter Wheat in Virginia

Practice	Previous Management	BMP Management	Environmental Advantages
Rotation	Sometimes	Always	Better pest control
Soil test	Unbalanced nutrients	Balanced nutrients	Better N use efficiency; quicker ground cover; increased crop residue
Seeding rate	1.5 bu/a	22 seeds/ft row	Quicker ground cover
N management	Single application	Use tissue test; split applications	Increased N efficiency; increased crop residue
Tramlines	None	Establish tramlines at planting	Apply N and other inputs with precision
Pest control	No integrated pest management; no scouting	Use integrated pest management; use scouting	Use pest control only when needed

Expected Results	Previous Management	BMP System
Yield	50 bu/a	85 bu/a
Total production (100 a)	5,000 bu	8,500 bu
Acres needed to produce 5,000 bu	100	59
N used	100 lb/a	130 lb/a
N use efficiency 0.50 bu/lb	0.65 bu/lb	
Total N applied (100 a)	10,000 lb	13,000 lb
(59 a)		7,670 lb
N remaining after harvest	40 lb/a	34 lb/a
Production costs, $/a	160	207
Production costs, $/bu	3.20	2.44

SOURCE: D. Brann, 1986, VPI.

Selected References

POWER, J. F. (Ed.). 1987. *The Role of Legumes in Conservation Tillage Systems.* Soil Conservation Society of America.

SPRAGUE, M. A., AND G. B. TRIPLETT (Eds.). 1986. *No-Tillage and Surface Tillage Agriculture: The Tillage Revolution.* John Wiley & Sons, New York.

TATE, R. L. 1987. *Soil Organic Matter: Biological and Ecological Effects.* John Wiley & Sons, New York.

TROEH, F. R., J. A. HOBBS, AND R. L. DONAHUE. 1991. *Soil and Water Conservation.* Prentice-Hall, Englewood Cliffs, N.J.

LAL, R., AND B. A. STEWART. 1994. *Soil Processes and Water Quality.* Advances in Soil Science Lewis Publ. Boca Raton, FL.

PAUL, E. A., K. PAUSTIAN, E. T. ELLIOT, AND C. V. COLE. 1997. *Soil Organic Matter in Temperate Agroecosystems.* CRC Press, Boca Raton, FL.

Index

A

Mg deficiency in corn. Interveinal yellowing or white discoloration beginning with lower leaves, as Mg is translocated from older to newer leaves. Can be confused with Fe deficiency.

Fe deficiency in grain sorghum. Severely stunted plant with interveinal chlorosis of entire leaf, occurring in newer leaves first. Leaves turn white under severe Fe stress.

Fe deficiency in strawberry. Interveinal chlorosis of newer leaves.

Zn deficiency in corn. Newer leaves exhibit bleached white or pale yellow discoloration in area between leaf edge and midrib.

Zn deficiency in corn. Severe stunting caused by shortening of internodes. Normal plant is on the right.

Mn deficiency in corn. Pale green to yellow discoloration between veins of newer leaves. Can be confused with Fe.